Industry 5.0 and Sustainable Development

Exploring the intersection of Industry 5.0 technologies and sustainable development, this book integrates newer technologies—such as artificial intelligence (AI), automation, the Internet of Things, and robotics—within sustainability frameworks to provide both theoretical foundations and practical applications.

Providing actionable insights for academics, policymakers, and industry professionals, this edited collection investigates the nexus among technology, social equity, and environmental responsibility in Industry 5.0 across a diverse range of chapters from authors based in Bangladesh, India, Indonesia, Lebanon, Morocco, Nigeria, Poland, South Africa, Sudan, and the United States, while it also includes case studies from developed and emerging economies, such as Germany, Canada, the United States, India, Brazil, China, and South Africa. Offering a balanced perspective on the challenges and opportunities in different regions, the global cases highlight the universal relevance of the discussions and address the pressing sustainability challenges faced by both developed and developing countries. By incorporating strategic recommendations for policymakers and business leaders, the book forges the connection between academic research and industry practices and serves as a comprehensive guide for aligning industrial progress with Sustainable Development Goals. It also looks to the future through a discussion of the transition toward Industry 6.0, offering valuable insight as technological advancements continue.

Industry 5.0 and Sustainable Development is intended for academics, researchers, policymakers, and industry professionals engaged in sustainable development, environmental studies, business management, and industrial engineering, as well as upper-level students exploring the integration of advanced technologies with sustainability principles.

Imran Hossain is a dedicated academician and researcher with a profound commitment to sustainable development and environmental governance. Currently, Mr. Imran serves as a lecturer in political science at Varendra University, Bangladesh. Mr. Imran has delved into extensive research on the SDGs, collaborating with both national and renowned international organizations. Mr. Imran's research portfolio is diverse and impactful, encompassing projects on various aspects of sustainable development, including environmental governance, waste management, climate change, water management, and sustainable cities. Mr. Imran actively engages in scholarly discourse by participating in national and international conferences. As a testament to his expertise, he serves as a reviewer for various scientific journals indexed in Scopus. Mr. Imran has published a number of research articles and book chapters indexed in Scopus and other renowned databases. Beyond academia, Imran is a prolific writer and columnist for national dailies, where he shares insights and perspectives on sustainable development, environmental governance, and smart technology.

Azrour Mourade currently works as a professor of computer science in the Department of Computer Science, Faculty of Sciences and Techniques, Moulay Ismail University of Meknès. He received his Ph.D. from the Faculty of Sciences and Techniques, Moulay Ismail University of Meknès, Morocco.

A.K.M. Mahmudul Haque is a professor of political science at the University of Rajshahi, Bangladesh. He attended his Ph.D. internship program under the Student for Development Program of the Centre for Asia-Pacific Initiatives (CAPI) at the University of Victoria (UVic), Canada.

S.M. Akram Ullah is a professor of political science at the University of Rajshahi, Bangladesh. He received his Ph.D. from the Institute of Bangladesh Studies, University of Rajshahi, Bangladesh. Apart from his full-time engagement in the Department of Political Science, Professor Akram provides teaching support to the Department of International Relations, University of Rajshahi, as an adjunct faculty member.

Routledge Studies in Sustainable Development

This series uniquely brings together original and cutting-edge research on sustainable development. The books in this series tackle difficult and important issues in sustainable development including values and ethics; sustainability in higher education; climate compatible development; resilience; capitalism and de-growth; sustainable urban development; gender and participation; and well-being.

Drawing on a wide range of disciplines, the series promotes interdisciplinary research for an international readership. The series was recommended in the *Guardian*'s suggested reads on development and the environment.

Inside the Sustainable Development Goals
From Fighting Poverty to Saving the World
Adele Orosz

Circular Economy Solutions for Sustainable Development
Katarzyna Nowicka

How to Achieve Sustainable Development
Local Actions for Global Change
Samuel T. Ledermann

Innovation Strategies and the Sustainable Development Goals
A Blueprint for Sustainable Transformation
Amit Singh Khokhar, Punita Duhan and Anshu Lochab

Industry 5.0 and Sustainable Development
Theoretical Foundations and Practical Applications
Edited by Imran Hossain, Azrour Mourade, A.K.M. Mahmudul Haque, and S.M. Akram Ullah

For more information about this series, please visit: www.routledge.com/Routledge-Studies-in-Sustainable-Development/book-series/RSSD

Industry 5.0 and Sustainable Development

Theoretical Foundations and Practical Applications

Edited by Imran Hossain, Azrour Mourade, A.K.M. Mahmudul Haque, and S.M. Akram Ullah

Routledge
Taylor & Francis Group
LONDON AND NEW YORK

earthscan
from Routledge

First published 2026
by Routledge
4 Park Square, Milton Park, Abingdon, Oxon OX14 4RN

and by Routledge
605 Third Avenue, New York, NY 10158

Routledge is an imprint of the Taylor & Francis Group, an informa business

British Library Cataloguing-in-Publication Data
A catalogue record for this book is available from the British Library

ISBN: 978-1-041-08927-8 (hbk)
ISBN: 978-1-041-08931-5 (pbk)
ISBN: 978-1-003-64762-1 (ebk)

DOI: 10.4324/9781003647621

Typeset in Times New Roman
by Apex CoVantage, LLC

Contents

About the Editors

Imran Hossain is a dedicated academician and researcher with a profound commitment to sustainable development and environmental governance. Currently, Mr. Imran serves as a lecturer in political science at Varendra University, Bangladesh. Mr. Imran has delved into extensive research on the SDGs, collaborating with both national and renowned international organizations. Mr. Imran's research portfolio is diverse and impactful, encompassing projects on various aspects of sustainable development, including environmental governance, waste management, climate change, water management, and sustainable cities. Mr. Imran actively engages in scholarly discourse by participating in national and international conferences. As a testament to his expertise, he serves as a reviewer for various scientific journals indexed in Scopus. Mr. Imran has published a number of research articles and book chapters indexed in Scopus and other renowned databases. Beyond academia, Imran is a prolific writer and columnist for national dailies, where he shares insights and perspectives on sustainable development, environmental governance, and smart technology.

Azrour Mourade received his Ph.D. from the Faculty of Sciences and Techniques, Moulay Ismail University of Meknès, Morocco. He currently works as a professor of computer science in the Department of Computer Science, Faculty of Sciences and Techniques, Moulay Ismail University of Meknès. His research interests include authentication protocol, computer security, the Internet of Things, smart systems, and machine learning.

A.K.M. Mahmudul Haque is a professor of political science at the University of Rajshahi, Bangladesh. He attended his Ph.D. internship program under the Student for Development Program of the Centre for Asia-Pacific Initiatives (CAPI) at the University of Victoria (UVic), Canada. Haque is a visiting scholar at UVic, Canada. During his decade-long teaching career, he has mostly taught courses on environmental governance and sustainable development, human rights and the civil society of Bangladesh, and local governance: policies and practices. He has already carried out a number of research projects funded by different national and international organizations. Dr. Haque has published over a hundred scientific papers in high-impact, Scopus-indexed journals and regularly contributes insightful columns to national dailies in Bengali and English.

S.M. Akram Ullah is a professor of political science at the University of Rajshahi, Bangladesh. Apart from his full-time engagement in the Department of Political Science, Professor Akram provides teaching support to the Department of International Relations, University of Rajshahi, as an adjunct faculty member. He received his Ph.D. from the Institute of Bangladesh Studies, University of Rajshahi, Bangladesh. Environmental governance, sustainable development, political theories, Bangladesh politics, international relations, good governance, and development discourse are the key areas of his teaching and research interests. He has published a number of research article indexed in Scopus and other renowned database. His research emphasizes the importance of integrating technology and participatory governance to improve service delivery and enhance the accountability of local government bodies. Through his publications and ongoing research, Dr. Akram Ullah contributes significantly to both the academic field of political science and the practical landscape of governance in Bangladesh.

Contributors

Abdullah Al Bassam is an emerging researcher in ML, AI, the Internet of Things (IoT), robotics, and full-stack engineering. In his final year of bachelor's degree in computer science and engineering, Bassam integrates his knowledge of data science, digital signal processing, and computational intelligence to take a multidisciplinary approach in completing his projects and research work. He is interested in smart systems, customer service automation using natural language processing (NLP), optoelectronics, and engine combustion analysis. Casually, Bassam writes code for full-stack software and contemporary robotics and IoT solutions for industrial automation. He is also an office and design engineer on funded projects and is writing a book in Bangla on JavaScript to help improve programming literacy. Fueled by ideas, he is eager to deliver meaningful innovations that harness AI, data science, and intelligent automation for technologies of the future.

Rezki Amalyadi pursued his bachelor's/vocational degree at the Malang Agricultural Development Polytechnic, majoring in animal husbandry extension and animal welfare, from 2015 to 2019. In 2020, he continued his education in the Department of Animal Nutrition, Faculty of Animal Science, Postgraduate School Gadjah Mada University. He has been actively teaching in the Animal Husbandry Study Programme at the Faculty of Life Sciences and Technology, Sumbawa University, from 2022 to 2024 and in the Animal Husbandry Study Programme at the Faculty of Animal Husbandry, Mataram University, from 2024 to the present. During his time as an educator, he has written several articles that have been published in reputable international journals (*Scopus* and *Web of Science*): "Effect of compost tea and harvest age on productivity, nutrient content, and in vitro digestibility *Cichorium Intybus*"; "From pasture to sea: Assessing the viability of ruminant forage species for dugong (*Dugong dugon*) domestication and welfare"; and "Screening and characterisation of lactic acid bacteria from the gastrointestinal tract of a native chicken (*Gallus gallus f. domesticus*) as probiotics."

P.R. Anisha is a faculty member in the Department of Computer Science and Engineering at Stanley College of Engineering and Technology for Women, India. Her teaching and research interests focus on computer networks, data science, and emerging technologies in computing. She is actively involved in mentoring students, publishing research, and contributing to academic development through innovative teaching practices and collaborative projects.

Nasir Assad is a distinguished researcher and educator with expertise in chemistry, nanotechnology, and materials science. He earned an M.Phil. in chemistry from the University of Sargodha, Pakistan, focusing on green synthesis, photocatalysis, and colorimetric sensors. Currently, he is pursuing postdoctoral research under the HEC Indigenous Fellowship, advancing his contributions to sustainable chemistry. Assad has earned recognition through several international publications and has gained industrial experience as an analyst at Biogen Pharma, Islamabad, Pakistan, and Nutrabiotics® Nutritional Life Science, Peshawar, Pakistan. He has actively participated in national and international conferences, presenting his research on green synthesis and sustainable development. Assad has received multiple scholarships, including the HEC Indigenous Ph.D. Fellowship and the PEEF Master Level Scholarship. Driven by a passion for sustainability, his research aims to develop eco-friendly materials for biomedical and environmental applications, bridging the gap between chemistry and real-world solutions.

Dr. Saumendra Das is currently working as a professor at the School of Management Studies, GIET University, Gunupur, Odisha. He has more than 22 years of teaching, research, and industry experience. He has published more than 116 articles in national and international journals and conference proceedings and as book chapters. He has also authored and edited 28 books of international and national repute. Dr. Das has participated in and presented many papers in seminars, conferences, and workshops in India and abroad. He has organized many international and national conferences, FDPs, and workshops during his career. He is an academician, author, and editor. He has also published two patents and two patent designs. He is an active member of various professional bodies in India. He was awarded the best teacher award by the Research Foundation of India.

Dr. Bilal Eneizan is an associate professor of marketing at Jadara University, Jordan, and an assistant professor at the College of Business Administration in Hotat Bani Tamim, Prince Sattam Bin Abdulaziz University (PSAU), Saudi Arabia. His research interests include digital marketing, artificial intelligence in marketing, consumer behavior, and green marketing. He has published in several peer-reviewed journals and actively contributes to advancing marketing knowledge, with a particular focus on technology-driven innovation and sustainability in business practices.

J. Smruthymol is an assistant professor of commerce at Karpagam Academy of Higher Education, Coimbatore. Her research interests lie in financial management, accounting, and business education. She is actively engaged in teaching, mentoring students, and contributing to departmental research initiatives. With a strong commitment to academic growth, she strives to bridge the gap between theoretical knowledge and practical applications in commerce and business studies.

Abdellah Jamali serves as a researcher at the Faculty of Sciences and Technologies, Hassan First University of Settat, contributing to the Research Laboratory on IT, Networks, Mobility, and Modeling. His academic work explores the intersection of information networks, modeling techniques, and mobile applications. Through publications and collaborative projects, he is advancing research in applied sciences and IT in Morocco and beyond.

Dr. M. Kalimuthu is a professor at Dr. N.G.P. Arts and Science College, Coimbatore. His academic and research expertise lies in management, economics, and social sciences, contributing to interdisciplinary research and higher education. He has published widely in reputed journals and presented papers at international conferences. Dr. Kalimuthu is known for his dedication to academic leadership, research excellence, and fostering innovation-driven education in Indian higher education institutions.

Rejaul Karim, Ph.D., is an associate professor of finance and banking in the Department of Business Administration at Varendra University, Rajshahi, Bangladesh, and a research fellow at INTI International University, Malaysia. He received his Ph.D. in finance from the Institute of Bangladesh Studies, University of Rajshahi, Bangladesh. He completed BBA and MBA in finance and banking from the Pabna University of Science and Technology, Pabna, Bangladesh, with outstanding academic performance. Dr. Karim has teaching and research experience for over 10 years at undergraduate and master's levels. He has a great passion for teaching and research in finance and economics, and his research focuses on financial performance, circular economy, financial structure, sustainable finance, and Sustainable Development Goals. Dr. Karim continuously contributes through his research and scholarly contributions and has published many papers in reputed national and international journals and as book sections and conference proceedings. Dr. Karim founded and headed BusinessXcel Lab, a center for quantitative business research. This voluntary organization aims to facilitate collaboration among domestic and international undergraduate, graduate, and Ph.D. students and professional researchers.

Munazzah Khan is a researcher in the Department of Computer Science and Engineering at Stanley College of Engineering and Technology for Women, India. She has been actively involved in teaching and academic research, focusing on data analytics, software development, and computational intelligence. Her contributions extend to student mentoring and participation in academic conferences.

Mahadih Kyambade is a distinguished academician and lecturer in the Department of Leadership and Governance at Makerere University Business School. He is pursuing a Ph.D. in responsible leadership and holds degrees of Master of Science in Leadership and Governance, Bachelor of Leadership and

Governance, and Bachelor of Laws from Makerere University. His research focuses on themes related to leadership, governance, psychological safety, gender diversity, higher education, sustainability, and environmental studies and policy.

Ayoub Louja is affiliated with the Faculty of Sciences and Technologies, Hassan First University of Settat, Morocco, where he contributes to the Research Laboratory on IT, Networks, Mobility, and Modeling. His scholarly interests include computer networks, mobile technologies, and applied computational modeling. He has been actively engaged in research and academic collaborations, contributing to the advancement of knowledge in IT and network systems.

Nasrullah Masud is currently serving as a lecturer in the Department of Electrical and Electronic Engineering at Varendra University, Bangladesh. He holds a BSc in Electrical and Electronic Engineering from Rajshahi University of Engineering and Technology (RUET) and is currently pursuing his MSc in the same discipline. His research interests span quantum computing, control system design, numerical analysis of electrical circuits, electrical machine fundamentals, power system operation and control, and applications of machine learning (ML) and deep learning in engineering. In addition to his academic role, Masud has actively contributed to scientific communication and outreach, writing for national science portals and engaging with international learning communities. He has also participated in professional training programs offered by global institutions, including Microsoft, Google, and MathWorks, covering areas such as digital journalism, project management, deep learning, and generative AI. With a strong commitment to bridging equity and accessibility in education, Masud envisions a future where learning becomes a lifestyle for global citizens. Through his teaching, research, and community engagement, he strives to integrate engineering innovations with societal needs while mentoring the next generation of learners and researchers.

Arnab Mitra, Ph.D. (Engg.), is an associate professor, an IEEE member, and a former Erasmus Mundus doctoral research fellow in mobility. Presently, he is associated with the Department of Information Technology, Techno International New Town, Kolkata, India. Along with his academic responsibilities, he holds several editorial board memberships with reputable international publication houses. His primary research interests include areas related to cellular automata and theoretical computer science. In addition to his primary research focus, his publications mainly focus on the following application areas: data security and privacy, eHealth, Industry 4.0, and social networks.

Md. Soleman Mollik is a lecturer in the Department of Business Administration in Marketing, Faculty of Business Studies, Bangladesh University of Professionals, Mirpur Cantonment, Dhaka, Bangladesh. He completed his MBA and BBA in marketing from the Department of Marketing, University of Rajshahi, Bangladesh. His research interests include contemporary issues in marketing, service marketing, digital tourism and hospitality marketing, Islamic marketing, circular economy, and sustainable marketing. He has published his research works in several journals, including the *International Journal of Religious Tourism and Pilgrimage*, Technological University, Dublin (Scopus indexed, Q1); the *International Journal of Services and Operations Management*, Inderscience (Scopus indexed, Q3); the *Geojournal of Tourism and Geosites* (Scopus indexed, Q3); the *Global Journal of Management and Business Research*; the *Journal of Business Studies* (Pabna University of Science and Technology); and the *Journal of Integrated Marketing Communications and Digital Marketing*. He received the "Dean's Award" from the University of Rajshahi in 2018.

Dr. Ganesan Muruganantham is working as a professor of management studies at the National Institute of Technology, Tiruchirappalli. His areas of interest in teaching, research, and consultancy include branding, consumer insights, and quality management. He has received the best paper awards for presenting his research paper in the United States, London, Australia, and India. His research works are available at Emerald, Inderscience, and Taylor and Francis publications.

N. Jaya Prakash is a Ph.D. research scholar in the Department of Commerce at Karpagam Academy of Higher Education, Coimbatore. His research interests lie in the fields of commerce, management, and sustainable business practices, with a focus on exploring emerging trends in corporate governance and financial management. He is dedicated to advancing academic knowledge through rigorous research and analytical inquiry, aiming to contribute valuable insights to the field of commerce and its role in socio-economic development.

Najib Naja is affiliated with the National Institute of Posts and Telecommunications, Rabat. His expertise lies in communication systems, networking, and IT management. He has actively participated in research initiatives focusing on technological development and its applications in modern telecommunications. Through his teaching and research, Dr. Naja contributes significantly to academic innovation and the development of advanced information technology (IT) solutions.

Afulah Namatovu is a renowned academician in the field of information management. She is a lecturer in the Department of Applied Computing and IT, Makerere University Business School, and holds a bachelor's degree in office and information management from Makerere University. Her research focuses on information management, information technology (IT), generative AI, and leadership and gender issues in organizations.

Amrita Nandy is a faculty member of the Department of Business Administration at BGC Trust University Bangladesh. Over the years, she has been instrumental in developing and implementing innovative teaching and works as a mentor for students. Lately, she has been pursuing her M.Phil. degree from the Bangladesh University of Professionals, Dhaka, Bangladesh. Her primary research interest lies in "sustainability issues" such as ethical consumption, eco-tourism, workplace cohesion, recycling, and promoting women entrepreneurship. Additionally, she has worked as a soft skills facilitator in several organizations. The author has published research works in multiple forms, including journal articles, book chapters, conference proceedings, and case studies.

Fitria Dewi Navisa is a faculty member in the Department of Law at Universitas Islam Malang (UNISMA), Indonesia. Her teaching and research span Indonesian legal systems, comparative law, and law-and-society perspectives. She is engaged in curriculum development and student mentoring and contributes to community outreach and legal literacy initiatives. Through classroom innovation and collaborative projects, she seeks to connect doctrinal analysis with real-world legal practice and public service.

Hafsa Nayeem is affiliated with the Department of Computer Science and Engineering at Stanley College of Engineering and Technology for Women, India. Her academic interests include programming, ML, and applied computing. With a passion for research and student engagement, she has been contributing to various academic initiatives and collaborative works in the field of computer science.

Udaya Sankar Patro is a Ph.D. scholar in the Department of Management Studies at the National Institute of Technology, Tiruchirappalli. Udaya Sankar Patro is the corresponding author.

Md. Sahidur Rahman (Ph.D., University of Chittagong; MScR, University of Edinburgh; MSc, University of Dundee; M.Com. and B.Com. (Hon.), University of Dhaka) is a professor in the Department of Management, University of Chittagong, Bangladesh. He is well known for his dynamic teaching tactics that persistently encourage students to excel in their performance. His prime focus of research lies in human resource management, such as organizational behavior, leadership, and emotional intelligence. The author has nearly 100 research articles and a dynamic presence in national and international conferences. In addition, he is an active reviewer of various international and national journals. Currently, he is supervising three Ph.D. and three M.Phil. researchers.

Dr. Mohammad Sahabuddin currently serves as an associate professor, Department of Finance and Banking, Faculty of Business and Entrepreneurial Sciences, University of Science and Technology Chittagong (USTC). Apart from this, he plays the role of the director of the USTC Research Cell (URC). His areas of research interest are financial markets and conventional and Islamic finance. His current projects are Islamic Fintech, halal tourism, entrepreneurship, blended learning, internationalization, and sustainable education system. He has presented several papers at national and international conferences and published more than 20 articles in Scopus, Web of Science (WoS), ADBC, and ABS-indexed journals. He completed his Ph.D. in business economics from Universiti Putra Malaysia (UPM) and a postdoctoral degree from Universiti Malaysia Perlis, Malaysia.

Dr. Phineas Sebopelo currently serves as the director of quality assurance (QA). In addition to his QA duties, he is a skilled professional who collaborates effectively with colleagues both locally and externally, ensuring clear communication. He is committed to reaching his goals while supporting the organization's vision and cultivating positive team relationships to encourage collective success. Dr. Sebopelo is also an aspiring academic with publications in quality management and artificial intelligence (AI), aiming to advance the field through his ongoing research. His recent AI-related work includes topics such as leveraging AI to improve quality in higher education institutions (HEIs), adaptive learning strategies in open and distance e-learning (ODeL)—highlighting opportunities and challenges for QA—AI integration and university service quality, and AI in ODeL institutions in sub-Saharan Africa, focusing on QA opportunities and challenges.

Dr. S. Senthilkumar is an associate professor of human resource management (HRM) and finance at Skyline University, Nigeria. His scholarly work bridges HRM and financial practices, emphasizing organizational development in emerging economies. He has published widely in peer-reviewed journals and presented papers at international conferences. With significant teaching and administrative experience, Dr. Senthilkumar actively engages in academic collaborations and knowledge dissemination across global higher education platforms.

Abdus Sobhan is an undergraduate student of computer science and engineering (CSE) at Varendra University, Rajshahi. His research interests include quantum computing, AI, ML, deep learning, and blockchain. He aspires to pursue higher studies and conduct advanced research in quantum computing to contribute to the development of next-generation computational systems. During his undergraduate studies, Abdus Sobhan has been actively engaged in research and academic activities. He has authored and co-authored publications in IEEE conferences and international book chapters, focusing on the applications of AI, ML, and blockchain to address real-world challenges.

Md. Tanvir Islam Sourav is a student at the Department of Computer Science and Engineering at a private university in Bangladesh. His focus is on IoT technologies, especially sensor-based systems for collecting real-time environmental and atmospheric data. He has a track record of IoT prototype designs that merge microcontrollers with advanced sensor modules to monitor air quality, temperature, humidity, and gas concentrations. His research focuses on IoT data acquisition and transmission techniques to build dependable, low-cost, and scalable environmental monitoring systems. Beyond hardware integration, he is interested in the application of IoT in smart cities, sustainable development, and public health research. He works on contributing intelligent sensing systems in the small and growing market of IoT-driven environmental solutions.

Dr. J. Sudarvel is an assistant professor of management at Karpagam Academy of Higher Education, Coimbatore. His research interests span organizational behavior, strategic management, and entrepreneurship, with a focus on innovative practices in higher education and business sustainability. He has contributed to several national and international conferences and journals, actively engaging in academic development and student mentoring. He is committed to integrating research insights with pedagogy to enhance management education in India.

Dr. Ravi Thirumalaisamy is a professor in the Department of Business and Economics at the Modern College of Business and Science, Muscat, Sultanate of Oman. His expertise covers economics, international business, and strategic management, with a strong focus on global market trends and sustainable economic growth. A prolific academician and researcher, Dr. Ravi has contributed extensively to scholarly publications and has been actively involved in academic leadership and curriculum innovation in international education.

Dr. R. Velmurugan is an associate professor of commerce at Karpagam Academy of Higher Education, Coimbatore. With extensive teaching and research experience, he specializes in accounting, finance, and corporate governance. His academic contributions include numerous publications in reputed journals and active participation in research forums. Dr. Velmurugan is dedicated to fostering academic excellence, guiding young scholars, and advancing research in commerce and management fields.

S. Vinodh is a professor in the Department of Production Engineering, National Institute of Technology, Tiruchirappalli, Tamil Nadu, India. He completed his Ph.D. under the AICTE National Doctoral Fellowship Scheme from PSG College of Technology, Coimbatore, India. He completed his master's in production engineering from PSG College of Technology, Coimbatore, India, and bachelor's in mechanical engineering from Government College of Technology, Coimbatore, India. He has published over 175 papers in international journals. He received the highly commended paper award from Emerald Publishers for the year 2016. His research interests include sustainable manufacturing, lean manufacturing, agile manufacturing, product development, and Industry 4.0.

Sridhar Reddy Vulapula is a faculty member in the Department of Information Technology at Vignana Bharathi Institute of Technology, India. His teaching and research interests include software engineering, information systems, and advanced computing applications. He has been actively contributing to research publications, curriculum development, and student guidance in higher education.

Hsing-Hao Wu is a faculty member in the Department of Economic and Financial Law at the National University of Kaohsiung, Taiwan. His academic interests include financial regulation, corporate governance, and the legal dimensions of fintech and capital markets. He teaches courses in economic and financial law and advises student research. His work emphasizes comparative perspectives and policy relevance, linking legal frameworks to market integrity and sustainable economic development.

Yassin Zaiouane is a researcher at the Faculty of Sciences and Technologies, Hassan First University of Settat, working with the Research Laboratory on IT, Networks, Mobility, and Modeling. His research focuses on network systems, mobile communications, and computational methods, with an emphasis on innovative technological solutions. He is committed to bridging theoretical advances with practical applications in the field of IT.

Preface

We are delighted to present the book *Industry 5.0 and Sustainable Development: Theoretical Foundations and Practical Applications*, a timely and comprehensive scholarly endeavor that explores the dynamic nexus between the Fifth Industrial Revolution and global sustainability imperatives. This edited collection emerged from a pressing recognition: while Industry 4.0 revolutionized automation, efficiency, and connectivity, the emerging paradigm of Industry 5.0 seeks to rehumanize industrial progress, embedding resilience, ethics, sustainability, and inclusivity into the very fabric of innovation.

This book is structured across three thematic sections that reflect both the theoretical evolution and applied relevance of Industry 5.0.

'Part I: Theoretical Perspectives and Foundations of Industry 5.0' provides the conceptual grounding for the volume. It begins with an exploration of the core tenets of sustainability within Industry 5.0 (Chapter 1), followed by a panoramic review of industrial revolutions from Industry 1.0 to 6.0 and their implications for societal development (Chapter 2). A scoping review of societal transformation through Industry 5.0 lenses (Chapter 3) further deepens our understanding of changing organizational behaviors and consumer expectations. Chapter 4 then critically assesses the sustainability-driven opportunities and challenges of Industry 5.0, laying the foundation for future discourse.

'Part II: Technology, Innovation, and Applications for Sustainable Development' shifts to the cutting-edge technological interventions that underpin Industry 5.0. Chapter 5 presents artificial intelligence (AI)-driven strategies aligned with the UN Sustainable Development Goals, while Chapter 6 evaluates the broader relationship between AI and sustainable development. Chapter 7 dives into AI applications in smart cities, energy, and climate action, offering a blueprint for urban resilience. Chapter 8 introduces an ethical AI approach to sustainability governance in healthcare using Internet of Things (IoT)-enabled deep learning, highlighting the convergence of smart technology and ethical responsibility.

'Part III: Industry 5.0 in Practice: Case Studies, Tools, and Strategic Insights' brings theory into the realm of practice. The case study from Taiwan (Chapter 9) evaluates corporate social responsibility (CSR) in sustainable agriculture, followed by a methodological application of the MCDM tool to assess Industry 5.0 drivers in manufacturing (Chapter 10). Chapter 11 employs advanced satellite simulation (ST-Vision Net) to study urban heat islands, linking AI with smart environmental planning. Chapters 12 and 13 elaborate on the concepts of circularity and eco-innovation, framing Industry 5.0 as a transformative platform for resilient, green industrial ecosystems. The final chapter (Chapter 14) presents an in-depth country case study on Uganda's policy and practice toward adopting Industry 5.0, highlighting its potential to reshape industrial futures in the Global South.

Collectively, the chapters in this book reflect a multidisciplinary and multi-sectoral approach. They bridge the gap between theory and application, offering critical insights for academics, policymakers, technologists, sustainability professionals, and forward-thinking industry leaders. The volume underscores that the true promise of Industry 5.0 lies not only in technological breakthroughs but also in fostering inclusive, equitable, and sustainable growth for our planet and its people.

We express our deepest gratitude to all contributing authors whose rigorous scholarship and visionary ideas have shaped this work. We are also thankful to Taylor & Francis for providing a platform that amplifies such vital and globally relevant discourse.

As we navigate a complex and uncertain future, we hope this book serves as both a guide and an inspiration for embracing Industry 5.0 as a force for sustainable transformation.

Imran Hossain
Varendra University, Rajshahi, Bangladesh

Part I

Theoretical Perspectives and Foundations of Industry 5.0

1 Sustainability in Industry 5.0

Rezki Amalyadi and Imran Hossain

1.1 Industry 5.0 Sustainability

Industry 5.0 (I5.0) signifies a model change in industrial operations, underscoring human-centric, sustainable, and also resilient strategies. This research examines the relationship between sustainability and I5.0, focusing on theoretical views, the circular economy (CE), and green manufacturing, along with ecological innovation and social and ethical factors. The aim of the research is to analyze how particular I5.0 principles can thoroughly drive sustainable industrial transformation while dealing with several ecological and social challenges. The methodology includes an exhaustive literature review. Additionally, it features a thorough case study analysis. The thorough review consolidates multiple fragmented understandings of I5.0's many sustainability dimensions, while multiple case studies highlight particular practical applications within CE and green manufacturing. Significant findings demonstrate how I5.0, which includes technologies like digital twins, blockchain (BC), and artificial intelligence (AI), fosters sustainability through human–machine collaboration. Green manufacturing relies heavily on CE principles like recycling and remanufacturing to reduce waste and dependency on resources. Ecological innovation, including energy-efficient processes coupled with renewable energy integration, improves green manufacturing practices, thereby contributing to diminished carbon footprints. Social and ethical considerations during implementation are important in I5.0. The research stresses upon the importance of ethical frameworks and value-sensitive design for promoting fair technology implementation. Dealing with data privacy, algorithmic bias, and worker well-being is vital to promoting trust and inclusivity within I5.0 ecosystems.

This research contributes considerably within the expanding body of knowledge on sustainable industrial practices by providing a thorough framework toward comprehensively integrating I5.0 principles alongside sustainability goals. It offers many actionable understandings to policymakers, as well as to industrialists. It also offers such understandings to researchers so they can navigate the overall transition toward a more sustainable industrial future. The results highlight how stakeholders must work together to address ethical, environmental, and technological issues in order to create a resilient and inclusive I5.0 future. The results emphasize how crucial it is to promote cooperation between various stakeholders, such as

DOI: 10.4324/9781003647621-2

legislators, business executives, scholars, and communities, in order to handle the complex issues raised by I5.0. These difficulties include ethical issues, environmental sustainability, and technical breakthroughs. Stakeholders may guarantee that I5.0 fosters inclusivity and sustainability in addition to increasing productivity by coordinating objectives and utilizing synergies. Coordinated tactics and creative solutions are needed to address problems including data security, algorithmic biases, and ecological footprints. A robust industrial ecosystem that strikes a balance among social responsibility, ethical frameworks, environmental stewardship, and technological advancement is made possible by this cooperative strategy.

A paradigm shift in industrial operations, I5.0 emphasizes resilient, sustainable, and human-centric approaches. In contrast to Industry 4.0, which prioritizes digitization and automation, I5.0 combines technology with sustainability and human–machine cooperation to reduce waste and improve resource efficiency (Rejeb et al., 2025). As environmental concerns grow, I5.0 provides a comprehensive framework for balancing industrial productivity with ecological responsibility.

Shafique et al. (2024) showed that technologies were divided into three different perspectives in the I5.0 literature review. The first is the continuation and expansion of current technologies such as enterprise systems, BC, Internet of Things (IoT), cybersecurity and encryption, big data, and industrial robots. The second has to do with cutting-edge technologies such as AI, humanoid robots, and the Internet of Medical Things. The third strategy focuses on intelligent energy management systems, the Internet of Everything, human identification technology, and adaptive (cognitive) robots. In addition, the three main I5.0 technologies are BC, big data analytics (BDA), and IoT. From a CE standpoint, technologies have also expanded to include cloud computing, big data, AI, machine learning (ML), IoT, and BC. Cloud/IoT, Cobot, cyber—physical systems, digital twins, radio frequency identification, and edge computing are all considered I5.0 technologies in the literature.

Only five well-known I5.0 technologies—AI, big data, IoT, ML, and BC—were the focus of this analysis. These were included because they are consistent with earlier research that relates certain technology to CE. The literature demonstrated how AI capabilities and circular business models are related. Furthermore, circular economy performance (CEPs) and sustainable supply chain performance were positively and significantly impacted by generative AI. The association between big data analytical skills and CE performance of the Czech manufacturing sector was examined in earlier research. Furthermore, the manufacturing sectors in Brazil and India were found to benefit from big data analytical capabilities in terms of CEPs.

A relationship framework between IoT and CE in the manufacturing sector was built by a bibliometric study. A thorough analysis of the literature also confirmed the connection between CE and IoT. Additionally, the literature discovered that Industrial IoT has a favorable effect on CE in Jordanian small and medium-sized enterprises (SMEs). Following the same strategy, a comprehensive literature review and bibliometric analysis were used to develop the link between ML and CE. Additionally, the plastic industry's CE programs used the ML method. Digital technologies, which are in line with the Sustainable Development Goals (SDGs), have an impact on CE in the Fourth Industrial Revolution. To close the gap between practice and research, a thorough literature analysis on the connection between BC and CE was carried out, building on the earlier reasoning. CE is impacted by the various facets of BC technology. BC also has an impact on CEPs.

Sustainability is at the core of I5.0, where technology is used to support greener industrial practices. One of the key aspects is waste reduction and resource optimization through CE-based approaches (Rejeb et al., 2025). In addition, green manufacturing is increasingly becoming a key focus in modern industry to reduce carbon footprint and improve energy efficiency (Shafique et al., 2024). Through the utilization of technologies like BC, digital twins, and AI, businesses may minimize adverse environmental effects and enhance manufacturing processes (Costa, 2024). In particular, we now emphasize how our work uniquely contributes to the I5.0 sustainability discourse by integrating human-centric design and CE principles into a unified framework. This addresses the current lack of comprehensive models that connect these elements in practical industrial contexts. In response, we have elaborated on the methodology section to include detailed descriptions of our literature search strategy, including databases (e.g., Scopus and Web of Science), keywords, inclusion and exclusion criteria, and publication date ranges. Furthermore, we now provide a clear rationale for the selected case studies and describe the analytical frameworks (e.g., thematic analysis and sustainability performance indicators) applied. These updates improve the transparency and reproducibility of our research and are presented in Section X.

1.2 Green Manufacturing

Specific approaches of green manufacturing are used for exploring sustainability. This section presents various features and approaches to green manufacturing practices (Figure 1.1).

The image illustrates the core approaches of green manufacturing that align closely with the principles of sustainability in I5.0, emphasizing not only technological advancement but also ecological responsibility and human-centric values. The cycle begins with recover–redesign–recycle, reflecting I5.0's focus on CE models where materials are reused and products are reimagined for longevity and minimal environmental impact. Reduction in energy use and decrease in water use resonate with I5.0's push for smart, energy-efficient technologies supported by AI and IoT to optimize resource consumption. The emphasis on emission reduction aligns with the broader goal of achieving carbon neutrality through intelligent manufacturing systems. Meanwhile, reducing waste generation supports the development of sustainable production chains where minimal waste is a design goal rather than a byproduct. Altogether, these green manufacturing approaches serve as foundational elements for implementing sustainable I5.0, where human–machine collaboration not only boosts productivity but also fosters environmental stewardship. This integration ensures that technological progress does not come at the cost of the planet, but instead contributes to long-term ecological balance and social well-being.

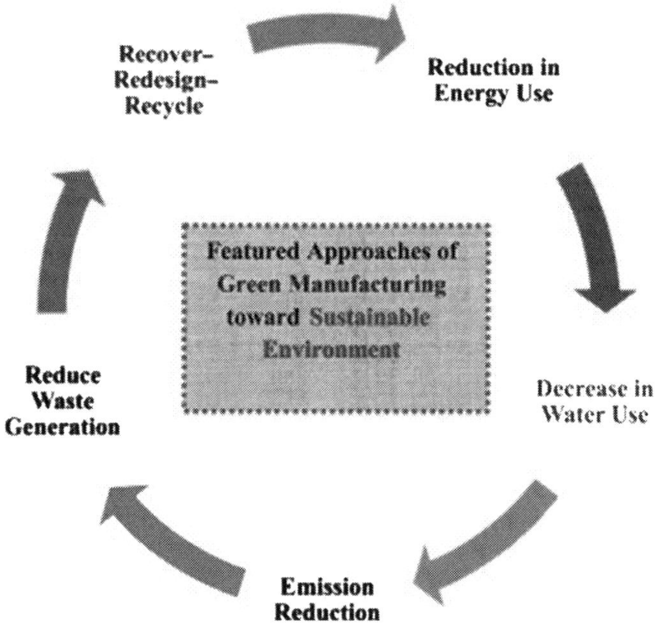

Figure 1.1 Important strategies for environmentally friendly, sustainable manufacturing (Haleem et al., 2023).

These strategies include many clever and energy-saving measures to help manufacturing activities become more environmentally friendly and compatible with the sustainability of the industry as a whole. Among the factors are lower energy and water use, lower levels of waste production during operations, and lower emissions. On the other hand, these methods for conceiving sustainable manufacturing operations are further enhanced by the problems or actions like recovering, redesigning, recycling, remanufacturing, and reprocessing (Javaid et al., 2022; Aminuddin et al., 2014).

Investing in new processes and technology is often motivated by resistance. The sector's traditional workers are afraid of losing their jobs; thus, it could be necessary to cease altering instances and make them more consistent and comprehensive. Work and conversations still have purpose. The transition to sustainable production and employee initiatives is about to happen (Milovanovic et al., 2022). Climate change is a symptom of a larger problem with modern overconsumption, which necessitates the mining of natural resources and the production of commodities to support our contemporary civilization. Greenhouse gas emissions are caused by transportation, energy consumption, industrial output, services, construction, and agriculture. People who desire cars, airplanes, electricity, large companies and homes, telecoms, technology, entertainment, etc., are at the heart of these extremely polluting and climate-related enterprises (Park et al., 2018; Yin et al., 2020).

Even the worst pollutants, such as those from chrome tanning, can be collected and recycled to decrease the quantity of direct waste in the water supply. As if the long-term environmental and health problems of direct dumping would not suffice, large sections of rainforest have been destroyed for farm usage. The rise in carbon emissions comes from the smoking stacks of plants and raw material farming. Agriculture is currently the main cause of rainforest degradation, as farmers use slash-and-brush practices to clear land for cattle grazing (Roschangar et al., 2018; Li et al., 2020). A consumer item is ultimately the culmination of all the inputs required to create a final product. The energy, water, materials, and equipment ultimately needed for manufacturing are examples of natural resources used in the production process. In order to achieve sustainable manufacturing, it is crucial that all inputs be taken into account and that sustainable business methods be used. Beyond environmental impact evaluations, truly sustainable production practices consider the living and working situations of those who toil diligently to create consumer goods (Mao & Wang, 2019).

This issue is made worse by the rapid advancement of technologies like computers, mobile phones, microwaves, and other devices, which society demands must be available at all times. Even while technology has improved production methods and efficiency, many products are made of untreated materials that are difficult to fix, leading to mountains of plastics and electronic garbage that fill landfills and release toxins into the ground. When robots perform risky tasks in sectors and companies that support sustainability goals, expenses are reduced. The same holds true when choosing not to take advantage of or behave in risky or unstable situations (Mafini & Loury-Okoumba, 2018; Hassan & Jaaron, 2021). Composite materials are commonly used in 3D printing, a developing technology that creates 3D items by building layers on top of layers. When creating intricate and lightweight products, this method is both economical and productive. A sustainable manufacturing process may be included in one or more of the aforementioned techniques. The most effective lean production technique will be green manufacturing.

1.3 Ethical Ecological and Technological Innovation through the Utilization of Artificial Intelligence Technology

Bimantoro et al. (2021) showed that, in the 5.0 era, it is planned that almost all aspects of life will be involved such as economic development, energy needs, food needs, education, health, housing, and industry. Technological developments will ultimately change the perspective, behavior, nature, and characteristics of social interaction with technological advances. The thing that is most affected by the evolution of information technology toward version 5.0 is human resources. Information technology significantly influences various aspects of life, especially the social dimensions of interpersonal communication. The Theory of Reasoned Action (1975) states that a person will use information technology if it is beneficial to improve their performance. Technology has been widely used and applied in real life, making it easier for people to carry out their daily activities. This certainly reduces interaction with one another in social life and the virtual world seems to be a place to interact that can be used easily across generations. However, this is followed by a decline in ethics and creates a paradox in the 5.0 era. The ethical values of today's society are a clear example of such a paradox and are very different from those of the past. The development of existing technology seems to take away their view of their surroundings. People's sensitivity to the social environment has decreased. Ethics such as friendly attitudes and greetings are quite difficult to find today. Technology has taken over the boundaries between one another so that ethics become underdeveloped. Before we go any further, we should recognize that ethics concerns how individuals act in accordance with moral standards. An ethical person is one who has good ethics, and an unethical person is one who violates ethics. Ethics is different from morals. Ethics is a science, while according to the dictionary of psychology, morals refer to principles related to social rules, laws, or customs that regulate behavior. Ethics is a science not a teaching. Development of technology influences human lives and ethics. Issues that often arise about ethics in information technology include health, privacy, intellectual, property rights, security, accuracy, and others. This was also previously mentioned in 1992 by the founder of the *Journal of Information Ethics*, Hauptmann, who discussed confidentiality, reliability, quality, and the use of information. One of the issues that can describe the ethics that everyone enjoys in this technological development is privacy. Privacy is an important point to live life even though everything is digital. Privacy is defined as the space of each individual that limits and regulates social life, which essentially focuses on individuals and privacy as a social value. It is undeniable that the current reality of violating information technology ethics has a lot to do with an individual's privacy being compromised. In Privacy in Context, sooner or later, disputes about privacy are fought hard, and the settlement benefits the strong parties. Therefore, privacy must be respected and even protected.

Through the usage of human–cyber–physical systems (HCPSs), AI in Society 5.0 gathers data from physical locations in cyberspace, facilitating the data analysis process in the future (Zhou et al., 2019), which is a new generation of AI that takes into account all human–machine interactions and is based on ontology. Additionally, this method makes it possible for industry and society to access previously unobtainable data on environmental and human biological factors. Scientists must apply the principles from biological evolution in order to achieve Society 5.0's goal of striking a balance between social problem resolution and economic advancement. I5.0 concentrates on the geostrategic changes brought about by synthetic biology, whereas I4.0 concentrated on the use of various digital/web-based technologies to gather and track data (Sachsenmeier, 2016). In order to achieve zero-failure processes for AI integration in smart cities, the concept calls for dynamic cities that are planned according to environmental conditions and that make use of bio-based goods, energy, and services. The objective is to continuously enhance process data, services, and products, as well as intelligent systems and associated infrastructure, in order to completely adapt the manufacturing process to the environment and human demands (Sachsenmeier, 2016). However, there are several obstacles to overcome in the implementation of Society 5.0, such as regulation, cybersecurity, human resource development, and social implications. The government must intervene in the establishment of supporting technology policies and infrastructure, cybersecurity protection, and education reform. The social implications between humans and robots must be normalized to achieve the full potential of Society 5.0. Society 5.0 is a step toward new digitalization and transformation that combines elements of AI with human cooperation to create a society that can enjoy life to the fullest. Technological and economic growth must be created for this purpose, not just for the benefit of the wealthy few.

AI refers to technology that assists people in carrying out tasks like learning, observing, solving issues, and getting notifications that typically call for human cognition. The goal of AI in computer science is to create technology that can intelligently solve challenging issues. In 1956, John McCarthy defined AI as the science and engineering involved in building intelligent machines. Since 1956, when the concept of AI was first discovered, this technology has continued to advance quickly.

In 1997, an AI machine first defeated a grandmaster in a game of chess, and in 2011, an AI machine won in a game of Jeopardy! against two expert human participants. Then in 2016, AlphaGo, a Google-made AI program, defeated the world Go champion, Lee Sedol. In 2021, GPT-3, an AI machine capable of writing human-like text, was launched by OpenAI. In recent decades, AI applications have spread to various fields, such as banking and financial services, healthcare, education, manufacturing, and automotive industries. In banking services, AI applications are used for credit risk analysis, fraud detection, and product marketing and development. In healthcare, AI is used to diagnose diseases, assist in drug design and development, and treatment of certain diseases. In education, AI is used to improve the efficiency and effectiveness of online and in-class learning. AI can be applied to a system with the aim of interpreting external data properly, and the data is later utilized to process results for a certain objective (Goralski & Tan, 2020). When AI is incorporated into a system, it can generally read images or voices as if a human were asking for something. The science of creating a system with multiple forms of intelligence is known as AI (Oktradiksa et al., 2021). AI offers humans services that maximize their existence for a variety of beneficial purposes. AI can analyze a lot of information and data in a reasonable amount of time. This capability enables AI to make astute and accurate predictions and judgments, and to discover and uncover patterns concealed in data. Because of this, AI is highly beneficial in industries like finance and medicine that demand quick and precise data processing and decision-making. The capacity of AI to process data and generate insightful predictions is one of its primary advantages. However, AI has some weaknesses in solving complex problems. One of them is its inability to make moral or ethical judgments. In addition, AI is also limited by the data it processes, so if the data used is unrepresentative or of low quality, the results produced by AI can also be inaccurate.

Taraya and Wibawa (2022) state that, the use of AI technology has made a significant contribution in realizing Society 5.0. By incorporating technology related to AI, society can benefit in various sectors such as industry, health, and education. However, challenges related to limited public understanding of AI technology, dependence on such technology, and potential unemployment due to automation must be overcome so that the utilization of AI technology can provide optimal benefits. Nevertheless, AI technology still has great opportunities to be used in the future. The development of AI technology can boost output and efficiency while raising people's standard of living, and provide solutions to social and environmental problems. Therefore, the development of AI technology must continue to be supported through research and development. All things considered, AI technology holds considerable promise for bringing about major changes in the realization of Society 5.0. However, challenges related to the development and utilization of AI technology must also be overcome so that the benefits of this technology can be felt optimally. Therefore, support from various parties, the government, industries, and the general public is needed to develop and utilize AI technology in an appropriate and responsible manner. AI combined with collaborative robotics is used in the automotive sector, and AI-enabled sustainability monitoring is transforming the manufacturing industry. These examples are now discussed in Section X and help bridge the gap between AI theory and practice, providing readers with a clearer understanding of how I5.0 principles are being implemented in real settings. Cybersecurity threats are amplified by increased connectivity, workforce skill gaps in AI and robotics, and potential economic barriers for SMEs. These considerations are now integrated into the Discussion section (pages X–Y) to provide a balanced view of I5.0 implementation. Algorithmic bias mitigation frameworks (e.g., the EU AI Act guidelines) and principles for safeguarding workers' rights, such as the use of transparency logs and participatory design in AI implementation, are essential for identifying and addressing biases in AI systems. These additions are presented in Section X and further enhance the manuscript's focus on human-centricity. To what extent does human–machine collaboration in I5.0 reduce operational carbon emissions in mid-sized enterprises? These additions aim to guide future empirical studies and are now located in Section X (pages Y–Z).

1.4 Circular Economy

Nugroho (2023) presents I5.0's effects on the Indonesian economy: the Indonesian economy is significantly impacted by the implementation of I5.0, according to the literature review and secondary data analysis. First, by utilizing cutting-edge technologies like AI, robotics, and the IoT, I5.0 has the potential to boost the industrial sector's production and efficiency in Indonesia. This makes it possible to reduce human error, optimize production processes, and boost productivity at a reduced cost. Second, implementing I5.0 can help Indonesian industries become more competitive. By applying more advanced technologies and utilizing data effectively, companies can improve product quality, production flexibility, and responsiveness to market needs, which can overall improve the competitiveness of Indonesian industries to market needs, in turn improving the industry's competitiveness at the global level. Challenges of I5.0 implementation in Indonesia: during the research, a number of challenges facing the implementation of I5.0 in Indonesia were identified. First, inadequate digital infrastructure is a major obstacle. While there has been significant progress in terms of Internet connectivity and penetration in Indonesia, there are still areas that are not connected to a stable and fast Internet network. This lack of digital infrastructure must be addressed to encourage the implementation of I5.0 evenly across the country. Second, the lack of relevant workforce skills is also a challenge. I5.0 requires a workforce that has understanding and skills in digital technology, data analysis, and programming. Intensive training and customized education programs need to be enhanced to improve the qualifications of the workforce in the I5.0 era. Finally, a conducive regulatory and policy environment also needs to be considered. To promote investment and innovation in the I5.0 sector, supportive frameworks and policies that ease the adoption of new technologies, safeguard privacy and data security, and offer financial incentives are required. I5.0's effect on employment: this study shows that I5.0 has a significant impact on employment in Indonesia. While there are concerns about the replacement of human labor by technology, it was found that the adoption of I5.0 also creates new job opportunities. Smart machines and technologies require high expertise in development, maintenance, and operation, creating demand for a highly skilled workforce. In addition, with increased productivity and efficiency, companies can expand their business activities, which in turn can create additional jobs in related sectors such as logistics, distribution, and customer service. It is crucial to remember, nevertheless, that in order for the workforce to stay relevant in the evolving labor market, I5.0 adoption also calls for ongoing skill development and adaptation. According to this analysis, implementing I5.0 in Indonesia has a significant chance of boosting industrial competitiveness, accelerating economic growth, and creating job possibilities. To realize the full potential of I5.0 in the nation, however, obstacles in the areas of worker skills, digital infrastructure, and the regulatory environment must be addressed. Therefore, it is important to develop policies and initiatives that support technology adoption, skills development, and collaboration between the public and private sectors to drive sustainable industrial transformation in Indonesia.

The figure illustrates how the integration of I5.0 technologies with CE practices contributes to achieving sustainable performance across environmental, social, and economic dimensions (Figure 1.2). At the foundation, advanced technologies such as AI, ML, BC, BDA, IoT, and eco-design function as key enablers that support intelligent, efficient, and human-centric manufacturing systems. These technologies are applied not in isolation but in synergy to optimize processes, enhance traceability, and improve resource management. Aligned with these capabilities, CE practices—such as eco-design, management systems, and investment recovery—are implemented to minimize waste, extend product lifecycle, and promote resource reuse. This strategic alignment results in measurable sustainable outcomes: improved environmental performance through reduced emissions and energy use; enhanced social performance by ensuring ethical practices and prioritizing human roles in production; and increased economic performance through efficiency gains and innovation. In essence, the diagram presents a holistic framework for sustainable I5.0, where technological advancement is harmonized with circular and ethical principles to create long-term value for businesses, society, and the environment.

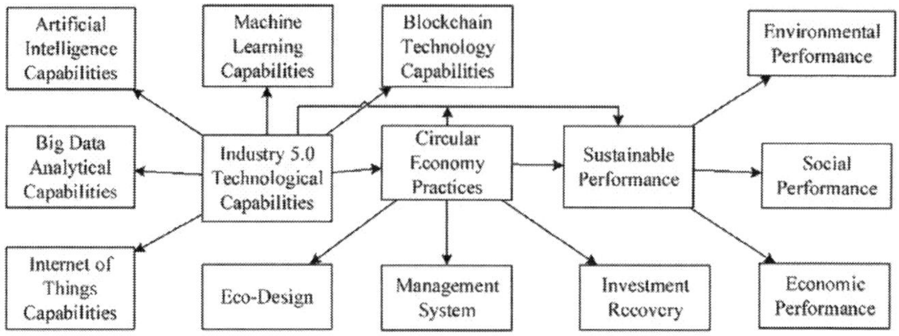

Figure 1.2 I5.0's conceptual framework for improving CE practices (Shafique et al., 2024).

Principles of CE, such as remanufacturing and recycling, play an important role in reducing dependence on natural resources and minimizing industrial waste (Rejeb et al., 2025). Green manufacturing, which prioritizes energy efficiency and renewable energy integration, is increasingly becoming a key focus in modern industry to reduce carbon footprint (Shafique et al., 2024). With this approach, companies can create a more sustainable production system and reduce negative impacts on the environment. I5.0's ecological innovation encompasses a range of sustainability-promoting technologies, including BC, AI, and digital twins (Costa, 2024). Businesses may increase production efficiency, cut waste, and maximize resource utilization, thanks to these technologies. Furthermore, the integration of renewable energy into the production process lowers carbon emissions and dependency on fossil fuels (Shafique et al., 2024). With a technology-driven approach, I5.0 can be a solution to the environmental challenges facing industries today. In addition to the environmental aspects, I5.0 also emphasizes the importance of technological ethics in industrial implementation. Social considerations such as data privacy, algorithmic bias, and worker welfare are crucial factors in ensuring that technology is used fairly and inclusively (Costa, 2024). With a values-based approach and ethics-sensitive design, I5.0 can foster a more responsible and sustainable application of technology.

According to Sinha (2022), since sustainable development aims to establish an economic model that decouples resource consumption, the creation of a CE is a positive step in that direction. Reusing leftover waste products from production and consumption as new input variables in the system helps cut the connection between resource use and natural resources. The production system's new input variables are production and consumption. The concept of sustainable development was first based on the notion that consumption and economic growth should be carefully planned to reduce adverse environmental effects. The Brundtland Commission Report essentially identifies three primary components of sustainable development, each of which is interconnected: economic, social, and environmental aspects. These three components are problems that have become global and urgent issues and threats to humanity around the world. Other issues that are part of the three components of sustainable development are poverty, AIDS, air pollution, biodiversity, water rights, birth control, and other aspects related to humanity and ecological footprint.

Circular economy is a model that links sustainable objectives with prudent levels of output and consumption (Kristianto & Nadapdap, 2021). According to the Ellen Macartur Foundation (2019), CE is a framework that may be used to develop solutions for global problems like pollution, waste, biodiversity loss, and climate change. The CE framework's design-led principles include reducing waste and pollution, conserving natural resources by prolonging the life cycle of goods, raw materials, and resources, and emphasizing regenerative and renewable natural resources. According to Beppenas (2022), the 9R framework encapsulates the CE's guiding principles, which center on lowering the consumption of materials and resources across the manufacturing chain. Making and using products more intelligently (refuse, rethink, and reduce), extending product life (reuse, repair, refurbish, remanufacture, and repurpose), and reaping additional benefits from existing materials (recycle and recover) are the three main components of the 9R framework, which is made up of ten CE principles ranked from 0 to 9 (Potting et al., 2017). The main principle of the CE concept is to minimize negative outputs, in the form of discarded waste from consumption and production (to be regenerated back into products with beneficial value). Utilizing a systems approach from product creation to consumption, minimizing resource consumption and waste, maintaining material use value, and being regenerative are the primary tenets of the CE model. In accordance with the Sustainable Development Goals, the adoption of a CE can enhance environmental resilience and community social welfare, lessen environmental harm, and boost the creation of new products with added value, and green economic growth (Lakshmi et al., 2020).

According to Subagyo et al. (2024), Society 5.0, through advanced technologies such as the IoT, AI, and smart grids, plays an important role in accelerating the growth of green economy. These technologies help improve resource efficiency, reduce carbon emissions, and optimize waste management. Research results show that the combination of IoT and AI in the agriculture and energy sectors can increase energy efficiency by 35% and reduce carbon emissions by 20%. This demonstrates how Society 5.0 can produce an environment that is more sustainable and support green economy goals in various sectors. The main challenges in implementing Society 5.0 for the green economy include high investment costs, infrastructure limitations, and resistance to change. However, there are great opportunities through government incentives, increased public awareness, and ongoing technological developments. This study shows that collaboration between government and industry, such as in the implementation of smart grids in Jakarta, can improve energy efficiency and strengthen the potential of Indonesia's green economy. Environmentally friendly technologies such as renewable energy are proving effective in supporting economic sustainability. Projects such as the Kupang solar power plant and Sidrap Wind Farm have reduced dependence on fossil fuels and significantly lowered carbon emissions. These findings demonstrate that Indonesia can use renewable energy to promote sustainable economic growth and lessen its negative effects on the environment if proper investments are made. Recent advancements that promote the fusion of sustainability and technology, particularly in bolstering the green economy, are consistent with this study. Society 5.0 has become a key strategy in addressing global environmental challenges, and Indonesia has a great opportunity to expand the adoption of these technologies in various sectors. Development prospects for future research could include a deeper exploration of the impact of Society 5.0 technologies on local communities, and case studies on specific sectors such as fisheries and forestry to strengthen the understanding of the effectiveness of technologies in the green economy.

Rapid change is occurring in I5.0, and there is a lot of talk and expectation about its necessity. The general consensus is that I5.0 has the capacity to solve enduring and worsening environmental and socioeconomic issues that endanger the welfare of coming generations. Given this perspective, the current study tackles a crucial question: how well has the I5.0 agenda been integrated into corporate structures, and has it produced positive results in terms of corporate sustainability and profitability? To answer this important question, the study used a systematic literature review (SLR) in conjunction with a concentric evaluation of relevant articles and an evidence-mapping exercise. The findings presented some intriguing and hotly contested facts on I5.0 that may have significant theoretical and practical ramifications.

Ghobakhloo et al.'s (2024), study provides insights into how the I5.0 agenda is developing and shows that it is still in its infancy. It is reasonable to assume that I5.0 as it exists today is still based on conjecture and theory. While this assertion may appear assertive, it is substantiated by certain factors. Notably, the very definition of I5.0 remains elusive, with diverse interpretations prevailing. It has been depicted as a framework for digital transformation, a regulatory initiative, or even an extensive industrial revolution. The truth is that there have been many different and perhaps conflicting viewpoints around I5.0. It was a pivotal moment when the European Commission stepped in and released policy agendas that brought people together and simplified the course of this phenomenon. Notably, though, the European Commission's involvement was akin to treading unknown seas, motivated by a recognition that the Industry 4.0 framework was insufficient for the future of European industries and that digital industrial change necessitated legislation. Although this regulatory initiative's general goal was evident, it was unclear how it would be carried out, who would be engaged, and how much involvement would take place.

The consequences of this directional and perceptual uncertainty are evident in the scholarly discourse shaped by the agendas of the European Commission. Many academics believe that I5.0 has the potential to improve many aspects of sustainability. Interestingly, they see I5.0's contributions to the social, environmental, and economic aspects of business performance as existing in a harmonious balance. However, there is clearly a lack of actual evidence supporting the adoption and execution of this agenda in business areas. As a result, it is still unknown how I5.0 will actually improve business profitability and sustainability performance. It is important to make clear that the aforementioned conclusion should not be interpreted as proof that I5.0 is ineffectual or unsuccessful. Rather, it represents the current stage of I5.0's development, which is distinguished by its conceptual advancement. Due mostly to the complexity of the I5.0 concept, this development is proceeding far more slowly than first projected. I5.0 encompasses more than just the application of new technologies. It is a comprehensive approach that calls for coordinated efforts to manage and supervise the full spectrum of technical progress, including development, integration, and application in the business and industrial sectors. Maintaining a longitudinal perspective that takes into account the possible long-term effects of technology on the environment and, more significantly, society is another requirement of I5.0.

Additionally, for I5.0 to succeed, several stakeholders, including businesses, technology companies, governments, social groups, labor unions, and regulators, must collaborate effectively. This partnership is essential to accomplishing I5.0's overarching objectives. Therefore, given the inherent complexity of coordinating collaborative efforts including societal goals, the slow speed at which I5.0 is developing is consistent with reasonable expectations. The difficulty is increased by the speed at which new digital technologies are developing. An example of this is the extraordinary emergence of generative AI, which is seriously upending socioeconomic systems. On the other hand, it appears that the I5.0 agenda and the concepts that go along with it in scholarly literature are lagging behind and failing to keep up with these revolutionary technological developments. In summary, this suggests that the quick rise of disruptive technologies or the more significant shifts in the market and business environment are outpacing the growth and maturation of I5.0. This may help explain why I5.0 seems to be in a theoretical and aspirational stage, with much rhetoric about attaining sustainable results but little real-world advancement. Despite the lack of hard data demonstrating I5.0's real-world benefits for the economy, society, and environment, we are adamant about making it a key regulatory framework. Major economies including the United States, Japan, and some European countries are making significant progress in regulating the creation, promotion, and broad use of technologies, especially AI. One plausible path for enabling such a regulatory shift is I5.0. The dedication of stakeholders to furthering the I5.0 goal is the crucial element. Their coordinated efforts ought to be accelerated in order to develop the framework and move it from a conceptual stage to real-world application.

The results of the SLR have indicated a number of interesting directions for further study, with the goal of filling in the current knowledge gaps on the contribution of I5.0 to sustainability goals. Of these, a longitudinal study of I5.0 adoption and its effects would be a particularly relevant and beneficial direction for future research. Future research could conduct a thorough longitudinal analysis to carefully track the development of I5.0 deployment across various industries. Such research projects would involve a long-term assessment of its effects on corporate sustainability and profitability. This may be accomplished by closely examining data from businesses that have fully adopted I5.0 approaches, which would enable a comparison of their sustainability and financial performance before and after implementing these cutting-edge techniques. It is crucial to emphasize that these case studies ought to go beyond simple technology incorporation. Rather, they ought to pay close attention to covering the strategic coordination of technology implementation that is purposefully in line with socio-environmental principles. This strategy ultimately adds to the larger conversation about the implications and possibilities of I5.0 by facilitating a deeper understanding of how to balance technological innovation, economic feasibility, and responsible ecological and societal stewardship.

Future studies could also look into the cooperative systems and alliances that should be formed among many stakeholders (such as governments, technology companies, corporations, and regulatory agencies) in order to support I5.0 deployment. An intriguing direction for further research would be to look at the difficulties, achievements, and lessons learned in creating and sustaining these kinds of alliances and how they affect business results. Furthermore, in the context of I5.0, it is imperative to compare the regulatory strategies used by various nations or regions to control the advancement and application of technologies like AI. It is critical to comprehend how these rules affect business sustainability and profitability.

It is necessary to reevaluate the widely held belief that implementing I5.0 in organizations exclusively has positive effects and no hazards. It stands to reason that each action involving strategic adjustments and technology investment includes a risk of negatively impacting a company's future. As a result, it is essential to carefully consider the potential risks involved in implementing the I5.0 framework. These include, among other things, ethical dilemmas, cybersecurity flaws, effects on staff, and interruptions to long-standing company procedures. Adopting I5.0 sensibly requires an understanding of these hazards. Therefore, in this situation, creating strict structures and strategies to address these risks becomes essential. Businesses can safeguard themselves against possible issues and strengthen their capacity to manage unforeseen difficulties by taking proactive measures to address these risks. It becomes crucial to take into account how these risk management techniques fit in with I5.0's primary objectives, which include boosting corporate profits and sustainability. This entails adopting a comprehensive strategy that blends advancements in technology with astute risk management techniques, thereby enhancing I5.0's ability to provide businesses with significant and long-lasting advantages. Notably, the academic literature, which offers a strong basis for comprehending the theoretical facets of I5.0 and its sustainability implications, was the primary focus of the current systematic study. Nonetheless, the gray literature and useful industry reports include a multitude of information and insightful observations.

This unexplored resource can provide a practical viewpoint on the state of I5.0 developments in the context of industry and policy. We suggest future study topics that make use of web crawling methods and AI-based tools in order to overcome this limitation and further our understanding. Scholars can explore the gray literature, which includes government publications, industry studies, and policy documents, by utilizing these cutting-edge methods. This strategy will make it possible to document the real-world applications and developments of I5.0, providing insight into how it is being embraced in actual situations. It will also aid in gaining an understanding of the difficulties and obstacles that businesses and legislators encounter when attempting to incorporate I5.0 practices.

A more thorough and comprehensive understanding of I5.0's current state, its effects on sustainability, and the issues that need to be resolved can be obtained by combining the results from the academic and gray literature. Our research will become more applicable and valuable for industry stakeholders, politicians, and researchers alike, thanks to this integrated methodology, which will help close the gap between theory and practice. The European Commission pilot study showed productivity gains of 18% and waste reduction by 12% in early I5.0 implementations. These references are now included in Section X, along with relevant citations to strengthen the credibility of our assertions.

References

Aminuddin, A. S. A., Nawawi, M. K. M., & Mohamed, N. M. Z. N. (2014, September). Analytic network process model for sustainable lean and green manufacturing performance indicator. In *AIP Conference Proceedings* (Vol. 1613, No. 1, pp. 32–38). American Institute of Physics.

Beppenas. (2022). *The Future Is Circular: Real Steps for Circular Economy Initiatives in Indonesia.* Jakarta: Cleanomic. www.undp.org/sites/g/files/zskgke326/files/2022-08/INS-The-Future-is-Circularfor%20upload.pdf

Bimantoro, A., Pramesti, W. A., Bakti, S. W., Samudra, M. A., & Amrozi, Y. (2021). The ethical paradox of utilizing information technology in the 5.0 era. *Jurnal Teknologi Informasi*, *7*(1), 58–68.

Costa, E. (2024). Industry 5.0 and SDG 9: A symbiotic dance towards sustainable transformation. *Sustainable Earth Reviews*, *7*(1), 4.

Ellen Macartur Foundation. (2019). *The Butterfly Diagram: Visualising the Circular Economy*. https://ellenmacarthurfoundation.org/circulareconomy-diagram

Ghobakhloo, M., Iranmanesh, M., Fathi, M., Rejeb, A., Foroughi, B., & Nikbin, D. (2024). Beyond Industry 4.0: A systematic review of Industry 5.0 technologies and implications for social, environmental and economic sustainability. *Asia-Pacific Journal of Business Administration*, Vol. ahead-of-print(No. ahead-of-print). https://doi.org/10.1108/APJBA-08-2023-0384

Goralski, M. A., & Tan, T. K. (2020). Artificial intelligence and sustainable development. *International Journal of Management Education*, *18*(1). https://doi.org/10.1016/j.ijme.2019.100330

Haleem, A., Javaid, M., Singh, R. P., Suman, R., & Qadri, M. A. (2023). A pervasive study on green manufacturing towards attaining sustainability. *Green Technologies and Sustainability*, *1*(2), 100018.

Hassan, A. S., & Jaaron, A. A. (2021). Total quality management for enhancing organizational performance: The mediating role of green manufacturing practices. *Journal of Cleaner Production*, *308*, 127366.

Javaid, M., Haleem, A., Singh, R. P., Suman, R., & Gonzalez, E. S. (2022). Understanding the adoption of Industry 4.0 technologies in improving environmental sustainability. *Sustainable Operations and Computers*, *3*, 203–217.

Kristianto, A. H., & Nadapdap, J. P. (2021). System dynamics community-based circular economy system. *Sebatik*, *25*(1), 59–67. https://doi.org/10.46984/sebatik.v25i1.1279

Lakshmi, V. V., Aruna, D. D., & Jhansi, R. K. P. (2020). *Wealth from Poultry Waste. Waste Management as Economic Industry towards Circular Economy*. Singapore: Springer.

Li, G., Lim, M. K., & Wang, Z. (2020). Stakeholders, green manufacturing, and practice performance: Empirical evidence from Chinese fashion businesses. *Annals of Operations Research*, *290*(1), 961–982.

Mafini, C., & Loury-Okoumba, W. V. (2018). Extending green supply chain management activities to manufacturing small and medium enterprises in a developing economy. *South African Journal of Economic and Management Sciences*, *21*(1), 1–12.

Mao, Y., & Wang, J. (2019). Is green manufacturing expensive? Empirical evidence from China. *International Journal of Production Research*, *57*(23), 7235–7247.

Milovanovic, S., Lukic, I., Kamiński, P., Dębczak, A., Klimkowska, K., Tyśkiewicz, K., & Konkol, M. (2022). Green manufacturing of high-value extracts from milk thistle seeds: Parameters that affect the supercritical CO2 extraction process. *Journal of CO2 Utilization*, *63*, 102134.

Nugroho, T. A. (2023). The development of Industry 5.0 for the Indonesian economy. *Manajemen Kreatif Jurnal (MAKREJU)*, *1*(3), 95–106.

Oktradiksa, A., Bhakti, C. P., Kurniawan, S. J., Rahman, F. A., & Ani. (2021). Utilization artificial intelligence to improve creativity skills in Society 5.0. *Journal of Physics: Conference Series*, *1760*(1). https://doi.org/10.1088/1742-6596/1760/1/012032

Park, J., Kim, S., & Lee, C. (2018). An analysis of pinned edge layer of slot-die coated film in roll-to-roll green manufacturing system. *International Journal of Precision Engineering and Manufacturing-Green Technology*, *5*(2), 247–254.

Potting, J., Hekkert, M., Worrell, E., & Hanemaaijer, A. (2017). *Circular Economy: Measuring Innovation in the Product Chain*. The Hague: PBL Netherlands Environmental AssessmentAgency. www.pbl.nl/en/publications/circulareconomy-measuring-innovation-in-productchains

Rejeb, A., Rejeb, K., Keogh, J. G., & Süle, E. (2025). When Industry 5.0 meets the circular economy: A systematic literature review. *Circular Economy and Sustainability*, 1–32.

Roschangar, F., Zhou, Y., Constable, D. J., Colberg, J., Dickson, D. P., Dunn, P. J., & Senanayake, C. H. (2018). Inspiring process innovation via an improved green manufacturing metric: iGAL. *Green Chemistry*, *20*(10), 2206–2211.

Sachsenmeier, P. (2016). Industry 5.0—the relevance and implications of bionics and synthetic biology. *Engineering*, *2*(2), 225–229. https://doi.org/10.1016/J.ENG.2016.02.015

Shafique, M. N., Adeel, U., & Rashid, A. (2024). The synergy between Industry 5.0 and circular economy for sustainable performance in the Chinese manufacturing industry. *Sustainability*, *16*(22), 9952.

Sinha, E. (2022). Circular economy—a way forward to sustainable development: Identifying conceptual overlaps and contingency factors at the microlevel. *Sustainable Development*, *30*(4), 771–783. https://doi.org/10.1002/sd

Subagyo, A. A., Perdana, I. J., Prayitno, E., & Iskandar, E. (2024). Green economy development in the era of Society 5.0: Technology and environment collaboration for the future. *Proceedings of the National Seminar on Science & Technology Applications (SNAST)*. http://dx.doi.org/10.34151/prosidingsnast.v1i1.4909

Taraya, P. C., & Wibawa, A. (2022). Mewujudkan Society 5.0 melalui pemanfaatan teknologi kecerdasan buatan. *Jurnal Inovasi Teknologi Dan Edukasi Teknik*, *2*(8), 378–385.

Yin, S., Zhang, N., & Li, B. (2020). Enhancing the competitiveness of multi-agent cooperation for green manufacturing in China: An empirical study of the measure of green technology innovation capabilities and their influencing factors. *Sustainable Production and Consumption*, *23*, 63–76.

Zhou, J., Zhou, Y., Wang, B., & Zang, J. (2019). Human–cyber–physical systems (HCPSs) in the context of new generation intelligent manufacturing. *Engineering*, *5*(4), 624–636. https://doi.org/10.1016/j.eng.2019.07.015

2 Industrial Revolutions

Revisiting Past, Present, and a Road Map to the Future and Its Impact on Healthcare, Agriculture, Education, and Society

Arnab Mitra

2.1 Introduction

The history of today's industrial automation process typically spans over two centuries, which specifies a variety from the early efforts for mechanization of textile mills to the modern era of smart assembly lines/smart production lines in smart factories, and several others. The different advancements concerning such industrial automation may be termed as Industry X, for

$X = 1.0, 2.0, 3.0, 4.0, 5.0$ and 6.0. It is challenging to mention all such characteristics for Industry X in a single literature. For this reason, the key efforts over time with reference to Industry X are briefly presented next.

- **Industry 1.0:** The early beginnings (in the 1800s) mainly focused on mechanization and pneumatic and hydraulic systems.
- **Industry 2.0:** The next revolution was due to the introduction of electrification; early automation, e.g., relay logic; and the development of feedback control of the industrial automation process (in the 1900s).
- **Industry 3.0:** The next stage of the industrial revolution, this was due to programmable logic controllers (PLCs) and thereafter digitized computerization (in the early 2000s).
- **Industry 4.0:** The next stage of the industrial revolution, this was due to the involvement of modern industrial automation, e.g., Industrial Internet of Things (IIoT), artificial intelligence and machine learning (AIML), and digital twins and simulation (in the 2010s).

In this regard, it might be noted that the term "Industry 4.0" (aka "Industrie 4.0" in German) was popularized in 2016 by K. Schwab, the founder and executive chairman of the World Economic Forum. The word originated in 2011 from a project in the high-tech strategy of the German government, which advanced the concept of cyber–physical systems (CPSs) into cyber–physical production systems (CPPSs) (Vogel-Heuser et al., 2012, 2017; Vogel-Heuser & Hess, 2016; Xu et al., 2021; Mitra, 2021).

- **Industry 5.0:** The next evolution over Industry 4.0, this was announced by M. Rada in 2015 (Rada, in "Industry 5.0 definition available at https://

DOI: 10.4324/9781003647621-3

michael-rada.medium.com/industry-5-0-definition-6a2f9922dc48") as an evolution led by humans involving the 6 R's (i.e., recognize, reconsider, realize, reuse, reduce, and recycle) and "Logistics Efficiency Design (LED) principles (i.e., transparency, profit sharing, efficiency) of industrial upcycling, a systematic waste prevention technique, and logistics efficiency design to evaluate life standards, innovative creations and produce high-quality custom products" (Rada, in "Industry 5.0 definition available at https://michael-rada.medium.com/industry-5-0-definition-6a2f9922dc48").

- **Industry 6.0:** The future is expected by 2050 and represents "the next phase of industrialization, and it's poised to transform entire industries through the integration of advanced technologies" (Bakla, 2024). Furthermore, it targets to produce fully integrated, intelligent manufacturing systems that can function with minimal human involvement, and are capable of combining human intelligence with several cutting-edge advanced technologies, e.g., artificial intelligence (AI), cloud computing, energy efficiency, and proper human–robot collaboration, in addition to being effective in handling big data issues and enabling quantum computing facility (Bakla, 2024).

In conclusion, Industry 6.0 is estimated to build upon these fundamentals, combining advancements in medical and bionic boosts, smart automation, and neo-critical industrial trends (Annanperä et al., 2021).

A diagrammatic representation of all Industrial revolutions with their key focus is presented in Figure 2.1.

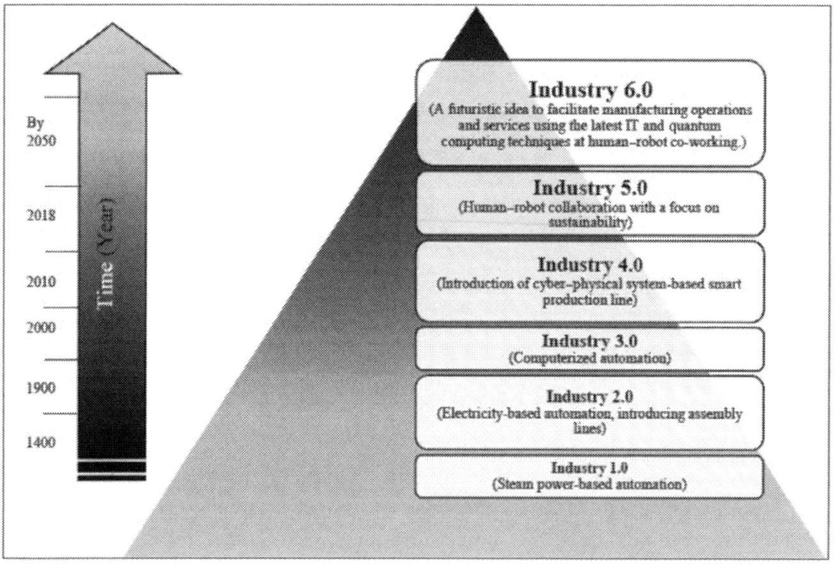

Figure 2.1 Evolution of industrial revolutions (partially inspired by Matuszak, 2024).

Figure 2.1 shows that the industrial revolutions resulted in major technological, socioeconomical, and cultural changes all over the world. It began in Britain in the late 1700s and is characterized by the conversion from hand production to machinery-based production techniques. The key aspects of the industrial revolutions included the mechanization of production: Industry 1.0 included the use of new power sources like steam, Industry 2.0 included electricity, Industry 3.0 included computerized automation and control, Industry 4.0 included CPSs, and thus we observed the rise of the smart and personalized factory systems in Industry 5.0 and Industry 6.0 (accessed from Encyclopedia Britannica, in "https://www.britannica.com/event/Industrial-Revolution", and "https://www.britannica.com/summary/Industrial-Revolution-Causes-and-Effects", accessed on July 04, 2025).

The major contributions of the presented chapter are as follows:

- It presents the key drivers, innovations, and impacts of each industrial revolution, highlighting their contributions (both positive and negative) to economic growth, societal development, and environmental sustainability.
- It presents a roadmap for the future of the Industrial Revolution, including the potential benefits and challenges.
- It presents valuable insights for business leaders, policymakers, and scholars seeking to investigate the complexities of industrial transformation and shape a sustainable future.

The rest of the chapter is organized as follows: a brief survey of related works is presented in Section 2.2; background works are briefly presented in Section 2.; the proposed work is presented in Section 2.4; results and associated discussions are presented in Section 2.3 and finally, the conclusions and a possible future research direction are presented in Section 2.4.

2.2 Related Works

Section 2.1 briefly discussed the important role of CPSs in the industrial revolutions. Next, a short discussion on CPSs and researchers' possible enhancement efforts follows.

As the evolution from Industry 1.0 to Industry 6.0 represents a significant upgradation in manufacturing, characterized by seamless human–machine collaboration enabling innovative and sustainable production methods, we sought to present the key contributions for each evolution level (Figure 2.1). It may be noted the term Industry 4.0 was first popularized in 2014, and thereafter earlier industrial revolution was categorized as Industry 1.0 to Industry 3.0, based on their characteristics (Figure 2.1). For this reason, we did not find literature focused specifically on Industry 1.0 to Industry 3.0.

It may further be noted that Industry 4.0 originated in 2011 from the Germany government's high-tech strategy project, which evolved the concept of CPSs into CPPSs (Vogel-Heuser et al., 2012, 2017; Vogel-Heuser & Hess, 2016; Xu et al., 2021; Mitra, 2021). For this reason, we began reviewing the literature with CPSs.

In different scientific databases and over the Internet, several research papers and reports might be found on the industrial revolutions, from Industry 4.0 to its transformation into the future Industry 6.0. For this reason, it is very hard to report each of them in a single effort. Hence, we carefully selected such references that help understand the industrial revolution processes starting from Industry 4.0 to Industry 6.0 along with their key driving features, and further present a conceptual synthesis to analyze their impact on several allied areas, e.g., healthcare, education, agriculture, and society.

Researchers discussed that the CPS typically integrates the physical and computational components to produce a new generation of engineering systems, which typically integrate several sensors, actuators, and communication networks to interact with the physical environment and respond to the changes in real time (Vogel-Heuser et al., 2012, 2017; Vogel-Heuser & Hess, 2016; Xu et al., 2021). Furthermore, Vogel-Heuser et al. (2012, 2017), Vogel-Heuser & Hess (2016), and Xu et al. (2021) showed that CPSs are beneficial with reference to the following points.

- Improved efficiency and productivity: CPSs are responsible for optimizing performance, enhancing decision-making processes, and reducing waste.
- Enhanced safety and security: CPSs are capable of detecting and responding to anomalies, and minimization of the risk of accidents and cyber threats.
- Increased autonomy and adaptability: CPSs are capable of being operated independently in dynamic scenarios/environments with improved responsiveness.

Although there are several benefits to implementing CPSs, researchers have pointed out a few drawbacks and challenges to implementing such CPSs (Vogel-Heuser et al., 2012, 2017; Vogel-Heuser & Hess, 2016; Xu et al., 2021). Among several others, the following three are particularly important.

1. Complexity and interoperability: Integrating the physical and computational components is often very complex, and further ensuring the interoperability between different systems is often challenging.
2. Security and privacy: CPSs are often found to be exposed to cyber threats. For this reason, enhancing the security and privacy of data in CPSs is vital.
3. Reliability and fault tolerance: CPSs must be designed to operate reliably and must be recoverable from faults or failures.

With the incorporation of CPSs, we observe researchers have presented several designs and possible enhancements in industry automation processes. As Numerous research papers on Industry 4.0 are available, and it is not possible to mention all of them in this study. Among several others, a detailed discussion on Industry 4.0 was presented in Lasi et al. (2014). In another attempt, M. Ghobakhloo concluded that Industry 4.0 and digitization together potentially offer significant opportunities for sustainability (Ghobakhloo, 2020). However, the author also highlighted the necessity for advanced research and investment to reap the benefits of Industry 4.0 and address potential limitations and challenges if any (Ghobakhloo, 2020). Furthermore, in Dalenogare et al. (2018), the authors concluded that Industry 4.0 technologies have the potential to significantly improve industrial performance, enabling companies to increase efficiency, productivity, and flexibility while reducing costs and improving quality. Frank et al. (2019) showed that Industry 4.0 technologies are being implemented in various ways in manufacturing companies, with a focus on improving efficiency and productivity; understanding such implementation patterns may help the companies develop effective strategies for adopting Industry 4.0 technologies. Vaidya et al. (2018) showed that Industry 4.0 has the potential to revolutionize manufacturing, enabling companies to improve efficiency, customer experience, and product quality. Besides the efforts from researchers on the scope, benefits, and challenges of Industry 4.0, we observe that several researchers have also focused on modeling different applications in Industry 4.0. Among several others, we particularly found interest in the MapReduce model.

Industry 4.0 scenario toward big data processing is presented in Mitra (2021, 2024), and enhanced data security and privacy involving blockchains in view of Industry 4.0 scenario is presented in Mitra (2022), as both models presented a cost-effective design with cellular automata (CAs), which have several advantages, e.g., capabilities for cheap physical implementation, inherent parallel computing compatibility, and support for very-large-scale integration (VLSI) (Mitra, 2021, 2022). Industry 4.0 primarily emphasizes on sustainability, human-centricity, and the integration of several advanced technologies. Several other sectors, e.g., healthcare, agriculture, education, and society, may draw advantages from adopting Industry 4.0 principles to enhance efficiency, personalization, and overall outcomes (Javaid et al., 2023; Rashid & Kausik, 2024; Samuels, 2025).

In Figure 2.2, some evolutions of other sectors with reference to Industry 4.0 are presented. The key features of Healthcare 4.0, Agriculture 4.0, Education 4.0, and Society 4.0 are briefly discussed as follows.

- Healthcare 4.0 denotes the fourth generation of healthcare, integrating progressive technologies such as AI, IoT, and robotics to provide personalized, predictive, and precise care.
- Agriculture 4.0 denotes the fourth generation of agricultural practices, which integrates latest technologies such as AI, IoT, drones, robotics, and precision agriculture to enhance crop production, reduce waste, and promote sustainable agricultural practices.
- Education 4.0 denotes the fourth generation of education, which integrates latest technologies such as AI, blockchain, IoT, and virtual and augmented reality (VR/AR) to produce a personalized, adaptive, and immersive learning experience.
- Society 4.0 refers to the fourth generation of societal development, characterized by the widespread adoption and integration of latest technologies, such as AI, blockchain, IoT, and robotics, into all aspects of life.

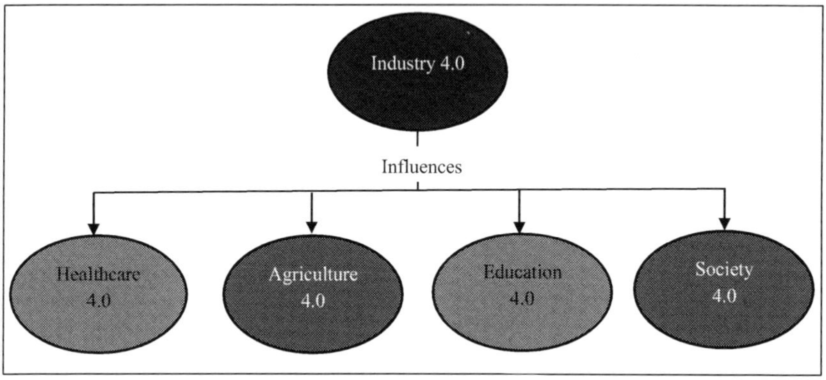

Figure 2.2 Industry 4.0-inspired development in other sectors.

On the other hand, we observe that during the implementation of Industry 4.0, in 2015, Michael Rada introduced the term Industry 5.0. According to M. Rada, Industry 5.0 "is future, but already penetrating trend, of change processes directing towards closer cooperation between man and machine, and systematic prevention of waste and wasting including industrial upcycling" (Rada, 2018, in "Industry 5.0 definition available at https://michael-rada.medium.com/industry-5-0-definition-6a2f9922dc48"), and its "priority is to utilize efficiently workforce of machines and people, in synergy environment. It goes back from virtuál environment to real one" (Rada, 2018, in "Industry 5.0 definition available at https://michael-rada.medium.com/industry-5-0-definition-6a2f9922dc48"). After the introduction of Industry 5.0, at present, a huge volume of literature targeting Industry 5.0 might be observed in different scientific databases. For this reason, we carefully selected a few important literature from such scientific databases. Among several others, a detailed survey on Industry 5.0 was presented in Maddikunta et al. (2022). The survey presented in Akundi et al. (2022) and Huang et al. (2022) concluded that Industry 5.0 has a true potential to change industries and societies, enabling proper human–machine collaboration, sustainability, and social responsibility. Like Industry 4.0, Industry 5.0 is also continuing to impact changes in other technology and utility sectors, e.g., Healthcare 5.0, Agriculture 5.0, and Society 5.0 (Akundi et al., 2022; Huang et al., 2022; Golovianko et al., 2023).

Figure 2.3 presents some evolvements of other sectors with reference to Industry 5.0. The key features for Healthcare 5.0, Agriculture 5.0, Education 5.0, and Society 5.0 are briefly discussed as follows.

- Healthcare 5.0 refers to the next generation of healthcare and integrates latest technologies such as artificial general intelligence (AGI), extended reality (XR), quantum computing, blockchain, and biotechnology to create a more personalized, predictive, and precise healthcare system.
- Agriculture 5.0 refers to the next generation of agricultural practices and integrates latest technologies such as AGI, XR, quantum computing, blockchain,

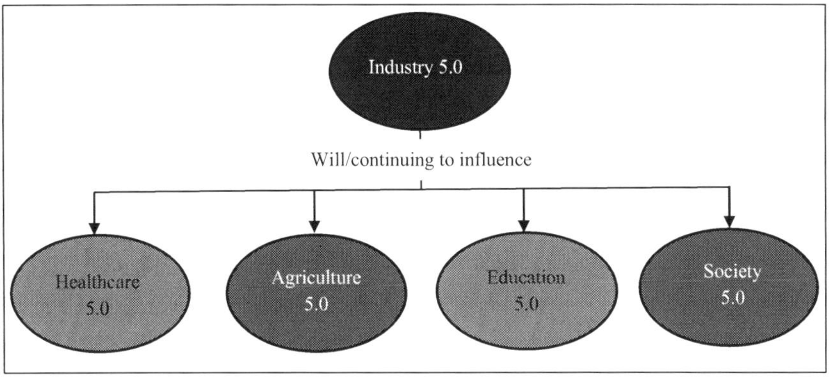

Figure 2.3 Industry 5.0-inspired development in other sectors.

and biotechnology to create a more sustainable, effective, and productive food system.

- Education 5.0 is a concept that refers to the next generation of education, which is expected to emerge in the mid-21st century. It is characterized by the integration of humans, machines, and the environment to create a personalized, adaptive, and immersive learning experience.
- Society 5.0 is a concept that refers to the next generation of societal development, which is expected to emerge in the mid-21st century. It is characterized by the integration of humans, machines, and the environment to create a more sustainable, equitable, and harmonious society.

Once again, the concept of Industry 6.0 was conceived during the rise of Industry 5.0 (Annanperä et al., 2021; Reddy et al., 2024). In Annanperä et al. (2021), Industry 6.0 was described as a paradigm shift toward antifragile manufacturing, which balances technological advancements with social and environmental responsibilities, ultimately creating a more sustainable and human-centered future. In Chourasia (2022), Industry 6.0 was described as a solution to the existing challenges and issues of Industry 5.0, e.g., integration difficulties, high implementation costs, skills gap, policy issues, and problems caused by social heterogeneity. In a different attempt in Duggal et al. (2022), Industry 6.0 was investigated and presented as a sequential roadmap that provides a structured approach for transitioning from the preceding version to Industry 6.0, enabling manufacturers to stay competitive and sustainable in the face of emerging technologies and trends. As the concept of Industry 6.0 is still not widely recognized or a globally established concept, researchers are still focusing on the refinement of the concept. One such recent effort is found in Singh et al. (2024), where the concept of Industry 6.0 has been reinvestigated, and in conclusion, it was discussed in Singh et al. (2024) that Industry 6.0 has the full potential to revolutionize modern society and manufacturing industries by providing a sustainable, efficient, and customer-centric pathway for the future. It is expected that Industry 6.0 will also continue to impact changes in other technology and utility sectors and may transform new models, e.g., Healthcare 6.0, Agriculture 6.0, and Society 6.0.

2.3 Discussions

Industrial evolutions for Industry X, for $X = 1.0, 2.0, 3.0, 4.0, 5.0$, and 6.0, represent a significant upgradation in manufacturing, characterized by seamless human–machine collaboration enabling innovative and sustainable production methods. This transformation is driven by intelligent automation with the help of AI and IoT, enabling efficient and adaptive production processes. The journey to the future of Industry 6.0 may be marked by the following points.

- Advancements in human–machine interaction: Evolving trends in collaboration between humans and machines.
- Advancements in materials and manufacturing: Materials influence manufacturing processes and product development.
- Advancements in automation: AI and IoT facilitate efficient and adaptive production processes.

The key characteristics of each version of Industry X, for $X = 1.0, 2.0, 3.0, 4.0, 5.0$ and 6.0, are briefly presented in Table 2.1, which mostly summarizes information from Hoque et al. (2018), Körner et al. (2019), Pilevari (2020), Pérez-Hernández et al. (2021), and Thaariq et al. (2024).

In Table 2.1, the key advantages and disadvantages with reference to technology involvement, economic benefits, societal effects, and environmental effects for several levels of Industry X, for $X = 1.0, 2.0, 3.0, 4.0, 5.0$, and 6.0, are presented. A diagrammatic representation of the successful evolution to Industry 6.0 is presented in Figure 2.4.

In conclusion, it might be noted that as Industry 6.0 is a hypothetical future in view of the future industrial revolution that is expected to be driven by emerging technologies, researchers are still trying to establish a solid foundation and regulation. For this reason, more investigations are required to fully uncover its true advantages and/or disadvantages.

It is expected that Industry 6.0 will also continue to impact changes in other technology and utility sectors like its predecessor versions (e.g., Industry 4.0 and Industry 5.0) and will bring out new models, e.g., Healthcare 6.0, Agriculture 6.0, and Society 6.0. A predicted future for such development is presented in Figure 2.5.

Figure 2.5 presents the role of Industry 6.0 in different sectors, a futuristic concept. It envisions a highly interconnected and intelligent manufacturing landscape, driven by several advancements (Figure 2.4) and human–machine collaboration, with a strong focus on sustainability and personalized production. It builds upon the foundations of Industry 4.0 and Industry 5.0 by emphasizing human-centric design and a more holistic approach to value creation, considering not only economic but also social and environmental well-being (Chourasia et al., 2022).

Table 2.1 Comparison of the Key Characteristics of Industry X, for $X = 1.0, 2.0, 3.0, 4.0, 5.0$, and 6.0.

Version	*Attribute*	Type	*Remarks*
Industry 1.0	Advantages	Technical	Focused on mechanization and standardization, and thus resulted in the division of labor and increased productivity
		Economical	Facilitated new job creation for skilled workers, increased economic growth, and increased trade
		Social	Caused new urbanization, improved transportation, and increased access to goods and services
		Environmental	Caused improved resource utilization
	Disadvantages	Technical	Lacks limited scalability and has limited automation facility, reliability issues, and safety concerns
		Economical	Caused environmental degradation, unequal distribution of wealth, and limited economic opportunities
		Social	Resulted in poor working environment and limited social mobility, and thus in a disruption of traditional ways of life
		Environmental	Resulted in air pollution, water pollution, deforestation, soil degradation, etc.
Industry 2.0	Advantages	Technical	Facilitated electrification and thus resulted in mass production, standardization, and improved transportation
		Economical	Facilitated increased productivity and thus resulted in new job creation for skilled people and a higher economic growth
		Social	Facilitated urbanization and increased mobility
		Environmental	Improved resource utilization and thus resulted in reduced volume of waste
	Disadvantages	Technical	Lacked flexibility as it was dependent on electricity; further high maintenance costs and safety concerns were also applicable
		Economical	Somehow suffered from unequal distribution of wealth and limited economic opportunities
		Social	Caused urban poverty, limited social mobility, and disruption of traditional ways of life
		Environmental	Resulted in air pollution, water pollution, deforestation, soil degradation, etc.
Industry 3.0	Advantages	Technical	Focused on automation and robotics, computerization and information technology (IT), increased use of electronics and sensors, and thus resulted in improved manufacturing processes

(*Continued*)

Table 2.1 (Continued)

Version	*Attribute*	Type	*Remarks*
		Economical	Achieved increased productivity and efficiency, improved product quality, along with increased customization and flexibility
		Social	Resulted in new job opportunities for skilled workers and increased mobility and flexibility
		Environmental	Achieved reduced amount of energy consumption and further achieved reduction in waste and emission; it also achieved improved resource utilization
	Disadvantages	Technical	Suffered from the existing dependence on technology. In addition, it also incorporates a high maintenance cost, limited flexibility, etc.
		Economical	Due to the high involvement of automation, job displacement increased. Furthermore, it suffered from the dependence on global supply chains
		Social	Due to the high participation of automation and technology, job security decreased. Thus, it resulted in an increased level of stress
		Environmental	Due to the high participation of automation and technology, it resulted in increased energy consumption by the system, thus, resulting in resource depletion, increased volume of electronic waste, and also climate change issues
Industry 4.0	Advantages	Technical	Achieved increased efficiency, improved accuracy, enhanced flexibility, and predictive maintenance. More might be achieved as researchers are still exploring
		Economical	Achieved cost efficiency, new business models, and improved supply chain management. More might be achieved as researchers are still exploring
		Social	Achieved improved working condition and product quality, and thus resulted in increased customer satisfaction. Furthermore, it resulted in new job opportunities for skilled workers. More might be achieved as researchers are still exploring
		Environmental	Achieved reduced energy consumption and reduced volume of waste, while it achieved improved resource allocation and somehow sustainable production. More might be achieved as researchers are still exploring

(Continued)

Table 2.1 (*Continued*)

Version	*Attribute*	Type	*Remarks*
	Disadvantages	Technical	Due to high dependency on the Internet, it invited cybersecurity risks. Besides, it suffers from dependence on technology, big data issues, and interoperability issues. More might be achieved as researchers are still exploring
		Economical	Involves high initial investment cost. Complexity is very high. More might be achieved as researchers are still exploring
		Social	Due to the high participation of automation and technology, job security decreased. Thus, it resulted in an increased level of stress. More might be achieved as researchers are still exploring
		Environmental	Due to the high participation of automation and technology, it resulted in an increased energy consumption by the system. Thus, it invited resource depletion and increased volume of electronic waste. More might be achieved as researchers are still exploring
Industry 5.0	Advantages	Technical	Achieves increased efficiency, improved accuracy, enhanced flexibility, and predictive maintenance along with proper human–machine collaboration. More might be achieved as researchers are still exploring
		Economical	Achieved cost efficiency, new business models, and improved supply chain management. More might be achieved as researchers are still exploring
		Social	Achieved improved working condition and product quality, and thus resulted in increased customer satisfaction. Furthermore, it resulted in new job opportunities for skilled workers. More might be achieved as researchers are still exploring
		Environmental	Achieved reduced energy consumption and reduced volume of waste (targeting a net zero waste), while it achieved improved resource allocation and sustainable production. More might be achieved as researchers are still exploring
	Disadvantages	Technical	Due to high dependency on the Internet, it invited cybersecurity risks. Besides, it suffers from dependence on cutting-edge technologies, and interoperability issues. More might be achieved as researchers are still exploring

(*Continued*)

Table 2.1 (*Continued*)

Version	*Attribute*	Type	*Remarks*
		Economical	Involves high initial investment cost. Complexity is very high. Furthermore, dependence on human–machine collaboration may result in uncertainty in return on investment. More might be achieved as researchers are still exploring
		Social	Due to the high participation of automation and technology, job security might be decreased. Besides, there exists a dependence on human–machine collaboration. More might be achieved as researchers are still exploring
		Environmental	Due to the high participation of automation and technology, it invited an increased energy consumption by the system. Thus, it resulted in resource depletion and increased volume of electronic waste. More might be achieved as researchers are still exploring
Industry 6.0	Advantages	Technical	Aims to achieve increased efficiency, improved accuracy, enhanced flexibility, and predictive maintenance along with proper human–machine collaboration through the involvement of autonomous systems, advanced AI and machine learning (ML), quantum computing, and extended reality. More might be achieved as researchers are still exploring
		Economical	Aims to achieve increased productivity, new business models, improved supply chain management, and increased revenue streams. More might be achieved as researchers are still exploring
		Social	Aims to achieve improved working conditions, new job opportunities, increased customer satisfaction, and improved product quality. More might be achieved as researchers are still exploring
		Environmental	Aims to achieve net zero waste, while it will achieve improved resource allocation and sustainable production. More might be achieved as researchers are still exploring
	Disadvantages	Technical	Complexity is expected to be high. Besides, there will be a high dependence on emerging and cutting-edge technologies, high cybersecurity risks, and big data management issues. More might be achieved as researchers are still exploring

(*Continued*)

Table 2.1 (*Continued*)

Version	*Attribute*	Type	*Remarks*
		Economical	It is expected that it will involve high initial investment cost. Furthermore, dependence on human–machine collaboration may result in uncertainty in return on investment and different regulatory challenges. More might be achieved as researchers are still exploring
		Social	It is expected that due to the high participation of automation and technology, job security might be decreased. Besides, there may exist dependence on human–machine collaboration. More might be achieved as researchers are still exploring
		Environmental	High participation of automation and technology has led to increased energy consumption by the system. Thus, it resulted in resource depletion and increased volume of electronic waste. Furthermore, the widespread adoption of Industry 6.0 technologies may have unintended environmental consequences that are not yet fully understood. More might be achieved as researchers are still exploring

The key features of Healthcare 6.0, Agriculture 6.0, Education 6.0, and Society 6.0 are briefly discussed as follows.

- Healthcare 6.0 is a hypothetical concept that represents the next generation of healthcare, beyond Healthcare 5.0. It is expected to be characterized by the integration of emerging technologies such as AGI and superintelligence, quantum computing and quantum AI, XR and brain–computer interfaces (BCIs), nanotechnology and synthetic biology, blockchain, and decentralized AI.

  It may be noted that this is a hypothetical timeline, and the actual development of Healthcare 6.0 may differ.
- Agriculture 6.0 is still a hypothetical concept, and its development and implementation will depend on the advancement of emerging technologies and the socioeconomical factors that shape the agriculture industry.

  It may be noted that this is a hypothetical timeline, and the actual development of Agriculture 6.0 may differ.
- Education 6.0 is still a hypothetical concept, and its development and implementation will depend on the advancement of emerging technologies and the socioeconomical factors that shape the education industry.

  It may be noted that this is a hypothetical timeline, and the actual development of Education 6.0 may differ.

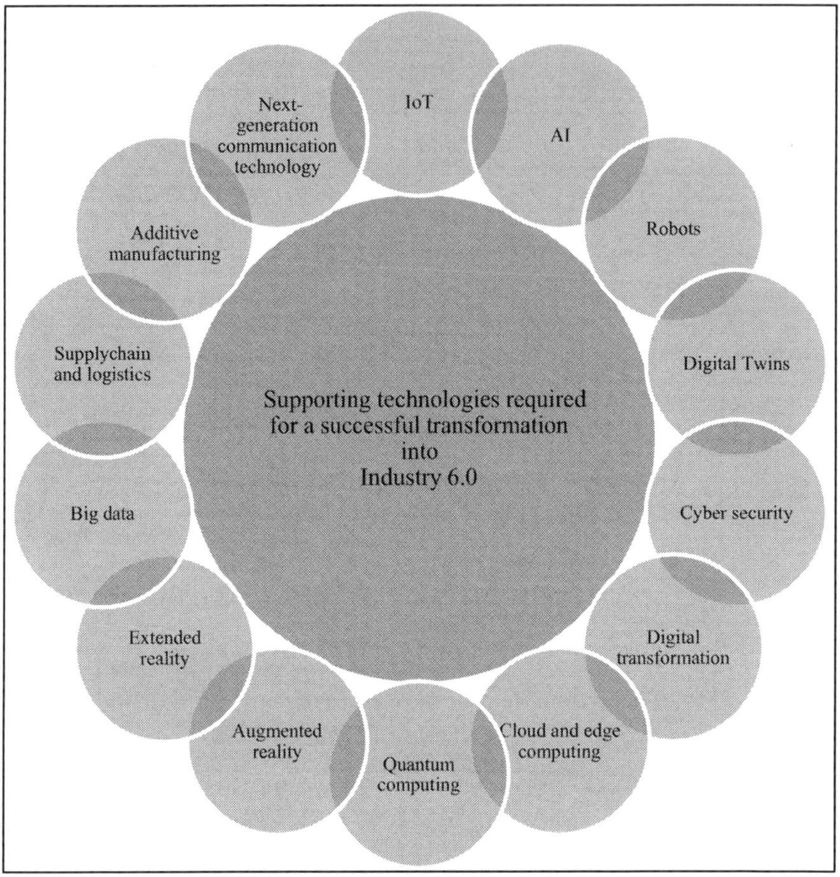

Figure 2.4 Diagram representing the components for the successful evolution to Industry 6.0.

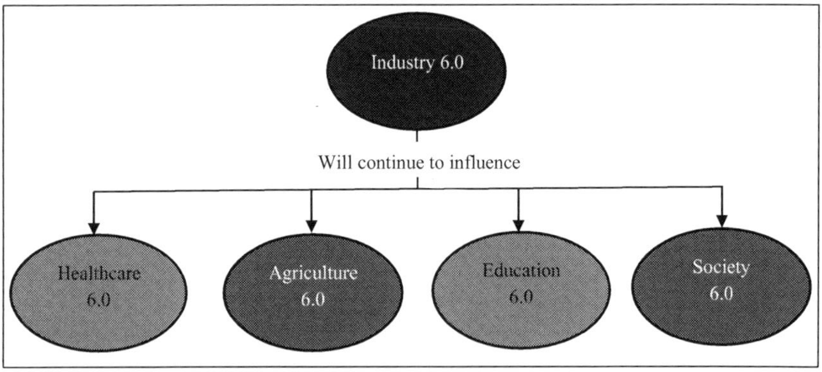

Figure 2.5 Industry 6.0-inspired development in other sectors.

- Society 6.0 is expected to be a post-scarcity society, where technology provides for all basic needs, and people are free to pursue their passions and interests.
  It may be noted that this is a hypothetical timeline, and the actual development of Society 6.0 may differ.

Although these presented concepts, i.e., Healthcare 6.0, Agriculture 6.0, Education 6.0, and Society 6.0, currently lack universally accepted definitions, for said reasons, a formal definition as studied from the state-of-the-art literature is presented next.

- Healthcare 6.0 denotes a vision of the future of personalized patient-centered healthcare, characterized by its emphasis on a data-driven nature enabled by several advanced computations and technologies to enhance efficiency, improve patient experience, and make healthcare more equitable and accessible (Ray & Bhattacharyya, 2025; Prabhakaran, 2025).
- Agriculture 6.0 denotes precision farming, integrating Internet of Things (IoT) technologies, big data analytics, and advanced automation, along with the use of quantum computing to optimize resource management, enhance efficiency, and promote sustainable practices toward a future in agriculture driven by sustainable technology and data-driven insights (Neves et al., 2023; Patil et al., 2024).
- Education 6.0 denotes personalized learning experiences in a rapidly changing, technology-driven world that focuses on developing social and emotional skills alongside technical skills for the future that emphasizes the integration of several emerging technologies (Moleka, 2023).
- Society 6.0 is also known as the Well-being Society 6.0 or the Sustainable Socially Responsible Society that imagines a future where real-world and digital solutions balance socioeconomical growth and environmental well-being (Rachmad, 2021; Žižek et al., 2021).

2.4 Conclusions and Future Scope

A brief discussion on the journey of Industrial Revolutions from Industry 1.0 to its future, i.e., Industry 6.0, along with its major characteristics at each level, is presented here. It might be observed that the journey from Industry 1.0 to Industry 6.0 demonstrates the importance of embracing change and innovation to drive progress and prosperity, and balancing the technological progress with environmental and social considerations for the design and development of a sustainable and equitable future. From the current study, we present the following key points toward a successful evolution into Industry 6.0.

- Transformational journey: The journey from Industry 1.0 to Industry 6.0 has been transformational, marked by significant technological, economic, social, and environmental changes.
- Increasing complexity: Each industrial revolution has introduced new technologies, processes, and systems, increasing complexity and interconnectivity.

- Improving efficiency and productivity: Each industrial revolution has focused on improving efficiency and productivity, driving economic growth and development.
- Shifting workforce skills: The journey may require the workers to adjust to new technologies and processes; thus, shifting the skills is required for employment.
- Growing environmental concerns: The increasing use of resources, energy consumption, and waste management and disposal have raised environmental concerns, emphasizing the need for sustainable practices.

In the future, we believe that there is scope for further investigations for the successful deployment of Industry 6.0. Among several others, the following areas need to be considered on a priority basis.

- Industry 6.0 and beyond: Emerging technologies like quantum computing, AI, and biotechnology will shape the future of industries.
- Sustainability and environmental considerations: Future industrial developments must prioritize sustainability, environmental stewardship, and social responsibility.

  The key aspects and frameworks for obtaining such sustainability and environmental considerations as studied from van Eldik et al. (2020), Chourasiya et al. (2024), Yadav et al. (2024), and Roshid et al. (2025) are briefly presented next.

  - **Sustainability integration:** This might be achieved through reducing waste, optimizing resource utilization, and minimizing energy consumption by leveraging hyperconnectivity, data analytics, and automation for efficient production and resource management.
  - **Human-centric automation:** This might be achieved by the integration of human expertise with advanced automation, creating a proper collaborative environment between humans and advanced machinery.
  - **Environmental impact assessment (EIA):** This might be achieved by assessing and mitigating the environmental impacts throughout the complete lifecycle of any product, from design to discarding.
  - **Circular economy principles:** These might be achieved by the acceptance of circular economy principles, aiming to optimize resource utilization, minimize waste generation, and promote recycling and upcycling, resulting in zero waste.
  - **Building information modeling (BIM)-based EIA:** This might be achieved by integrating BIM with EIA for infrastructure projects, enabling automated environmental impact assessments during the design phase, and facilitating complete data-driven decision-making.

- Human–machine collaboration: The future work will encompass better collaboration between humans and machines and will need workers to upgrade their skills and to adapt to the changing job responsibilities.
- Global cooperation and knowledge sharing: Future industries will require global cooperation, knowledge sharing, and collaboration to address common challenges and opportunities.

Declaration of Competing Interest

The author announces that he has no known or unknown opposing financial activities that might sway the work reported in this chapter. Hence, the author declares no conflicts.

Declaration of generative AI and AI-assisted technologies in the writing process

During the preparation of the chapter manuscript, MetaAI (Llama 3.2) was used to enhance the readability of the content. After using this tool/service, the author reviewed and edited the content as needed and takes full responsibility for the content of the publication.

References

Akundi, A., Euresti, D., Luna, S., Ankobiah, W., Lopes, A., & Edinbarough, I. (2022). State of Industry 5.0—analysis and identification of current research trends. *Applied System Innovation*, *5*(1), 27.

Annanperä, E., Jurmu, M., Kaivo-oja, J., Kettunen, P., Knudsen, M., Lauraéus, T., Jukka, M., & Porras, J. (2021). *From Industry x to Industry 6.0: Antifragile Manufacturing for People, Planet, and Profit with Passion*. https://cris.vtt.fi/en/publications/from-industry-x-to-industry-60-antifragile-manufacturing-for-peop, accessed on December 20, 2024.

Bakla, M. (2024). *The Future: Industry 6.0*. https://medium.com/@mustafabakla/the-future-industry-6-0-4ed653ec897f, accessed on December 20, 2024.

Chourasia, S., Tyagi, A., Pandey, S. M., Walia, R. S., & Murtaza, Q. (2022). Sustainability of Industry 6.0 in global perspective: Benefits and challenges, *Journal of Metrology Society of India*, *37*(2), 443–452.

Chourasiya, R., Pandey, S., Malviya, R. K., & Pujara, A. A. (2024). Towards sustainable success: A framework for assessing performance of sustainable manufacturing adoption in Indian textile industry. *Sustainable Futures*, *7*, 100216.

Dalenogare, L. S., Benitez, G. B., Ayala, N. F., & Frank, A. G. (2018). The expected contribution of Industry 4.0 technologies for industrial performance. *International Journal of Production Economics*, *204*, 383–394.

Duggal, A. S., Malik, P. K., Gehlot, A., Singh, R., Gaba, G. S., Masud, M., & Al-Amri, J. F. (2022). A sequential roadmap to Industry 6.0: Exploring future manufacturing trends. *IET Communications*, *16*(5), 521–531.

The Editors of Encyclopaedia Britannica. (2025, June 11). Industrial revolution. *Encyclopedia Britannica*. https://www.britannica.com/event/Industrial-Revolution, accessed on July 4, 2025.

Frank, A. G., Dalenogare, L. S., & Ayala, N. F. (2019). Industry 4.0 technologies: Implementation patterns in manufacturing companies. *International Journal of Production Economics*, *210*, 15–26.

Ghobakhloo, M. (2020). Industry 4.0, digitization, and opportunities for sustainability. *Journal of Cleaner Production*, *252*, 119869.

Golovianko, M., Terziyan, V., Branytskyi, V., & Malyk, D. (2023). Industry 4.0 vs. Industry 5.0: Co-existence, transition, or a hybrid. *Procedia Computer Science*, *217*, 102–113.

Hoque, A., Mohiuddin, M., & Su, Z. (2018). Effects of industrial operations on socio-environmental and public health degradation: Evidence from a least developing country (LDC). *Sustainability*, *10*(11), 3948.

Huang, S., Wang, B., Li, X., Zheng, P., Mourtzis, D., & Wang, L. (2022). Industry 5.0 and Society 5.0—comparison, complementation and co-evolution. *Journal of Manufacturing Systems*, *64*, 424–428.

Javaid, M., Haleem, A., Khan, I. H., & Suman, R. (2023). Understanding the potential applications of artificial intelligence in agriculture sector. *Advanced Agrochem*, *2*(1), 15–30.

Körner, U., Müller-Thur, K., Lunau, T., Dragano, N., Angerer, P., & Buchner, A. (2019). Perceived stress in human–machine interaction in modern manufacturing environments—results of a qualitative interview study. *Stress and Health*, *35*(2), 187–199.

Lasi, H., Fettke, P., Kemper, H. G., Feld, T., & Hoffmann, M. (2014). Industry 4.0. *Business & Information Systems Engineering*, *6*, 239–242.

Maddikunta, P. K. R., Pham, Q. V., Prabadevi, B., Deepa, N., Dev, K., Gadekallu, T. R., . . . Liyanage, M. (2022). Industry 5.0: A survey on enabling technologies and potential applications. *Journal of Industrial Information Integration*, *26*, 100257.

Matuszak, J. (2024). *Is Your Business Ready for Industry 5.0?* https://knowhow.distrelec.com/manufacturing/is-your-business-ready-for-industry-5-0/, accessed on December 20, 2024.

Mitra, A. (2021). On the capabilities of cellular automata-based MapReduce model in Industry 4.0. *Journal of Industrial Information Integration*, *21*, 100195.

Mitra, A. (2022). How can we enhance reputation in blockchain consensus for Industry 4.0–a proposed approach by extending the PageRank algorithm. *International Journal of Information Management Data Insights*, *2*(2), 100138.

Mitra, A. (2024). Cellular automata-based MapReduce design: Migrating a big data processing model from Industry 4.0 to Industry 5.0. *e-Prime-Advances in Electrical Engineering, Electronics and Energy*, *8*, 100603.

Moleka, P. (2023). Exploring the role of artificial intelligence in Education 6.0: Enhancing personalized learning and adaptive pedagogy. https://www.preprints.org/frontend/manuscript/160be85f5d69a44df018763e6f746c04/download_pub, accessed on December 20, 2024.

Neves, M. F., Casagrande, B. P., Cambaúva, V., de Oliveira Teixeira, G., & Toledo, P. J. F. (2023). Agriculture 6.0: A new proposal for the future of agribusiness. *Revista de Gestão Social e Ambiental*, *17*(9), 1–16.

Patil, Y. H., Patil, R. Y., & Ozer, O. (2024). Next-gen Agriculture 6.0: A smart solution pioneering the future of farming. In *Evolution and Advances in Computing Technologies for Industry 6.0* (pp. 167–187). CRC Press.

Pérez-Hernández, H., Pérez-Moreno, A., Sarabia-Castillo, C. R., García-Mayagoitia, S., Medina-Pérez, G., López-Valdez, F., . . . Fernández-Luqueño, F. (2021). Ecological drawbacks of nanomaterials produced on an industrial scale: Collateral effect on human and environmental health. *Water, Air, & Soil Pollution*, *232*, 1–33.

Pilevari, N. (2020). Industry revolutions development from Industry 1.0 to Industry 5.0 in manufacturing. *Journal of Industrial Strategic Management*, *5*(2), 44.

Prabhakaran, P. (2025). Healthcare 6.0 promising the era of smart and intelligent hospitals: A paradigm shift. In *Exploration of Transformative Technologies in Healthcare 6.0* (pp. 39–70). IGI Global Scientific Publishing.

Rachmad, Y. E. (2021). *Life in the New Normal: The Role of Society 6.0 Management in Leading Global Recovery and Sustainability Post-COVID-19*. The Academic Professor Board Protocol 2021, United Nations Global Compact 137635 UIPM.

Rada, M. (2018). *Industry 5.0 Definition*. Retrieved December 20, 2024, from https://michael-rada.medium.com/industry-5-0-definition-6a2f9922dc48

Rashid, A. B., & Kausik, A. K. (2024). AI revolutionizing industries worldwide: A comprehensive overview of its diverse applications. *Hybrid Advances*, 100277.

Ray, S., & Bhattacharyya, S. (2025). Roadmap for healthcare transformation: Leveraging Industry 6.0. In *Exploration of Transformative Technologies in Healthcare 6.0* (pp. 1–38). IGI Global Scientific Publishing.

Reddy, C. K. K., Doss, S., Khan, S. B., & Alqhatani, A. (Eds.). (2024). *Evolution and Advances in Computing Technologies for Industry 6.0: Technology, Practices, and Challenges*. CRC Press.

Roshid, M. M., Waaje, A., & Karim, R. (2025). Industry 6.0 as an emerging field of research: A systematic and bibliometric analysis. In *EMIDWORLD 2nd International Congress on Economics Public Finance Business & Social Sciences* (pp. 784–802).

EMIDWORLD. https://www.researchgate.net/publication/387873513_Industry_60_as_an_Emerging_Field_of_Research_A_Systematic_and_Bibliometric_Analysis/link/68559d6e93040b17338c4c80/download?_tp=eyJjb250ZXh0Ijp7ImZpcnN0UGFnZSI6InB1YmxpY2F0aW9uIiwicGFnZSI6InB1YmxpY2F0aW9uIn19, accessed on December 20, 2024.

Samuels, A. (2025). Examining the integration of artificial intelligence in supply chain management from Industry 4.0 to 6.0: A systematic literature review. *Frontiers in Artificial Intelligence*, *7*, 1477044.

Singh, R., Tyagi, A. K., & Arumugam, S. K. (2024). Imagining the sustainable future with Industry 6.0: A smarter pathway for modern society and manufacturing industries. In *Machine Learning Algorithms Using Scikit and TensorFlow Environments* (pp. 318–331). IGI Global.

Thaariq, Z. Z. A., Utami, W. B., Haryono, M. B., & Vaz, B. L. (2024). Educational technology in Industry 6.0. *Wacana Akademika: Majalah Ilmiah Kependidikan*, *8*(1), 134–141.

Vaidya, S., Ambad, P., & Bhosle, S. (2018). Industry 4.0–a glimpse. *Procedia Manufacturing*, *20*, 233–238.

van Eldik, M. A., Vahdatikhaki, F., dos Santos, J. M. O., Visser, M., & Doree, A. (2020). BIM-based environmental impact assessment for infrastructure design projects. *Automation in Construction*, *120*, 103379.

Vogel-Heuser, B., Bauernhansl, T., & Ten Hompel, M. (2017). Handbuch industrie 4.0 Bd. 4. *Allgemeine Grundlagen*, *2*.

Vogel-Heuser, B., Bayrak, G., & Frank, U. (2012). *Forschungsfragen in "Produktionsautomatisierung der Zukunft": Diskussionspapier für die acatech Projektgruppe "ProCPS-Production CPS";[im Rahmen der Erarbeitung der Forschungsagenda Cyber-Physical Systems (AgendaCPS)]*. acatech, Dt. Akad. der Technikwiss.

Vogel-Heuser, B., & Hess, D. (2016). Guest editorial Industry 4.0—prerequisites and visions. *IEEE Transactions on Automation Science and Engineering*, *13*(2), 411–413.

Xu, X., Lu, Y., Vogel-Heuser, B., & Wang, L. (2021). Industry 4.0 and Industry 5.0—inception, conception and perception. *Journal of Manufacturing Systems*, *61*, 530–535.

Yadav, S., Rab, S., & Wan, M. (2024). Metrology and sustainability in Industry 6.0: Navigating a new paradigm. In *Handbook of Quality System, Accreditation and Conformity Assessment* (pp. 855–885). Singapore: Springer Nature.

Žižek, S. Š., Mulej, M., & Potočnik, A. (2021). The sustainable socially responsible society: Well-being society 6.0. *Sustainability*, *13*(16), 9186.

3 Industry 5.0 and the New Society

A Scoping Review

Amrita Nandy and Md. Sahidur Rahman

3.1 Introduction

Historians often debate on the precise commencement of the era of industrial revolutions (IRs) that sublimely pronounced multihued economic and social changes across divergent terrains of the world, yet they accede on earlier four major shifts: IR1.0, IR2.0, IR3.0, and IR4.0. All these ancient origins have incorporated outstanding technologies and acknowledged the intra- and interdependence of diverse ecological agents (Knowles et al., 2024). Along with radical operational improvements, each version of the industrial wave has steered major societal progression in terms of workplace communication, consumerism, and ethical drills, all of which pave the way for imminent industrial evolutions (Islam et al., 2025). Particularly, the repercussion of IR4.0 or I4.0 has extended toward environmental responsibility and sustainability prioritizing optimum resource utilization with minimal waste at each stage of production (Fuertes et al., 2024), which is being welcomed globally as the philosophy of Industry 5.0 or I5.0 (Neumann et al., 2021). Some schools of doctrines view I5.0 as a successive upgradation of I4.0, whereas others acknowledge it as an equivalent progress after the Industry 4.0 paradigm (Bakator et al., 2024). Scientists repeatedly argue that the adoption of human-centric notion, upscaling human capabilities, and manifesting symbiotic relationships between humans and machines are the core deviation of I5.0 from all prior industrial revolutions (Gamberini & Pluchino, 2024; Grabowska et al., 2022).

Once again, being a refreshing notion, the submerged influence of I5.0 gradually strengthens sustainability concerns (like green manufacturing, recycling, reusing, and reduced carbon footprints) worldwide along with creating awareness toward ethical factors (such as workplace safety, supreme quality, and compliance). Likewise, system supremacy of I5.0 such as cyber–physical cognitive mechanism, adaptive manufacturing treats, and combined physical elements with intelligent software has majestically contributed to real-time resource optimization (Alawag et al., 2023). Yet research by Bazel et al. (2024) revealed that certain businesses and stakeholders of various societies are struggling to adopt this innovative vision in full capacity. The aim of this study is to exterminate the conceptual mismatch relating to manufacturing and social needs and to capture how strategically this novel attempt is revolving the social practices.

DOI: 10.4324/9781003647621-4

The chapter holds threefold objectives: first, to develop an inclusive overview of I5.0; second, to introduce how new social principles are created under this development concept; and third, to showcase its underlined significance with Sustainable Development Goals (SDGs). In addition, this chapter will contribute valuable insights to academics, executives, practitioners, and decision-makers from copious societal wings alike.

3.2 Literature Review

From the characteristics of the market economy, global economic expansion has been transformed to the extreme definition of the knowledge-based principle of I4.0 in the 20th century, where technology has remained the single most influential driver (Kolade & Owoseni, 2022). Academics frequently stress that such smart, digital, connected, and self-controlling manufacturing (I4.0) has drastically transformed all aspects of human existence, work, and way of interaction. For instance, collaborative robots (cobots) assist assembly lines, exoskeletons empower production and logistics support, while various cloud-based software advocates enterprises to tackle evolving competition (Cagno et al., 2024). On the contrary, erudite works explored several human-oriented negative impacts of I4.0 such as exaggerated employment, work complexity, and high layoff rate (Islam et al., 2025; Kolade & Owoseni, 2022), justifying that automation should comprise such habits where collectively both individual and technology work to abolish the stressful post-work utopia and to form a work-less society in near future (Spencer, 2018). Similarly, an assessment by Antonaci et al. (2024) revealed that the future of the I4.0 ideology is inseparable from the progression of a sustainable, resilient, and human-adjusted production logit, given the paramount priority to the human aspect. Acknowledging akin scholarly demand and exercising interest, global citizens are witnessing a new techno reality where the urge is set to avail technology for humans and to protect the planet rather than the previous focus of rapid automation (Grybauskas et al., 2022; Sarkar et al., 2024). Nation-wide this wave of new paradigm is identified as Industry 5.0. Simply, with technological underpinnings, I5.0 harmonizes economic growth, social well-being, and environmental preservation (Rame et al., 2024). In an investigation, Rijwani et al. (2024) pointed out that by embodying humans' adaptability and cognitive abilities, and deliberating ethical priorities in industrial systems, this new paradigm focuses on human welfare with augmented technology. Another critical comparison by Maddikunta et al. (2022) streamlined that rather than I4.0 envision, I5.0 markets the establishment of more skilled jobs by strengthening both people and machines, where individuals are more necessitated to focus on critical thinking and improving creativity.

Learning of Gamberini and Pluchino (2024) further highlighted that I5.0 encounters social challenges by restoring people-centered perspectives, and thus it strongly adopts the principles of social sustainability such as capitalizing on workforce needs and respecting human rights. An extended inspection of Ruppert et al. (2022) identified three fundamental pillars of I5.0: i) sustainability, ii) resilience, and iii) human-centricity. Additionally, the findings of Rame et al. (2024) testified the alignment of I5.0 objectives with promoting democracy and advocating political discussions in order to curtail uncertainties and industrial disputes by fostering scientific advancements side-by-side; hence, it requires interdisciplinary connectivity and collaboration with multiple sectors of an economy that ultimately generate far-reaching benefits beyond business maneuvers.

Through a comprehensive investigation, Bazel et al. (2024) illuminated how ample cutting-edge technologies are pronouncing the success of I5.0 in multiple industries, and a glimpse of this is presented in Table 3.1. For simplification, the findings of Xu et al. (2021) categorized two branches of I5.0 technologies: i) individualized human–machine interaction technologies and ii) bio-inspired technologies. Another consequential analysis of Ivanov (2023) revealed three spears of I5.0, namely, automation technology, infrastructure, and communication, that play a pivotal impression in improving work–life quality, mounting efficiency, and fostering innovation enterprise-wide.

On the whole, I5.0 implies future-oriented and cross-sectoral technologies (Barata & Kayser, 2024) that prioritize social goals to provide robust prosperity for the sustainable heightening of all humanity (Leng et al., 2022). Nonetheless, controversy exists regarding the unrivaled contribution of I5.0. According to Neumann et al. (2021), irrespective of the prevalent role of I5.0 technologies in terms of production and logistics, its people-centric notions are yet underrepresented. Yet again, the investigation of Antonaci et al. (2024) questioned the realistic justification of techno-driven and human-centric coexistence. Besides, Leng et al. (2022) admitted that the current exploration of I5.0 is relatively scarce, and widening systematic analysis will enrich the content beyond its usual lens. This inquiry thus contributes to the current debate by analyzing the concept through relative understanding. Expectedly, the discussion will shed light on multidisciplinary insights relevant to I5.0 that are going to shape the world toward perseverance and prosperity.

3.3 Research Methods

A scoping review was performed according to the PRISMA Extension for Scoping Reviews (PRISMA-ScR) standards, concentrating on the following research questions:

R1: Is the I5.0 era contributing to new societal practices?
R1a: Is the I5.0 era improving organizational practices?
R1b: Is the I5.0 era promoting green consumerism?
R2: Is the I5.0 era contributing to the SDGs?

A scoping review is an emerging body of knowledge synthesis that maps or represents the existing literature focusing only on a specific topic when the subject is complex and heterogeneous (Galway et al., 2022). Such a promising review technique is replacing the traditional narrative style of literature review with

Table 3.1 Industry-Wise Major Enablers of I5.0 and Its Contributions.

Sector	*I5.0 Enablers*	*Key Service Outcomes*
Intelligent healthcare	Artificial intelligence (AI), robotics, Internet of Things (IoT), and blockchain	Patient database, personalized medicine, remote patient assistance, precise diagnostics, AI-assisted smart medical devices
Supply chain management	IoT, blockchain, and data analytics	Product tracking and monitoring in real time, transparency, optimize logistics and flowless inventory management
Energy and utilities	IoT apparatus plus data analytics	Real-time monitoring of energy usage, minimized downtime, optimized energy consumption with reduced costs, and analytical maintenance practices
Cloud manufacturing	Cloud computing	Shared manufacturing resources, digital marketplace on demand service, efficient resource distribution with customization, increased agility, low cost, and superior production efficiency
Manufacturing production	Collaborative robots, digital twins, and predictive analytics	Agile and intelligent manufacturing processes with better quality, and customization capabilities
Education	Virtual and augmented reality technologies, smart classrooms equipped with IoT devices, real-time data analytics, and life-long learning with digital resources	Personalized plus immersive learning environments, intelligent tutoring systems, seamless communication among students and teachers. Enhanced student performance patterns with timely interventions
Transportation and logistics	Intelligent systems, incorporating smart warehouses, and self-sufficient vehicles with real-time tracking systems	Boost up safety, efficiency, and sustainability
Agriculture and food production	IoT sensors, data analytics, and AI algorithms	Farming accuracy with proficient food supply chain management, augment resource consumption, crop health observation, and predictive analysis of harvest optimization

Source: Bazel et al. (2024)

more precise and transparent insights. According to Marshall et al. (2018), the utmost usefulness of a scoping review rests in examining new and interdisciplinary wings of literature. Following Arksey and O'Malley (2005), scoping review data was accumulated in five stages: 1) identifying the research question/s; ii) recognizing relevant studies; iii) selecting the studies; iv) mapping, arranging, and summarizing; and v) reporting the results (Knowles et al., 2024). An extensive argument by Galway et al. (2022), however, suggested that the stages often operate iteratively instead of following a linear pattern.

3.3.1 Identifying Relevant Studies

In the preliminary search stage, we accessed seven well-indexed and peer-reviewed databases, namely, Springer, ScienceDirect, MDPI, RePEc, Emerald Insight, Wiley Online Library, and Sage, because of their enabled scale and comprehensiveness in terms of journals and disciplinary views. The proportion of each database is shown in Figure 3.1. The majority of our papers (37%) were repossessed from the Emerald Insight database, and 35% of our research work were retrieved from Elsevier journals. Apparently, this also signifies the amplified focus of those databases toward such an emerging phenomenon.

For constructing the search equation, thesaurus descriptors were consulted additionally with thorough insights. A 2-month period (April to May 2025) was allocated to select intellectual papers based on the research questions. Due to our interest in synthesizing current and emerging fields, publications beyond the year 2018 were not included. A detailed note of the central search equation, database, and implied search strategy is presented in Table 3.2.

The primary database searches identified a total of 163 papers. After the search execution, all accrued documents were filtered based on the screening criterion. Eliminating all duplication, irrational, and irrelevant studies, finally 62 papers (n = 62) were transferred to a supplementary Excel sheet for extensive analysis. Table 3.3 presents the inclusion and exclusion criteria that were strictly monitored as part of study selection.

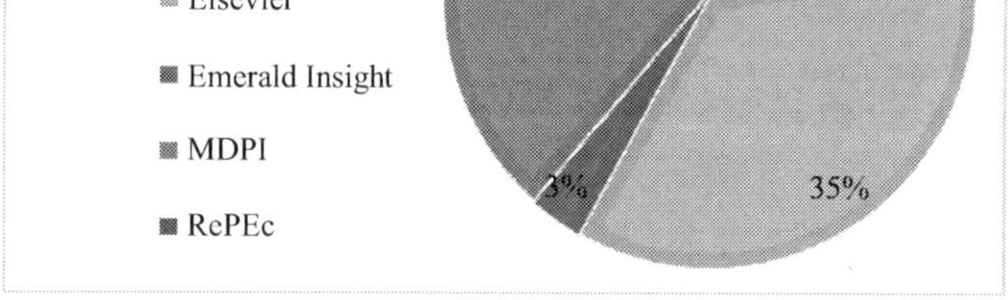

Figure 3.1 Usage of various databases.

Source: Authors' contribution (2025).

Table 3.2 List of Central Search Equation, Database, and Implied Search Strategy.

Central Search Equation	
((("Industry 5.0") OR ("Industry 4.0 & Industry 5.0") OR ("IR 4.0, IR 5.0") OR ("Industrial Revolution 5.0") OR ("Industry 5.0 and Workplace") OR ("Industry 5.0 and Employee well-being") OR ("Industry 5.0 and Work Safety") OR ("Industry 5.0 and Consumerism") OR ("Industry 5.0 and Workplace & Green consumerism") OR ("Industry 5.0 and Green consumerism") OR ("Industry 5.0 and Emerging trend of consumerism") ("Industry 5.0 and New Marketing") OR ("Future of Industry 5.0) OR ("Challenges of Industry 5.0") OR ("Industry 5.0 and Sustainability") OR ("Industry 5.0, Ethics") OR ("Industry 5.0 and SDG") OR ("Industry 5.0 and Cobot") OR ("Industry 5.0 and Impact on SDGs")))	
Database	**Search Strategy**
Springer, ScienceDirect, MDPI, and Elsevier	All fields Article title, abstract, and keywords

Source: Authors' contribution (2025).

Table 3.3 Record of Inclusion and Exclusion Conditions.

Inclusion Condition	*Exclusion Condition*
Article is published in an academic journal	Conference abstracts or proceedings, theses, and pre-print
Article published in English	Article published in other languages than English
Peer-reviewed papers	Non-peer-reviewed papers
Paper visibly focuses on I5.0 and its impact	Paper does *not* visibly focus on I5.0 and its impact
Research explicitly driven by sustainability concern as an endpoint/ objective	Research explicitly *not* driven by sustainability concern as an endpoint/ objective
Published after 2018	Published prior to 2018

Source: Authors' contribution (2025).

3.3.2 PRISMA Chart

Subsequently, the final sample comprised only 62 research papers. The PRISMA-ScR chart in Figure 3.2 shows the detailed process of findings and finalizing the scope of this examination.

The 62 core articles were validated as shown in the PRISMA diagram, and three reports were ultimately identified using other methods (websites). The inclusion of both peer-reviewed journals and other sources is in line with Peters et al. (2021), who approved that incorporating both empirical and non-empirical evidence as well as policy documents can be utilized in the scoping review to address broader subjects or research questions (Maraveas & Hahladakis, 2025). Additionally, using the iterated snowballing process, the appropriateness of all selected articles was

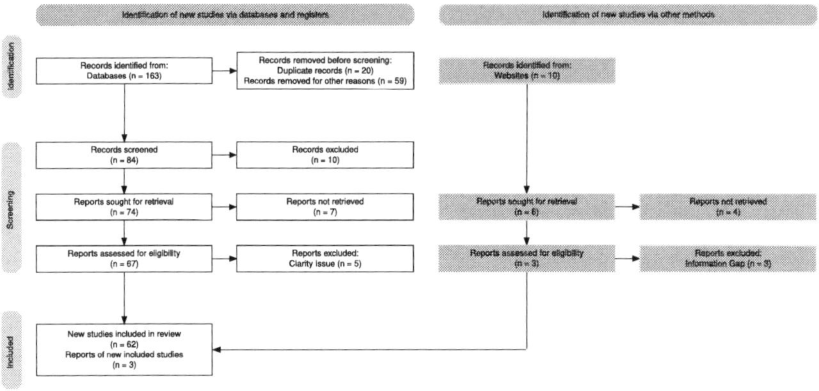

Figure 3.2 PRISMA-ScR diagram illustrating the article selection process.
Source: Authors' (2025) and Haddaway et al. (2022).

verified. An academic investigation by Wohlin et al. (2022) described that such a forward and backward snowballing technique is useful when the target population holds rare characteristics, when it is difficult to reach deprived communities, or when the topic is heterogeneous.

3.3.3 Data Extraction, Synthesis, and Research Bias

The data extraction and synthesis process was carried out without any external support. The data was extracted and recorded in an Excel document to facilitate an effective review process. Various descriptive information, such as the title, year of publication, author/authors' information, the journal or database, unit of analysis, and the type of research, were separated for fact analysis.

For data synthesis, the argumentative synthesis technique was used, where selected arguments were gathered from complementary and contradictory sources and later integrated to form a greater context than tradition (Cuevas et al., 2024). Systematically, the researchers screened the title, abstract, and full text of each inquiry to resolve the present research questions and arguments. Because of the nature of the scoping review, we prioritized the use of diverse academic databases, which ensures the methodological excellence of the selected studies. Again, the main aim of this study was to identify new social practices associated with I5.0 rather than justifying the aggregated evidence.

Moreover, to eradicate the risk of bias at each stage of data collection and synthesis, it has been reiterated as in Maraveas and Hahladakis (2025) that the researchers have no vested interest in any predefined inference. The core purpose of this investigation was to contribute to an added form of knowledge, insight, or guidance.

3.3.4 Theme Analysis

A thematic analysis was carried out to unveil the multidimensional significance of this undertaking, which is depicted in Table 3.4. We classified the stipulated

Table 3.4 Theme Analysis and Literature Distribution.

Content	f	P	CP	
Theme 1 I5.0 from I4.0	30	48	48	
I4	6	10	10	
I4 and I5	7	11	21	
I5	17	27	49	
Theme 2 Human centricity	21	34	82	
Human machine collaboration	4	6	55	
Cobot (collaborative robot)	3	5	60	
Ethical practice	3	5	65	
Health and safety	4	6	71	*n* = 62
Marketing 5.0	1	2	73	
Mass automation	1	2	75	
TQM	1	2	76	
Workplace well-being	4	6	83	
Theme 3 Sustainable approaches	11	18	100	
Sustainability	7	11	94	
Circular economy	4	6	100	
n = 62				
**f* = Frequency or count	**P* = Percent		*CP = Cumulative percentage	

Source: Authors' contribution (2025).

literary work into three broad categories (themes). Here, nearly half of the selected articles (48%) scrutinized the evolutionary phase of "I5.0 from I4.0," where a proportion of 27 precise documents emphasized on I5.0, which again indicates the universal acceptance of such emerging technologies. After that, various sustainability approaches covered almost 18% of the research discipline. To understand the social transition in detail, substantial discussion (34 percentile) revealed the human-centric notion of I5.0 digital sophistication on diverse humanistic attributes like ethical practices, workplace well-being, health and safety measures, and human–machine collaboration among the total sample ($n = 62$) of this research.

Additionally, Figure 3.3 provides a critical understanding of the emergence of several themes over our research timeframe (2018–2025). We can find an increasing focus on divergent themes related to I5.0 and I4.0, human-centricity, and sustainability issues annually. The diagram again signifies researchers' growing interest in I5.0-related subjects worldwide from 2022, as well as the demand of such inspections over time.

3.3.5 Ethical Considerations

To complement ethical aspects, the authors acknowledge the fulfillment of all ethical recommendations in selecting articles, and a good scientific process adhering to honesty, integrity, objectivity, and responsibility (Valencia-Contrera et al., 2025). Besides, we declare proper attribution of authorship in this research paper.

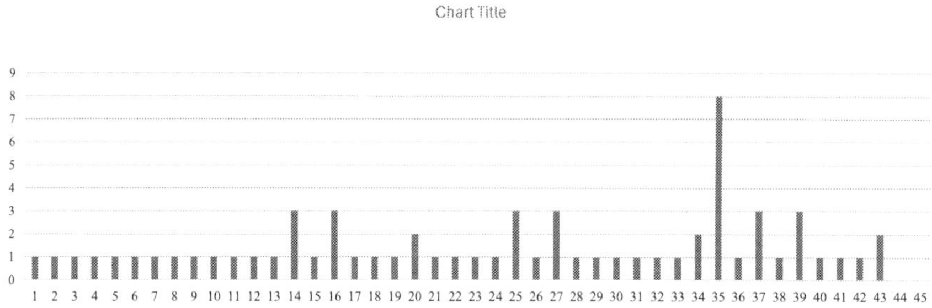

Figure 3.3 Critical assessment of theme retrieval over the research timeline.

Source: Authors' contribution (2025).

3.4 Results

Analyzing the end PRISMA outputs, existing pieces of the literature highlight several uprising trends, while the regime of I5.0 is culminating prior organizational customs from a traditional to a smart, novel, and digitally connected industry with progressive consumerism pragmatically (Antonaci et al., 2024). For generalization, the investigation findings are classified into two broad categories: i) improved organizational practices and ii) change in consumerism.

3.4.1 Improved Organizational Practices

Numerous intellectual testimonies verified that implementation of I5.0 sophistications pay back firms far beyond its operational profitability (Neri et al., 2023). Preferably the human-centricity of such technological supremacy adjusts human–machine interactions (Lu et al., 2022), resulting in better organizational and social traits in operation. The study by Cagno et al. (2024) prescribed that irrespective of the business size, firms that rely heavily on I5.0 have achieved momentous advances in the field of work culture as employees face less stress with "0" (zero) error and altogether enjoy a certain degree of autonomy in job performance (Rahman et al., 2024).

As the context of I5.0 underlines a combination of technical, soft skills, and multidisciplinary knowledge, it is vested in lifelong learning to drive innovation in the ever-changing industrial landscape (Pinto et al., 2024). One unrecognized value of such interaction lies in promoting employee empowerment via holistic training (Gamberini & Pluchino, 2024). Therewithal, academics (Cónego et al., 2024; Rijwani et al., 2024) repetitively suggested that the human–machine partnership of I5.0 leads to the development of vivid collaborative skills through flexibility, encourages critical thinking, and prompts agility to embrace change in dynamic environments along with ethical consideration on implicit social trends. Through an elaborative experiment, Leon (2023) proved that several upskilling and re-skilling techniques of I5.0 delicacies significantly influence employees' multiple psychological attributes such as interpersonal trust, self-efficacy, inclusiveness, and job satisfaction.

Again, the analysis of Gamberini and Pluchino (2024) supported that the I5.0 era has majestically redefined work dynamics, where workforce well-being such as talent management, innovation, and the drift of continuous learning are advocates of priority that cater to equitable opportunity and sustainable career paths to diverse people. Ample scholarly works are evident that the versatility of I5.0 helps control workers' behaviors such as using safety measures or availing specific authorization for valuable equipment or documents (Cagno et al., 2024). Likewise, the exploration of Rahman et al. (2024) justified that by using motion capture (MoCap) technology of I5.0, a number of occupational hazards can be regulated drastically, and it also largely works to ensure workplace health and safety measures. Relatedly, an examination by Alves et al. (2023) revealed that I5.0 technologies render improved safety performance in particularly hazardous situations because of their clarity and trustworthiness in each platform.

Empirical evidence by Grabowski et al. (2021) also elucidated how virtual reality techniques (I5.0) are utilizing collaborative mobile robot movement to control cost and product quality in extreme conditions like life-threatening temperature, high dust concentration, or explosion hazards. Further research by Cagno et al. (2024) scrutinized how several extraordinary capabilities of I5.0 can help managers handle emergencies like aiding evacuations or tracking employees' positions in crisis, which vastly ensures their well-being and secures the business from impending loss.

3.4.2 Change in Consumerism

With persistent technological push from I4.0, diverse pull forces of I5.0 are aiming to drive modern consumerism to a new height (Frank et al., 2019). Akundi et al. (2022) documented that with equal concern on mass personalization, I5.0 alternatively provides consumers a platform or urge to express themselves and allows them to pay a premium for doing so. One such fascinating backing is the resolution toward circular practices, which involves circular design, recycling, reusing, and repairing any product as long as possible (Shafique et al., 2024). In addition, I5.0 underlines reducing resource usage and controlling production up to the essential level with elimination of waste at each point (Shafique et al., 2023; Fuertes et al., 2024) and, thus, prescribes circularity as a combined effort of governmental planning, economic initiatives, product development, managemental support, and knowledge from academia (Riggs et al., 2023), where each social representative needs to act innovatively and responsibly to attain larger benefits. Furthermore, in an inquiry, Neumann et al. (2021) stressed that since the fulfillment of individualized demands without increasing the costs (frequently referred to as 'mass customization' or 'product individualization') is one of the superordinate goals of I5.0, consumer demand is concurrently evolving in various aspects such as supreme quality, improved safety, and sustainable manufacturing (Sharma,

2025). For instance, as collaborative robots (cobots) are undertaking repetitive and hazardous tasks, the workforce is at present more entitled to value-added responsibilities (Alves et al., 2023) like creative problem-solving, prompt innovation, and rising diversity-oriented mindset, which ultimately results in an urge for lesser operational cost with supreme quality and minimalized waste (Islam et al., 2025). By the same token, an analysis by Bakator et al. (2024) identified a number of customers' changing priorities like facilitating on-demand production and flexible delivery schedules that are aligned with I.50 intelligence. One classified list of rising social primacies (Table 3.5) in this latest industrial revolution further supports the circumstances.

Abruptly, with radical changes in individual methods and development scenarios, this smart cyber–physical socio-technical interphase is gradually shaping a new biome not only for work culture but also for the entire host ecosystem.

3.5 Discussions

The available scientific research has attested that concentrating solely on production efficiency, I5.0 instantaneously strengthens human cognitive abilities and agreement; ensures ethical submission at the workplace and in artifacts (Islam et al., 2025); and later establishes balanced synergy by linking people and technology (Adel, 2022). A comprehensive evolutionary vein of I5.0 proposed by Leng et al. (2022) revealed the additional tribute of industry I.50 in society through a tridimensional system architecture, where the scientists have identified three subsequent scales: i) technical, ii) reality, and iii) application dimensions. Over rationalization, the researchers explained how interconnectivity (through knowledge empowerment and value-driven measures at respective phases) imparts supreme precedence on science, technology, and innovation where human–machine alliance befits paramount. Again, considering the current and future boundary constraints, numerous scholarships affirmed the direct interrelation between the technological capacities of I5.0 and sustainability (Shafique et al., 2024), while in several circumstances, I5.0 digital solutions prioritize social objectives beyond productivity and development (Akundi et al., 2022). Moreover, with changing consumer demands (like eco-design, green manufacturing, Internet environment management, and investment recovery), sustainable practices positively affect such connotations (Shafique et al., 2024). Dynamic exploration of Mourtzis et al. (2022) further established the reciprocity between each emerging I5.0 proficiency and individual SDG achievement, where the authors once again rationalize the transition toward an advanced society through the execution of all the SDGs.

It is important to note that the wave of I5.0 commands a synchronized alignment among technology adoption, human capital development, and ethical governance (Bakator et al., 2024), where managers also need to be alert regarding new and emerging risks arise due to cultural transformations and constant digitalization (Ávila-Gutiérrez et al., 2022). The prospect of the research findings is illustrated in Figure 3.4.

Table 3.5 Catalog of Advancing Social Primacies.

Preventive Specialization	*Labor Conditions*	*Production*	*Waste Management*	*Secondary Raw Material*	*Material Health*	*Competitiveness and Innovation*
Occupational medicine	Labor risks Human rights	Self-sufficiency for raw materials	Recycling rates	Contribution of recycled materials to raw material demand	Nutrient type	Private investment, jobs, and gross value
Occupational safety	Security and health	Green public procurement	Recycling recovery for specific Waste	Trade in recyclable raw materials	Material characteristics	Number of patents
Ergonomic	Innovative social projects (diversity)					
Hygiene	Continuous improvement	Waste generation			Volume of carbon emissions	Number of patents related to secondary raw materials

Source: Ávila-Gutiérrez et al. (2022).

Figure 3.4 Summarized prospect of the research.

Source: Authors' contribution (2025).

3.6 Conclusion

The beginning of I5.0 (the socio-technical determinants) does not mark the end of I4.0 (the technology-driven force) in industrial production (Adel, 2022), both are likely to mature with time (Barata & Kayser, 2024). The fundamental aim of this research is to generate a panoramic understanding of the emerging body of knowledge for societal contribution. Theoretically, this study envisages broader literacy gaps that lead to further sustainable progression. Stakeholders from the divergent arena will find this investigation insightful by realizing how a harmonized adaptation of human capital and technological sophistication can ultimately be beneficial for a business and for society. A development path aligning the SDGs and I5.0 tech ventures has also been presented to point out the underscored significance of a changing society that can equally contribute to designing precise theory, and policy formulation plus successful execution by experts in multiple fields. On the practical ground, this learning will assist managers to understand the extensive scenario of I5.0 and encourage more to prioritize the human and environmental attributes. The knowledge foremost will guide policymakers and practitioners of diverse fields to design, facilitate, and follow human-centric, resilient, and sustainable values for a radical progression with emerging technological advances. Due to time constraints, the search strategy used in this research can be a possible limitation as it might have missed a few important research cases. Likewise, the use of only English keywords is another shortcoming, since a handful of perceptive studies are conducted locally and published in native repositories.

Although human–machine symbiosis has not been completely exploited in the industrial context yet, value-oriented I5.0 know-how will stimulate an interplay between "human-centric design" and "ethical and responsible innovation" of future workplaces or work processes beyond imagination (Longo et al., 2020). Thus, it is expected that future researchers should prioritize producing a multimodal framework that combines all these aspects with precise practical use. Empirical investigations to reveal in-depth human-centric notions along with progression through I5.0 technologies will enrich the research outcome further. Moreover, being an incipient concept, successful adoption of the I5.0 philosophy requires adjustment with industrial laws and regulations that can ensure better human–robot collaborations (cobots) in reality. Prominent scholars thus need to analyze and propose governance structures to contemplate the ethical implications of cyber security, data privacy, and personnel well-being, to safeguard conscientious and equitable deployment of I5.0.

References

Adel, A. (2022). Future of Industry 5.0 in society: Human-centric solutions, challenges and pros prospective research areas. *Journal of Cloud Computing*, *11*(40). https://doi.org/10.1186/s13677-022-00314-5

Akundi, A., Euresti, D., Luna, S., Ankobiah, W., Lopes, A., & Edinbarough, I. (2022). State of Industry 5.0—analysis and identification of current research trends. *Applied System Innovation*, *5*(1), 27. https://doi.org/10.3390/asi5010027

Alawag, A. M., Alaloul, W. S., Liew, M. S., Baarimah, A. O., Musarat, M. A., & Al-Mekhlafi, A. B. A. (2023). The role of the total-quality-management (TQM) drivers in overcoming the challenges of implementing TQM in industrialized-building-system (IBS) projects in Malaysia: Experts' perspectives. *Sustainability*, *15*(6607). https://doi.org/10.3390/su15086607

Alves, J., Lima, T. M., & Gaspar, P. D. (2023). Is Industry 5.0 a human-centred approach? A systematic review. *Processes*, *11*(1), 193. https://doi.org/10.3390/pr11010193

Antonaci, F. G., Olivetti, E. C., Marcolin, F., Castiblanco Jimenez, I. A., Eynard, B., Vezzetti, E., & Moos, S. (2024). Workplace well-being in Industry 5.0: A worker-centered systematic review. *Sensors*, *24*(17), 5473. https://doi.org/10.3390/s24175473

Arksey, H., & O'Malley, L. (2005). Scoping studies: Towards a methodological framework. *International Journal of Social Research Methodology*, *8*(1), 19–32. https://doi.org/10.1080/1364557032000119616

Ávila-Gutiérrez, M. J., Suarez-Fernandez de M., S., & Aguayo-González, F. (2022). Occupational safety and Health 5.0—a model for multilevel strategic deployment aligned with the sustainable development goals of agenda 2030. *Sustainability*, *14*(11), 6741. https://doi.org/10.3390/su14116741

Bakator, M., Ćoćkalo, D., Makitan, V., Stanisavljev, S., & Nikolić, M. (2024). The three pillars of tomorrow: How Marketing 5.0 builds on Industry 5.0 and impacts Society 5.0? *Heliyon*, *10*(17), https://doi.org/10.1016/j.heliyon.2024.e36543

Barata, J., & Kayser, I. (2024). How will the digital twin shape the future of Industry 5.0? *Technovation*, *134*, https://doi.org/10.1016/j.technovation.2024.103025

Bazel, M. A., Mohammed, F., Baarimah, A. O., Alawi, G., Al-Mekhlafi, A. B. A., & Almuhaya, B. (2024). The era of Industry 5.0: An overview of technologies, applications, and challenges. In Saeed, F., Mohammed, F., & Fazea, Y. (Eds.), *Advances in Intelligent Computing Techniques and Applications. IRICT 2023. Lecture Notes on Data Engineering and Communications Technologies* (pp. 274–284). Springer. https://doi.org/10.1007/978-3-031-597 07-7_24

Cagno, E., Accordini, D., Neri, A., Negri, E., & Macchi, M. (2024). Digital solutions for workplace safety: An empirical study on their adoption in Italian metalworking SMEs. *Safety Science*, *177*, https://doi.org/10.1016/j.ssci.2024.106598

Cónego, L., Pinto, R., Pinto, J., & Gonçalves, G. (2024). Leveraging gamification in Industry 5.0: Tailored solutions for workplace' employees. *Procedia Computer Science*, *232*, 1769–1778. https://doi.org/10.1016/j.procs.2024.01.175

Cuevas, I., Mateos, M., Casado-Ledesma, L., Olmos, R., Granado-Peinado, M., Luna, M., Núñez, J. A., & Martín, E. (2024). How to improve argumentative syntheses written by undergraduates using guides and instructional rubrics. *European Journal of Psychology of Education*, *39*, 4573–4596. https://doi.org/10.1007/s10212-024-00890-x

Frank, A. G., Dalenogare, L. S., & Ayala, N. F. (2019). Industry 4.0 technologies: Implementation patterns in manufacturing companies. *International Journal of Production Economics*, *210*, 15–26, https://doi.org/10.1016/j.ijpe.2019.01.004

Fuertes, G., Zamorano, J., Alfaro, M., Vargas, M., Sabattin, J., Duran, C., Ternero, R., & Rivera, R. (2024). Opportunities of the technological trends linked to Industry 4.0 for achieve sustainable manufacturing objectives. *Sustainability*, *14*.

Galway, L. P., Levkoe, C. Z., Portinga, R. L. W., & Milun, K. (2022). A scoping review examining governance, co-creation, and social and ecological justice in living labs literature. *Challenges*, *13*(1). https://doi.org/10.3390/challe13010001

Gamberini, L., & Pluchino, P. (2024). Industry 5.0: A comprehensive insight into the future of work, social sustainability, sustainable development, and career. *Australian Journal of Career Development*, *33*(1), 5–14. https://doi.org/10.1177/10384162241231118

Grabowska, S., Saniuk, S., & Gajdzik, B. (2022). Industry 5.0: Improving humanization and sustainability of Industry 4.0. *Scientometrics*, *127*(6), 3117–3144. https://doi.org/10.1007/s11192-022-04370-1

Grabowski, A., Jankowski, J., & Wodzyński, M. (2021). Teleoperated mobile robot with two arms: The influence of a human–machine interface, VR training and operator age. *International Journal of Human–Computer Studies*, *156*. https://doi.org/10.1016/j.ijhcs.2021.102707

Grybauskas, A., Stefanini, A., & Ghobakhloo, M. (2022). Social sustainability in the age of digitalization: A systematic literature review on the social implications of Industry 4.0, *Technology in Society*, *70*. https://doi.org/10.1016/j.techsoc.2022.101997

Haddaway, N. R., Page, M. J., Pritchard, C. C., & McGuinness, L. A. (2022). PRISMA2020: An R package and Shiny app for producing PRISMA 2020-compliant flow diagrams, with interactivity for optimised digital transparency and open synthesis. *Campbell Systematic Reviews*, *18*, e1230. https://doi.org/10.1002/cl2.1230

Islam, M. T., Sepanloo, K., Woo, S., Woo, S. H., & Son, Y.-J. (2025). A review of the Industry 4.0 to 5.0 transition: Exploring the intersection, challenges, and opportunities of technology and human–machine collaboration. *Machines*, *13*(4), 267. https://doi.org/10.3390/machines13040267

Ivanov, D. (2023). The Industry 5.0 framework: Viability-based integration of the resilience, sustainability, and humancentricity perspectives. *International Journal of Production Research*, *61*(5), 1683–1695. https://doi.org/10.1080/00207543.2022.2118892

Knowles, B., Marsh, P., Prehn, J., & Maxwell, H. (2024). Addressing ethical issues in outdoor health practice: A scoping review. *Journal of Outdoor and Environmental Education*, *27*, 7–35 https://doi.org/10.1007/s42322-024-00160-w

Kolade, I., & Owoseni, A. (2022). Employment 5.0: The work of the future and the future of work. *Technology in Society*, *71*. https://doi.org/10.1016/j.techsoc.2022.102086

Leng, J., Sha, W., Wang, B., Zheng, P., Zhuang, C., Liu, Q., Wuest, T., Mourtzis, D., & Wang, L. (2022). Industry 5.0: Prospect and retrospect. *Journal of Manufacturing Systems*, *65*, 279–295. https://doi.org/10.1016/j.jmsy.2022.09.017

Leon, R. D. (2023). Employees' reskilling and upskilling for Industry 5.0: Selecting the best professional development programmes. *Technology in Society*, *75*. https://doi.org/10.1016/j.tetechsoc.2023.102393

Longo, F., Padovano, A., & Umbrello, S. (2020). Value-oriented and ethical technology engineering in Industry 5.0: A human-centric perspective for the design of the factory of the future. *Applied Sciences*. *10*, 4182. https://doi.org/10.3390/app10124182

Lu, Y., Zheng, H., Chand, S., Xia, W., Liu, Z., Xu, X., Wang, L., Qin, Z., & Bao, J. (2022). Outlook on human-centric manufacturing towards Industry 5.0. *Journal of Manufacturing Systems*, *62*, 612–627. https://doi.org/10.1016/j.jmsy.2022.02.001

Maddikunta, P. K. R., Pham, Q. V., Prabadevi, B., Deepa, N., Dev, K., Gadekallu, T. R., & Liyanage, M. (2022). Industry 5.0: A survey on enabling technologies and potential applications. *Journal of Industrial Information Integration*, *26*. https://doi.org/10.1016/j.jii.2021.100257

Maraveas, C., & Hahladakis, J. N. (2025). A scoping review on the European Union agricultural plastic waste management strategies: Focusing on liquefaction, *Journal of Hazardous Materials Advances*, *18*, https://doi.org/10.1016/j.hazadv.2025.100727

Marshall, R. E., Levison, J. K., McBean, E. A., Brown, E., & Harper, S. L. (2018). Source water protection programs and indigenous communities in Canada and the United States: A scoping review. *Journal of Hydrology*, *562*, 358–370.

Mourtzis, D., Angelopoulos, J., & Panopoulos, J. (2022, September 7). *Transition from Industry 4.0 to Society 5.0*. https://encyclopedia.pub/entry/26927

Neri, A., Negri, M., Cagno, E., Kumar, V., & Garza-Reyes, J. A. (2023). What digital-enabled dynamic capabilities support the circular economy? A multiple case study approach. *Bus Strategy Environment*. *32*, 5083–5101. https://doi.org/10.1002/bse.3409

Neumann, W. P., Winkelhaus, S., Grosse, E. H., & Glock, C. H. (2021). Industry 4.0 and the human factor—a systems framework and analysis methodology for successful development. *International Journal of Production Economics*, *233*.

Peters, M. D. J., Marnie, C., Colquhoun, H., Garritty, C. M., Hempel, S., Horsley, T., Langlois, E. V., Lillie, E., O'Brien, K. K., Tunçalp, Ö., Wilson, M. G., Zarin, W., & Tricco, A. C. (2021). Scoping reviews: Reinforcing and advancing the methodology and application. *Systematic Reviews*, *10*(1). https://doi.org/10.1186/s13643-021-01821-3

Pinto, R., Žilka, M., Zanoli, T., Kolesnikov, M. V., & Gonçalves, G. (2024). Enabling professionals for Industry 5.0: The self-made programme. *Procedia Computer Science*, *232*. https://doi.org/10.1016/j.procs.2024.02.107

Rahman, M. H., Hasan, M. R., Chowdhury, N. I., Syed, M. A. B., & Farah, M. U. (2024). Predictive health analysis in industry 5.0: A scientometric and systematic review of motion capture in construction. *Digital Engineering*, *1*. https://doi.org/10.1016/j.dte.2024.100002

Rame, R., Purwanto, P., & Sudarno, S. (2024). Industry 5.0 and sustainability: An overview of emerging trends and challenges for a green future. *Innovation and Green Development*, *3*(4). https://doi.org/10.1016/j.igd.2024.100173

Riggs, R., Roldán, J. L., Real, J. C., & Felipe, C. M. (2023). Opening the black box of big data sustainable value creation: The mediating role of supply chain management capabilities and circular economy practices. *International Journal of Physical Distribution & Logistics Management*, *53*(7), 762–788. https://doi.org/10.1108/IJPDLM-03-2022-0098

Rijwani, T., Kumari, S., Srinivas, R., Abhishek, K., Iyer, G., Vara, H., Dubey, S., Revathi, V., & Gupta, M. (2024). Industry 5.0: A review of emerging trends and transformative technologies in the next industrial revolution. *International Journal on Interactive Design and Manufacturing*, *19*, 667–679.

Ruppert, T., Darányi, A., Medvegy, T., Csereklei, D., & Abonyi, J. (2022). Demonstration laboratory of industry 4.0 retrofitting and operator 4.0 solutions: Education towards industry 5.0. *Sensors*, *23*(1), 283.

Sarkar, B. D., Vipulesh, S., Dwivedi, A., & Pamucar, D. (2024). Digital transition from Industry 4.0 to Industry 5.0 in smart manufacturing: A framework for sustainable future. *Technology in Society*, *78*. https://doi.org/10.1016/j.techsoc.2024.102649

Shafique, M. N., Adeel, U., & Rashid, A. (2024). The synergy between Industry 5.0 and circular economy for sustainable performance in the Chinese manufacturing industry. *Sustainability*, *16*(22), 9952. https://doi.org/10.3390/su16229952

Shafique, M. N., Rashid, A., Yeo, S. F., & Adeel, U. (2023). Transforming supply chains: Powering circular economy with analytics, integration and flexibility using dual theory and deep learning with PLS-SEM-ANN analysis. *Sustainability*, *15*.

Sharma, N. (2025, May 21). *Industry 5.0-Revolutionizing the Factory Floor with Human-Centric Manufacturing*. https://appinventiv.com/blog/industry-5-0-manufacturing/

Spencer, D. A. (2018). Fear and hope in an age of mass automation: Debating the future of work. *New Technology, Work and Employment*, *33*(1), 1–12. https://doi.org/10.1111/ntwe.12105

Valencia-Contrera, M., Rivera-Rojas, F., Villa-Velásquez, J., Cancino-Jiménez, D., Vallejos-Vergara, S., & Febré, N. (2025). Scoping review on ethical considerations in research on the work–family interaction process. *Nursing Reports*, *15*(2). https://doi.org/10.3390/nursrep15020031

Wohlin, C., Kalinowski, M., Romero Felizardo, K., & Mendes, E. (2022). Successful combination of database search and snowballing for identification of primary studies in systematic literature studies. *Information and Software Technology*, *147*, 106908. https://doi.org/10.1016/j.infsof.2022.106908

Xu, X., Lu, Y., Vogel-Heuser, B., & Wang, L. (2021). Industry 4.0 and Industry 5.0—inception, conception and perception. *Journal of Manufacturing Systems*, *61*, 530–535. https://doi.org/10.1016/j.jmsy.2021.10.006

4 Industry 5.0 as a Driver of Sustainability

Opportunities, Challenges, and Future Directions

Imran Hossain, A.K.M. Mahmudul Haque, and Rezki Amalyadi

4.1 Introduction

The world economy, social structures, and technological landscapes have all undergone significant changes as a result of the fast development of industrialization over the last 200 years (Castells, 1986). From steam power and mechanized manufacturing to electrification, mass production, and digitalization, industries have gradually adopted transformative technologies since the First Industrial Revolution in the late 18th century (Kumar et al., 2025; Xu et al., 2018). Incorporating cyber–physical systems, the Internet of Things (IoT), artificial intelligence (AI), and big data analytics into production processes, Industry 4.0, which was introduced in the early 21st century, marked a significant milestone in industrial advancement (Liao et al., 2017). Unprecedented automation, interconnection, and efficiency within industrial ecosystems were promised by this digital revolution. Although Industry 4.0 greatly improved operational performance, it frequently ignored important human and environmental factors in favor of focusing primarily on technology and financial benefits.

A paradigm change toward a more comprehensive industrial approach that specifically incorporates resilience, sustainability, and human-centric values alongside technology advancement is represented by the rise of Industry 5.0 (Sheikh et al., 2024). Developed as a follow-up and enhancement to Industry 4.0, Industry 5.0 promotes cooperation between intelligent systems and humans by repositioning the human workforce as a key participant in manufacturing and industrial processes (Demir et al., 2023). Industry 5.0 prioritizes ethical considerations, sustainable development, and customized production over automation and efficiency. The European Commission (2021) defined Industry 5.0 as an industry that "moves beyond efficiency and productivity as the sole goals" to prioritize "human-centricity, sustainability, and resilience." This new industrial revolution recognizes that future production systems need to be planned for environmental stewardship and societal well-being in addition to economic competitiveness.

A key global agenda today is sustainable development, which the Brundtland Commission defined as development that "meets the needs of the present without compromising the ability of future generations to meet their own needs" (WCED, 1987). A comprehensive framework for tackling important global issues, such as

DOI: 10.4324/9781003647621-5

poverty eradication, climate change mitigation, responsible consumerism, and equitable economic growth, was made available by the United Nations' (UN's) approval of the Sustainable Development Goals (SDGs) in 2015 (United Nations, 2015). Given their enormous impact on economic systems, employment, resource consumption, and environmental effects, industries are essential to reaching these objectives (Jie et al., 2023). However, there are significant risks to social cohesion and environmental sustainability associated with the traditional industrial growth model, which is frequently typified by resource-intensive production, linear value chains, and a disregard for social equity (Fernández Ocamica et al., 2025).

Growing regulatory restrictions, societal expectations, and commercial demands for morally and environmentally responsible manufacturing have all contributed to the recent acceleration of the integration of sustainability principles into industrial practice (Berry & Rondinelli, 1998). However, attaining sustainability in industrial settings calls for more than small-scale increases in resource efficiency or decreases in emissions. It calls for a fundamental shift in industrial paradigms, one that harmonizes technological advancement with socio-environmental goals. Given this, Industry 5.0 has promise as a driving force behind integrating sustainability into industrial strategies. Industry 5.0 promises to improve industrial adaptability, encourage circular economy practices, and enable more robust supply chains by promoting synergies between cutting-edge technologies and human creativity (Nahavandi, 2019).

The idea of human–machine collaboration is one of the hallmarks of Industry 5.0 (Pizoń & Gola, 2023). Industry 5.0 envisions a future where human intelligence, creativity, and emotional insight complement the precision and computational power of machines, in contrast to the automation-centric approach of Industry 4.0, which frequently aimed to minimize human intervention (Demir et al., 2023). It is expected that this mutually beneficial relationship will result in creative production methods, personalized goods, and flexible manufacturing systems that are more in line with customer demands and environmental limitations. Technological enablers of this human-centric industrial model include the emergence of digital twin technologies, advanced AI interfaces, and collaborative robotics (cobots) (Maddikunta et al., 2022). In addition to increasing productivity, these technologies open doors for skill development, fulfilling work, and increased job satisfaction.

Additionally, circular economy, a regenerative economic model that aims to minimize waste, maximize resource efficiency, and extend product lifecycles, is closely aligned with Industry 5.0 (Geissdoerfer et al., 2017). Industry 5.0 can help create sustainable production models that lessen environmental impact and aid in the shift to low-carbon economies by incorporating circularity into manufacturing processes. For example, remanufacturing, additive manufacturing, and AI-driven predictive maintenance can minimize material waste and maximize resource use, promoting economic efficiency and environmental preservation (Jabbour et al., 2018). Furthermore, local manufacturing should be encouraged, and the carbon intensity linked to global supply chains can be decreased by implementing decentralized production models made possible by digital technologies.

In the context of Industry 5.0, the social aspect of sustainability is just as important. Concerns regarding job displacement, digital inequality, and the ethical consequences of AI have gained traction as sectors become more automated and data-driven (Martini et al., 2024). In order to address these issues, Industry 5.0 places a strong emphasis on empowerment, inclusivity, and human well-being. Instead of replacing them with technology, it promotes the development of high-quality professions that make use of human abilities in creativity, invention, and problem-solving (Nahavandi, 2019). Furthermore, societal ambitions for justice, accountability, and human rights in the digital age are reflected in the development of ethical AI, data transparency, and participatory governance models within industrial contexts.

Industry 5.0 has a lot of potential, but there are a number of obstacles in the way of its actualization. These include technological challenges pertaining to cybersecurity, interoperability, and system integration; financial obstacles related to investment costs and market uptake; and regulatory and policy deficiencies that could impede the smooth adoption of sustainable and human-centered industrial practices (Jabbour et al., 2018). In order to address these issues, governments, business executives, researchers, and civil society must work together to create frameworks that are supportive, cultivate innovative ecosystems, and encourage inclusive discussions about the direction of industry.

The purpose of this study is to investigate the new field of Industry 5.0 as a sustainability driver, critically analyzing its advantages, disadvantages, and potential avenues for further study. The study aims to give a thorough grasp of how Industry 5.0 might support SDGs by combining recent scholarly research, governmental documents, and industry reports. The study emphasizes how important it is to combine technological innovation with socio-environmental factors in order to make sure that industrial development benefits both human welfare and the health of the planet. By doing this, it hopes to influence scholarly discussions, direct business operations, and aid in the formulation of laws intended to promote an inclusive and sustainable industrial future.

4.2 Conceptual Framework of Industry 5.0

In the past, technological advancements and their significant effects on production methods, productivity, and social structure have propelled the development of industrial revolutions. Every industrial revolution added new technology capabilities and redefined industrial paradigms, building on the one before it. Factory-based manufacturing was pioneered by Industry 1.0, which introduced mechanization using steam and water power (Kumar et al., 2025). Automation, electronics, and information technology became more prevalent in manufacturing processes in Industry 3.0, while electricity and assembly lines enabled mass production in Industry 2.0 (Yin et al., 2018). Big data analytics, AI, machine learning, digitalization, cyber–physical systems, and the IoT changed industrial processes with the advent of Industry 4.0 in the early 21st century (Xu et al., 2018). Although Industry 4.0 has the potential to revolutionize the industry, it has been criticized for prioritizing automation, operational efficiency, and technological optimization over social and human factors (Liao et al., 2017).

In light of this, Industry 5.0, which prioritizes human-centricity, sustainability, and resilience in addition to ongoing technological advancement, rises not as a replacement for Industry 4.0 but rather as its natural progression. Technology advancement, human ingenuity, social responsibility, and environmental stewardship are all aimed at striking a balance in Industry 5.0, which is a change to an industrial paradigm that "goes beyond efficiency and productivity as the sole goals" to embrace more general societal objectives, according to the European Commission (2021). Because it addresses growing concerns about the social, ethical, and environmental effects of industrial development and recognizes the limitations of a purely techno-economic focus, this shift is significant. Instead of just automating or digitizing processes, as its predecessors did, Industry 5.0 aims to create a cooperative relationship between advanced technologies and human intelligence (Table 4.1).

Human–machine collaboration is the fundamental tenet of Industry 5.0, which describes a synergistic relationship where human cognitive abilities, creativity, emotional intelligence, and problem-solving skills complement the accuracy, efficiency, and computational capabilities of machines and AI systems. This contrasts with the automation-driven goals of Industry 4.0, which frequently reduced human workers to supervisory roles or attempted to completely eliminate human intervention (Demir et al., 2023). Using collaborative robots, or "cobots," which are made to function securely alongside people in shared workspaces, is a prime example of this collaborative concept. With sensors, AI-driven control systems, and adaptive algorithms, cobots can interact with human operators in real time, improving workplace safety and productivity in contrast to traditional industrial robots that operate independently in remote areas (Maddikunta et al., 2022).

According to Tallat et al. (2023), Industry 5.0's enabling technologies include a wide range of digital tools and intelligent systems. In particular, AI is essential for improving decision-making, streamlining production lines, and facilitating predictive maintenance. AI-driven analytics enable businesses to make data-driven decisions that are sustainable and efficient by providing real-time insights into market trends, supply chain dynamics, and operational performance. A virtual depiction of actual assets, systems, or processes that enables real-time monitoring, modeling, and optimization is known as a "digital twin." By using digital twins, manufacturers can reduce material waste and energy consumption by improving product design through virtual prototyping, optimizing resource use, and predicting system behavior (Tao et al., 2019). In addition, human–machine interfaces (HMIs) have developed to facilitate smooth and intuitive communication between operators and intricate systems, frequently utilizing virtual reality (VR) and augmented reality (AR) technologies to improve maintenance, training, and operational effectiveness.

The focus Industry 5.0 places on customization and personalization in production is one of its most notable characteristics. With Industry 5.0, mass customization tactics that satisfy individual consumer preferences while preserving scalability and efficiency are supported, departing from the mass production models typical

Table 4.1 Conceptual Framework of Industry 5.0 in the Context of Sustainable Development.

Core Elements of Industry 5.0	*Enabling Technologies*	*Associated Sustainability Dimension*	*Strategic Contributions to Sustainable Development*	*Potential Challenges*
Human–machine collaboration	Cobots, HMIs, and AI-driven decision systems	Social and economic	Enhances productivity with human creativity, fosters meaningful employment, and promotes safe and ergonomic workplaces	Workforce skill gaps, resistance to adoption, and need for re-skilling
Advanced digitalization and automation	AI, IoT, big data analytics, and digital twins	Economic and environmental	Optimizes production efficiency, enables mass customization, supports predictive maintenance, and minimizes resource waste	High integration costs, interoperability issues, and cybersecurity risks
Circular economy integration	Blockchain for supply chain transparency, material tracking systems, and green IoT	Environmental and Economic	Promotes resource efficiency, reduces waste, and supports closed-loop manufacturing and sustainable supply chains	Supply chain complexity, initial investment barriers, and technological maturity
Personalization and mass customization	Additive manufacturing (3D printing) and AI-driven design tools	Economic and social	Meets individual customer needs, reduces overproduction, and supports small and medium-sized enterprise (SME) competitiveness and innovation	High cost of customization, and balancing scalability and personalization
Sustainable energy management	Smart grids, renewable energy integration systems, and AI-based energy optimization	Environmental	Reduces carbon footprint, enhances energy efficiency, and supports low-carbon industrial transitions	Integration with existing grids, regulatory hurdles, and high implementation cost
Ethical governance and human-centric values	Data governance frameworks, ethical AI, and transparency tools	Social	Ensures data privacy, promotes trust, supports fair labor practices, and fosters corporate responsibility	Ethical ambiguities, regulatory compliance challenges, and need for stakeholder alignment

of earlier industrial eras (Table 4.1). With the use of AI-driven design tools and advanced manufacturing technologies like additive manufacturing (3D printing), highly customized items can be produced with less loss of resources (Maddikunta et al., 2022). In line with changing consumer demands for distinctive, environmentally friendly, and ethically made products, this trend toward personalization strengthens the bond between producers and consumers.

Additionally, Industry 5.0 incorporates the circular economy and sustainability concepts into its conceptual framework (Table 4.1). Traditional linear production models, which are predicated on the "take–make–dispose" methodology, have resulted in waste production, resource depletion, and environmental degradation. Industry 5.0, on the other hand, stresses closed-loop supply chains, remanufacturing, resource efficiency, and product life extension (Geissdoerfer et al., 2017). Technologies like AI and the IoT make it possible to precisely follow the movements of materials, goods, and energy, supporting circular production models that reduce waste and maximize the use of renewable resources. Industry 5.0 also helps achieve global sustainability goals, like the SDGs of the UN, and facilitates the shift to low-carbon economies by integrating sustainability considerations into design, manufacturing, and distribution processes (United Nations, 2015).

Industry 5.0's human-centric orientation is further highlighted by its social dimension. Unlike other industrial revolutions, which frequently placed a higher priority on automation and labor cost reduction, Industry 5.0 promotes the development of worthwhile job possibilities that take advantage of what makes people uniquely human. It places a strong emphasis on acquiring new skills, especially in areas like moral decision-making, system design, digital literacy, and creative problem-solving (Nahavandi, 2019). Industry 5.0 also tackles issues of digital exclusion and employment displacement by encouraging ethical technology governance, inclusive working practices, and chances for ongoing learning. Workplace ergonomics, mental health, and safety are all included in the emphasis on human well-being, which reflects a corporate culture and workforce development strategy that is comprehensive.

Governance structures and ethical issues are essential to the idea of Industry 5.0. There are growing worries about data privacy, algorithmic bias, transparency, and accountability as new technologies are incorporated into industrial and societal systems. Workers, consumers, legislators, and civil society are just a few of the stakeholders where Industry 5.0 promotes the adoption of ethical AI principles, responsible data management techniques, and participatory governance models (Demir et al., 2023). This focus on ethics and governance promotes social acceptance of Industry 5.0 initiatives and builds public trust while also reducing the risks of technological misuse.

4.3 Sustainability Dimensions in Industry 5.0

4.3.1 Economic Sustainability

In the context of Industry 5.0, economic sustainability is the ability of industrial systems to provide economic value in a way that is inclusive, resilient, and consistent with long-term environmental and societal objectives (Ivanov, 2023). Historically, economies of scale, cost reduction, and productivity maximization have propelled industrial economic growth, frequently at the price of environmental integrity and social justice (Islam et al., 2003). However, the Industry 5.0 paradigm reinterprets economic sustainability by integrating technological innovation, human-centricity, and resilience as key facilitators of sustainable value generation. Industry 5.0 promotes a symbiotic relationship between human creativity and cutting-edge technological systems to foster new forms of economic productivity, in contrast to the automation-centric Industry 4.0 model, which primarily focused on minimizing labor costs and maximizing operational efficiency through machine-driven processes (Sun & Song, 2025; Demir et al., 2023). In addition to improving operational flexibility, this cooperative approach creates opportunities for innovation-driven economic models that can be adjusted to quickly shifting market conditions.

Increasing productivity through human–machine cooperation is one of the most important components of Industry 5.0's economic sustainability. According to Rahman et al. (2024), the integration of intelligent manufacturing platforms, AI-driven decision support systems, and cobots enables the optimization of production processes while utilizing human cognitive abilities like creativity, problem-solving, and ethical judgment. By fusing the accuracy and durability of machines with the flexibility and inventiveness of human workers, cobots, which are made to operate securely alongside human operators, increase productivity (Maddikunta et al., 2022). In addition to increasing production efficiency, this human–machine symbiosis also helps lower operational errors, improve product quality, and promote a continuous improvement culture in industrial settings.

The importance of mass personalization and customization in manufacturing techniques is also emphasized by the Industry 5.0 model's emphasis on economic sustainability (Saniuk et al., 2023). Through advanced manufacturing technologies like additive manufacturing, AI-driven design, and digital twin systems, Industry 5.0 allows manufacturers to accommodate individualized consumer preferences, in contrast to the mass production models of earlier industrial revolutions that relied on standardized products for global markets (Tao et al., 2019). By producing distinctive, value-added products, mass customization not only boosts consumer happiness and loyalty but also helps increase profit margins. Digital supply chain integration, which enables responsive production planning, effective resource allocation, and real-time monitoring of supply and demand dynamics, facilitates the trend toward personalization. As a result, businesses can strengthen their economic resilience by cutting inventory costs, reducing waste, and reacting quickly to changes in the market.

By simplifying processes and cutting down on inefficiencies, Industry 5.0's spread of smart factories and intelligent production systems also promotes economic sustainability. Real-time process optimization, autonomous quality control, and predictive maintenance are made possible by smart factories, which are driven by IoT devices, AI algorithms, and data analytics (Jabbour et al., 2018). These features improve equipment longevity, decrease maintenance expenses, and decrease downtime, all of which increase total cost effectiveness. Additionally, the use of digital twins makes it easier to simulate and test production processes virtually, which enables manufacturers to find possible problems, streamline operations, and create novel product ideas without having to pay for physical prototyping. A lean production mindset is supported by this digitization of industrial processes, which maximizes output and value creation while consuming the fewest resources possible.

Industry 5.0 promotes the creation of new business models that are consistent with the concepts of economic sustainability in addition to offering operational benefits. Manufacturers are shifting from traditional product sales to value-added services and customer-centric solutions, as demonstrated by the emergence of platform-based economies, product-service systems (PSS), and servitization strategies (Geissdoerfer et al., 2017). Businesses can provide integrated services like lifecycle management, performance monitoring, and predictive maintenance by utilizing digital platforms, which will increase revenue and improve client relations. By encouraging product longevity, reuse, and remanufacturing, the servitization movement not only encourages company innovation but also advances the circular economy. These business models increase the economic resilience of companies operating in volatile markets by fostering more consistent and varied revenue streams.

The empowerment of SMEs is a crucial aspect of economic sustainability in Industry 5.0. The democratization of digital technology in Industry 5.0 gives SMEs the chance to compete on an even playing field, whereas prior industrial revolutions frequently favored huge firms with substantial capital investment capacities. SMEs can engage in global value chains, personalize their products, and reach niche markets with creative solutions when they have access to cloud-based platforms, collaborative networks, and reasonably priced digital tools (Xu et al., 2018). Through encouraging entrepreneurship, job creation, and regional industrial diversification, this inclusivity not only promotes local economic development but also advances socioeconomic fairness on a larger scale.

Furthermore, the economic aspect of Industry 5.0 is becoming more and more impacted by the incorporation of sustainability considerations into financial and investment strategies. A growing understanding of the financial materiality of sustainability performance is reflected in the increasing use of environmental, social, and governance (ESG) metrics in investment decisions. Businesses that embrace Industry 5.0 concepts show their dedication to environmental stewardship, worker well-being, and responsible production, which puts them in a better position to achieve ESG standards (Jabbour et al., 2018). Long-term financial performance, corporate reputation, and access to finance can all be improved by this alignment with investor expectations. The economic feasibility of Industry 5.0 adoption is further supported by legislative frameworks and governmental incentives that encourage sustainable industrial practices.

But achieving Industry 5.0's potential for economic sustainability also means overcoming some obstacles, mainly related to workforce preparedness, technology integration, and investment costs. Particularly for SMEs and businesses in developing nations, the upfront capital costs necessary to deploy advanced manufacturing systems, digital infrastructures, and upskilling initiatives can be a major obstacle. The effective implementation of Industry 5.0 initiatives depends on a number of crucial factors, including managing technological complexity, ensuring interoperability among heterogeneous systems, and protecting data security (Maddikunta et al., 2022). Furthermore, significant expenditures in education, training, and ongoing learning are required to develop a trained workforce that can function and innovate in human–machine collaborative situations.

4.3.2 Environmental Sustainability

Environmental sustainability has emerged as a key issue in industrial development, especially in light of the rapid deterioration of the environment, resource depletion, and climate change. Large-scale environmental effects, high carbon emissions, and intensive resource consumption have all historically been linked to industrial growth, which frequently puts financial gain ahead of environmental stewardship. By integrating environmental sustainability into its fundamental framework, Industry 5.0 offers a revolutionary approach to industrial practice and encourages production systems that are sustainable and technologically sophisticated. In contrast to earlier industrial paradigms that prioritized efficiency and output maximization, Industry 5.0 promotes manufacturing procedures and industrial operations that minimize environmental damage, maximize resource use, and support more general ecological objectives like biodiversity preservation, climate resilience, and the shift to low-carbon economies (European Commission, 2021).

Regarding environmental sustainability, Industry 5.0's incorporation of the circular economy's tenets is one of its distinguishing features. Depletion of resources and pollution of the environment have been greatly exacerbated by the conventional linear economic model of "take, make, dispose." By eliminating waste from the system and separating economic activity from the use of limited resources, the circular economy, on the other hand, aims to redefine growth. Industry 5.0 makes use of cutting-edge technologies like AI, big data analytics, and the IoT to support closed-loop production systems, optimize resource flows, and enable circular manufacturing processes (Geissdoerfer et al., 2017). Manufacturers may expand product lifecycles and lessen their environmental impact by using recycling and remanufacturing techniques, improving resource efficiency, and reducing material waste by employing real-time data and predictive analytics.

The potential of Industry 5.0 to be environmentally sustainable is further enhanced by digital twin technology. Real-time monitoring, simulation, and optimization of industrial processes are made possible by digital twins, which create virtual versions of physical assets, production lines, or entire manufacturing systems (Tao et al., 2019). With this capability, manufacturers can test alternative production scenarios, anticipate possible inefficiencies, and optimize system performance without requiring the resource-intensive trials that are usually associated with physical prototyping. As a result, using digital twins helps make manufacturing processes more sustainable by lowering material waste, energy use, and production downtime. Digital twins also facilitate the use of eco-design concepts, which limit the environmental impact of processes and products throughout their lifecycle by designing them from the ground up.

In Industry 5.0, AI is also essential for improving environmental sustainability through intelligent process control, energy management, and waste reduction. AI-driven solutions can lower the carbon footprint of industrial activities by tracking patterns in energy usage, spotting inefficiencies, and suggesting ideal operating conditions (Jabbour et al., 2018). AI-powered predictive maintenance increases the lifespan of machinery, lowers the chance of unscheduled downtime, and lowers the risk of equipment failure, all of which further reduce energy consumption and resource depletion. AI can also be used in supply chain optimization, which enables businesses to minimize transportation emissions, streamline logistics, and encourage the localization of production processes, all of which together lessen their negative effects on the environment.

Industry 5.0's contribution to environmental sustainability is also characterized by the use of additive manufacturing technology, or 3D printing. According to Maddikunta et al. (2022), additive manufacturing is a more efficient way to produce complicated components than standard subtractive manufacturing techniques. Additive manufacturing, which builds things layer by layer, lowers the amount of raw materials and scrap while enabling product customization, which lowers overproduction and unsold inventories. Additionally, localized manufacturing made possible by additive manufacturing lowers the environmental costs of global supply chains and long-distance transportation by supporting decentralized production models. These decentralized methods support local economic resilience and lessen the environmental impact of industrial operations, which is in line with sustainable development aims.

The development and incorporation of renewable energy sources into industrial operations is another benefit of Industry 5.0. Innovations in energy management that include incorporation of renewable energy sources like solar, wind, and biomass into manufacturing processes are made possible by the growing digitalization and connectivity of production systems (Nahavandi, 2019). IoT-enabled energy systems and smart grids make it easier to monitor and manage energy use in real time, maximizing energy use and effectively utilizing renewable energy sources. By decreasing dependence on fossil fuels and improving energy efficiency, Industry 5.0 directly supports international initiatives to mitigate climate change and advance sustainable energy transitions.

Responsible supply chain management is also closely related to environmental sustainability in Industry 5.0. Companies can track environmental performance from the sourcing of raw materials to the delivery of finished goods, thanks to advanced data analytics, blockchain technology, and IoT devices that improve transparency, traceability, and accountability across supply chains (Demir et al., 2023). This enhanced visibility makes it possible to spot and stop unsustainable behaviors, encourage ethical sourcing, and implement more environmentally friendly logistical solutions. In markets that are becoming more and more concerned with sustainability, supply chain transparency not only helps businesses adhere to environmental laws and norms but also enhances customer confidence and brand recognition.

Industry 5.0 has the potential to achieve environmental sustainability, but there are a number of obstacles in the way. Technical obstacles pertaining to data security, system compatibility, and the integration of legacy systems with new digital infrastructures are among them. For SMEs in emerging nations in particular, the substantial upfront costs associated with sustainable process redesign and sophisticated manufacturing technology can be a major obstacle. Additionally, the ecological footprints of digital technologies, such as data center energy consumption, electronic trash creation, and the sourcing of rare materials for advanced hardware, must be carefully weighed against the environmental benefits of these technologies (Jabbour et al., 2018). To overcome these obstacles, a comprehensive strategy that incorporates industry standards, cross-sectoral cooperation, supportive legislative frameworks, and technological innovation is needed to guarantee that the implementation of Industry 5.0 results in real and quantifiable environmental benefits.

4.3.3 Social Sustainability

In the discussion of sustainable development, social sustainability has become a crucial component that highlights the significance of human well-being, equity, inclusion, and societal resilience in industrial and economic systems. As the paradigm aims to realign industrial advancement with human-centric ideals, social sustainability assumes increasing significance in the context of Industry 5.0. Industry 5.0 introduces a shift toward collaborative intelligence, where human creativity, intuition, and emotional intelligence complement machine efficiency and computational capabilities. This contrasts with Industry 4.0, which was primarily defined by task automation and efficiency optimization through digital technologies (Demir et al., 2023). This collaborative vision places people back at the heart of industrial processes, striving not only to boost productivity but also to ensure that technical breakthroughs contribute positively to societal well-being and the creation of meaningful, dignified work.

The loss of human labor as a result of automation and AI has been one of the most urgent issues surrounding earlier industrial revolutions, especially Industry 4.0. Although there is no denying that automation has increased productivity, it has also increased concerns about job losses, growing skills gaps, and the decline of traditional employment sectors. In order to solve these issues, Industry 5.0 advocates for the idea of human–machine collaboration, in which cutting-edge technologies are intended to enhance human skills rather than replace human labor (Nahavandi, 2019). Industry 5.0 envisions a future where workers are empowered to engage in creative problem-solving, innovation, and decision-making, tasks that machines alone cannot effectively perform, by utilizing the strengths of both humans and machines. This change promotes employment stability and worker resilience by assisting in the development of high-quality, socially relevant, cognitively challenging, and less automatable jobs.

Industry 5.0's focus on human-centricity also emphasizes the necessity of lifelong learning and ongoing skill development. Digital literacy, systems thinking, and interdisciplinary problem-solving are among the new skill sets that are becoming increasingly necessary as industrial systems grow more complex and dependent on cutting-edge technologies. Industry 5.0 promotes funding for educational and training initiatives that give employees the skills they need to succeed in cutting-edge workplaces (Maddikunta et al., 2022). In addition to improving individual employability, this emphasis on human capital development fortifies societies' ability to adapt to the unpredictabilities of technological change. Furthermore, lowering social inequality and guaranteeing that the advantages of technological advancement are fairly shared among various demographic and socioeconomic groups depend on promoting equitable access to education and upskilling opportunities.

In the context of Industry 5.0, workplace well-being and the standard of working life are essential components of social sustainability. Intelligent systems, ergonomic design, and collaborative robotics can all be combined to create safer, healthier, and more satisfying workplaces. Industry 5.0 can lower occupational health risks and enhance overall workplace safety by automating repetitive, physically taxing, and hazardous tasks (Jabbour et al., 2018). At the same time, assistive technologies and HMIs can improve job satisfaction by allowing employees to concentrate on tasks that call for creativity, critical thinking, and interpersonal skills. The larger objectives of social sustainability, which place a high value on justice, dignity, and the development of human potential in the workplace, are in line with this human-centric strategy.

A key component of Industry 5.0's social sustainability agenda is ethical considerations. Concerns regarding algorithmic fairness, data security, privacy, and the possibility of technical abuse are brought up by the extensive usage of AI, data-driven decision-making, and linked systems. According to Demir et al. (2023), Industry 5.0 encourages the use of ethical frameworks and governance models that guarantee responsibility, transparency, and respect for human rights in the creation and application of industrial technology. This entails putting in place regulations that protect private information, stop discriminatory behavior, and encourage inclusive involvement in technical governance procedures. Industry 5.0 promotes the development of relationships based on trust between industry stakeholders, employees, customers, and society at large by integrating ethical considerations into industrial innovation.

Another essential component of Industry 5.0's social sustainability dimension is social inclusion. Technology democratization, or opening up cutting-edge tools and platforms to a wider range of people and organizations, can aid in closing the digital divide and advancing fair economic participation. In industrial ecosystems, Industry 5.0 promotes the active participation of a variety of stakeholders, such as local economies, SMEs, and marginalized communities (European Commission, 2021). Industry 5.0 can enable underrepresented groups to participate in innovation, entrepreneurship, and sustainable economic activities by creating cooperative networks and knowledge-sharing platforms. In addition to fostering social cohesion, this inclusivity helps industrial systems become more resilient and adaptable to global issues like demographic changes, economic inequality, and geopolitical tensions.

The Industry 5.0 paradigm significantly improves corporate social responsibility (CSR) and community participation. Stronger ties between businesses and the communities where they operate are promoted by the human-centric approach. Businesses are urged to embrace socially conscious operations that put the welfare of the community, environmental conservation, and moral labor standards first. These activities can include educational partnerships, social innovation assistance, local employment programs, and community development projects (Geissdoerfer et al., 2017). Industry 5.0 promotes an economically and socially responsible industrial growth model by coordinating corporate strategies with social sustainability goals.

But there are obstacles in the way of Industry 5.0's social sustainability realization. Resolving issues with workforce preparedness, organizational culture, and technology accessibility is necessary to achieve meaningful human–machine collaboration. Economically disadvantaged areas and smaller businesses may be disproportionately impacted by the upfront costs of upskilling, infrastructure adaptation, and technology deployment. Additionally, proactive approaches that put an emphasis on open communication, stakeholder engagement, and inclusive policy development are required to manage the socio-psychological effects of technological change, such as fears of job displacement, resistance to innovation, and cultural adaptation (Jabbour et al., 2018). To guarantee that the shift to Industry 5.0 does not worsen already existing socioeconomic inequalities or give rise to new kinds of exclusion, these issues must be addressed.

4.4 Opportunities of Industry 5.0 for Sustainable Development

Through the use of advanced digital technology, human–machine collaboration, and a renewed emphasis on environmental and societal values, Industry 5.0 offers a revolutionary chance to incorporate sustainable development ideas into industrial processes (Adel & HS Alani, 2024). In contrast to its predecessor, Industry 4.0, which placed a higher priority on automation and technological efficiency, Industry 5.0 introduces a paradigm that equally emphasizes ecological stewardship, social responsibility, and human-centricity (Sheikh et al., 2024). This transition fosters innovation that tackles urgent global issues like social inequity, resource scarcity, and climate change in addition to increasing industrial productivity. Industry 5.0 creates new opportunities for industries to actively contribute to sustainable economic, social, and environmental transformation by coordinating technological advancement with the SDGs of the UN (European Commission, 2021).

Personalized production and mass customization are two of Industry 5.0's most important prospects for sustainable development (Martini et al., 2024). Industry 5.0 enables manufacturers to transition from the inflexible mass production models of the past to highly flexible, customer-oriented manufacturing systems, made possible by advanced manufacturing technologies like digital twins, additive manufacturing, and AI (Maddikunta et al., 2022). This move toward customization reduces overproduction, minimizes inventory waste, and improves product lifecycle value, all of which contribute to sustainable consumption patterns in addition to meeting the growing demand from consumers for personalized products. More precise and resource-efficient production on demand immediately supports more sustainable manufacturing methods and strengthens the bonds between producers and consumers (Hegab et al., 2023).

According to Lewandowska et al. (2023), Industry 5.0 presents a noteworthy prospect for empowering SMEs and promoting regional economic growth. Industry 5.0 lowers entry barriers for SMEs and empowers them to compete successfully in domestic and international markets by democratizing digital tools, cloud-based platforms, and collaborative networks (Xu et al., 2018). The empowerment of SMEs encourages entrepreneurship, job creation, and regional industry diversification, all of which contribute to inclusive economic growth. Additionally, decentralized manufacturing technologies enable production to be localized, which improves supply chain resilience and lessens the environmental impact of centralized production models and long-distance transportation (Nandi et al., 2021). Industry 5.0 is in line with SDGs that prioritize environmental responsibility and economic inclusivity by encouraging distributed manufacturing networks.

According to Varriale et al. (2023), the Industry 5.0 paradigm presents exceptional prospects for improving sustainability through the incorporation of smart technologies into supply chain management. All supply chains can be monitored in real time, transparently, and traceably, thanks to the use of IoT devices, blockchain technology, and AI-driven analytics (Jabbour et al., 2018). Accountability and trust among stakeholders are fostered by this degree of visibility, which encourages the adoption of ethical labor standards, environmental compliance, and sustainable sourcing methods. According to Liu et al. (2024), intelligent supply chain management may also improve energy efficiency, minimize waste, and optimize logistics, all of which help create value chains that are more resilient and environmentally friendly. Customers are also better equipped to make educated selections because of the increased transparency, which increases demand for eco-friendly goods and ethical corporate practices.

In the realm of smart cities and sustainable urban development, Industry 5.0 technologies hold promise for establishing human-centered, integrated, and efficient urban systems (Adel & HS Alani, 2024). Intelligent transportation systems, smart utilities, and sustainable housing options can all result from the fusion of industrial technologies and urban infrastructure (Demir et al., 2023). In urban settings, for instance, AI-powered energy management systems and IoT-enabled smart grids can maximize the distribution and usage of renewable energy sources, lowering greenhouse gas emissions and boosting energy resilience (Mishra & Singh, 2023). Urban circular economies can be promoted and reliance on external supply chains can be decreased by supporting localized production and services using collaborative robotics and digital platforms (Sarc et al., 2019). Industry 5.0 helps achieve sustainable cities and communities as defined by SDG 11 by promoting mutually beneficial relationships between industrial ecosystems and urban communities.

The potential for Industry 5.0 to help mitigate climate change and improve environmental resilience is another important possibility (Mourtzis et al., 2022). The implementation of energy-efficient production systems, predictive maintenance, and modern manufacturing techniques can help enterprises drastically reduce their carbon emissions and environmental impact (Geissdoerfer et al., 2017). Real-time monitoring and predictive analytics make it possible to manage emissions, waste production, and resource consumption proactively, enabling industries to match their operations with international climate action targets like those established by the Paris Agreement. Furthermore, smart energy management systems facilitate the integration of renewable energy sources into industrial processes, which aids in the shift to low-carbon industrial models (Zheng, 2024). In light of growing climate-related risks and uncertainties, Industry 5.0's ability to promote resilient and adaptive industrial practices is especially important (Table 4.2).

Additionally, Industry 5.0 presents special chances to encourage social innovation and community involvement. According to Nahavandi (2019), Industry 5.0 promotes the adoption of inclusive business models that give priority to stakeholder participation, community well-being, and ethical governance by highlighting human-centric principles. Through partnerships among businesses, universities, governments, and civil society, collaborative platforms, digital learning environments, and open innovation ecosystems stimulate social innovation that tackles regional issues and advances sustainable development programs (Costa & Matias, 2020). In addition to improving the societal impact of industrial operations, this cooperative method fosters social cohesiveness, trust, and the development of shared value. Building robust, inclusive, and sustainable communities is a goal that Industry 5.0 pursues by integrating CSR into fundamental industrial objectives.

Industry 5.0 has the potential to significantly contribute to sustainable development through the improvement of risk management and disaster resilience (Table 4.2) (Kasinathan et al., 2022). According to Tao et al. (2019), the incorporation of AI, the IoT, and predictive analytics into industrial and infrastructure systems facilitates early warning, quick response, and flexible risk management techniques. In reducing the effects of supply chain interruptions, industrial catastrophes, and natural disasters, these capabilities are very helpful. Building resilient infrastructure, promoting inclusive and sustainable industrialization, and fostering innovation are the three main goals of SDG 9, which is supported by Industry 5.0 through the development of resilient infrastructure and adaptive manufacturing systems (Costa, 2024). Furthermore, resilient industrial systems help maintain the stability and sustainability of economies and societies by being better equipped to endure global shocks like pandemics and economic crises (Table 4.2).

However, a range of stakeholders, including legislators, company executives, academic researchers, and civil society organizations, must collaborate in order for these prospects to be fulfilled. Effective integration of Industry 5.0 into sustainable development strategies requires investments in digital and human capital, the establishment of supportive policy frameworks, and the promotion of cross-sectoral collaboration. Regulations that encourage the deployment of ethical technology, encourage responsible innovation, and offer incentives for sustainable practices

Table 4.2 Opportunities of Industry 5.0 for Sustainable Development.

Opportunity Area	*Industry 5.0 Mechanism*	*Aligned SDGs*	*Anticipated Positive Impacts*	*Potential Stakeholders*
Human–machine collaboration for productivity and job enrichment	Integration of cobots, AI-assisted tools, and human-centric design	SDG 8 (Decent Work and Economic Growth), SDG 9 (Industry, Innovation, and Infrastructure)	Increases productivity, enhances job quality, reduces routine task burdens, and fosters creative and high-value roles	Industry leaders, workforce, training institutions, and labor unions
Mass personalization and sustainable consumption	AI-driven design, additive manufacturing, and flexible production systems	SDG 12 (Responsible Consumption and Production), SDG 9 (Industry, Innovation, and Infrastructure)	Reduces overproduction, minimizes waste, aligns production with consumer needs, and supports sustainable market dynamics	Manufacturers, consumers, SMEs, and digital platform providers
Empowerment of SMEs and local economies	Access to decentralized manufacturing, digital tools, cloud services, and collaborative platforms	SDG 8 (Decent Work and Economic Growth) and SDG 10 (Reduced Inequalities)	Enhances SME competitiveness, fosters entrepreneurship, strengthens regional economies, and supports equitable growth	SMEs, policymakers, financial institutions, and local governments
Circular economy integration and resource efficiency	Blockchain-enabled supply chain transparency, AI-based resource management, and digital twins	SDG 12 (Responsible Consumption and Production) and SDG 13 (Climate Action)	Promotes closed-loop systems, enhances material traceability, and reduces resource consumption and environmental impact	Supply chain partners, environmental agencies, and industrial consortia
Sustainable supply chain and ethical sourcing	IoT-enabled supply chain monitoring, AI-driven logistics optimization, and blockchain for ethical sourcing	SDG 12 (Responsible Consumption and Production) and SDG 16 (Peace, Justice, and Strong Institutions)	Increases transparency, promotes ethical labor practices, reduces environmental footprint, and enhances stakeholder trust	Global supply chain actors, NGOs, regulatory bodies, and consumers
Smart cities and urban–industrial symbiosis	Integration of industrial systems with smart urban infrastructure, localized manufacturing, and AI-enabled resource sharing	SDG 11 (Sustainable Cities and Communities) and SDG 9 (Industry, Innovation, and Infrastructure)	Enhances urban sustainability, supports local production, optimizes resource usage, and reduces transport emissions	Urban planners, municipal governments, industry players, and technology providers

(*Continued*)

Table 4.2 (Continued)

Opportunity Area	*Industry 5.0 Mechanism*	*Aligned SDGs*	*Anticipated Positive Impacts*	*Potential Stakeholders*
Climate change mitigation and environmental resilience	Adoption of renewable energy, AI-driven energy management, and predictive maintenance for resource conservation	SDG 7 (Affordable and Clean Energy) and SDG 13 (Climate Action)	Reduces greenhouse gas emissions, enhances energy efficiency, and promotes sustainable industrial operations	Energy providers, environmental organizations, industries, and governments
Social innovation and community engagement	Co-creation platforms, participatory innovation ecosystems, and CSR initiatives	SDG 17 (Partnerships for the Goals) and SDG 8 (Decent Work and Economic Growth)	Fosters innovation for social good, enhances community well-being, and promotes collaborative industrial models	Civil society organizations, industries, academic institutions, and local communities
Disaster resilience and risk management	Predictive analytics, AI-enabled risk assessment, and resilient supply chain systems	SDG 9 (Industry, Innovation, and Infrastructure) and SDG 13 (Climate Action)	Enhances industrial adaptability, improves disaster preparedness, and strengthens infrastructure resilience	Emergency management agencies, industries, policymakers, and research institutions

are necessary to fully realize Industry 5.0's potential for sustainable development (European Commission, 2021). To ensure that the transition to Industry 5.0 significantly contributes to global sustainability goals, industrial businesses must also foster a culture of ethical responsibility, inclusivity, and continuous learning.

4.5 Challenges and Barriers

Industry 5.0 offers promising opportunities for long-term expansion, but its actual application is fraught with a wide range of organizational, legal, technological, and economic barriers and difficulties (Table 4.3). The transition to a human-centered, resilient, and sustainable industrial paradigm requires substantial adjustments in industrial ecosystems. There are difficulties with these adjustments. Technical integration is one of the main challenges, particularly with regard to platform and system compatibility. As opposed to the relatively modular frameworks of Industry 4.0, Industry 5.0 seeks to facilitate human–machine collaboration, which calls for high levels of system integration, real-time data exchange, and harmonization of heterogeneous technologies (Ostrom et al., 1999). Such interoperability is technically challenging to accomplish and often requires significant investments in digital infrastructure, software standards, and cross-platform compatibility. Moreover, the advanced technologies promoted by Industry 5.0 might not work with more traditional systems in well-established sectors, necessitating costly upgrades or whole system redesigns. More widespread adoption may be deterred by this (Table 4.3).

Implementing Industry 5.0 technologies is made more difficult by risks related to data security and cybersecurity. Industrial operations are more susceptible to cyber threats, data breaches, and malicious attacks as a result of their growing reliance on cloud-based platforms, AI-driven processes, and interconnected systems (Alcaraz & Zeadally, 2015). Strong cybersecurity procedures are essential as human-machine collaboration increases and sensitive operational data is transferred across digital networks. But a recurring obstacle is the quickly changing nature of cyberthreats and the disparities in cybersecurity maturity among enterprises. Furthermore, strict data governance procedures and adherence to global laws like the General Data Protection Regulation (GDPR) are required due to worries about data privacy, especially in situations involving human-centric applications and customized production (Voigt & Von dem Bussche, 2017). These issues underline the necessity for comprehensive cybersecurity plans that go beyond technical safeguards to include corporate awareness, employee training, and stakeholder involvement.

Adopting Industry 5.0 is also fraught with financial challenges, particularly when it comes to the upfront expenditures of labor upskilling, digital transformation, and new manufacturing technologies. SMEs frequently face financial limitations that restrict their ability to fully adopt Industry 5.0 principles, whereas large multinational corporations may have the financial resources to invest in cutting-edge technologies and training programs (Mittal et al., 2018). Purchasing cobots, putting AI-driven systems in place, and incorporating digital twins might be prohibitively expensive, especially for businesses that operate in developing

Table 4.3 Challenges and Barriers to Industry 5.0 Implementation for Sustainable Development.

Challenge Category	*Specific Challenge*	*Affected Domain(s)*	*Implications for Sustainable Development*	*Potential Mitigation Strategies*
Technological integration	Lack of interoperability between legacy and advanced systems	Manufacturing and IT infrastructure	Limits digital transformation and increases resource inefficiency	Standardization protocols, modular systems, and government-supported R&D
	Complexity in HMI design and cognitive ergonomics	Workforce productivity and safety	Reduces effectiveness of collaboration and risks operator burnout or misuse	Inclusive User Experience (UX)/urban index (UI) research, adaptive HMI systems, and participatory design
	Insufficient maturity of enabling technologies (e.g., digital twins and cobots)	Industrial operations and R&D	Slows implementation and hinders scaling of sustainable production models	Investment in pilot programs, and cross-sector innovation consortia
Cybersecurity and data privacy	Increased vulnerability to cyberattacks and data breaches	Digital supply chains and data-driven manufacturing	Threatens business continuity, and undermines consumer trust and ethical standards	AI-based threat detection, zero-trust architectures, and General Data Protection Regulation (GDPR)-compliant protocols
	Ethical risks of AI-driven decision-making and algorithmic bias	Governance, human rights, and workplace equity	Undermines social trust and exacerbates inequalities	Transparent AI models, algorithm audits, and regulatory oversight
Economic constraints	High initial investment costs for SMEs	Financial access and SME participation	Excludes small firms from transformation and deepens digital divide	Public–private partnerships, tax incentives, and accessible funding mechanisms
	Uncertain ROI in sustainability-focused technologies	Business strategy and capital markets	Reduces motivation for sustainable innovation	Lifecycle costing models and ESG-linked financial instruments

(*Continued*)

Table 4.3 (Continued)

Challenge Category	*Specific Challenge*	*Affected Domain(s)*	*Implications for Sustainable Development*	*Potential Mitigation Strategies*
Workforce readiness	Skills gap in emerging technologies and collaborative systems	Education and labor market	Limits workforce adaptability and risks displacement or under-employment	Reskilling/ upskilling programs, vocational training, and university–industry collaboration
	Resistance to change due to automation anxiety and job insecurity	Organizational culture and employee engagement	Slows the adoption of human–machine collaboration and undermines morale	Change management, participative decision-making, and human-centered leadership
Organizational culture	Misalignment between existing business models and Industry 5.0 values	Strategic planning and operations	Limits organizational agility and sustainability adoption	Cultural transformation, sustainability KPIs (key performance indicators), and leadership alignment
	Siloed innovation processes lacking inter-departmental coordination	R&D, production, and CSR	Prevents holistic implementation of sustainability and collaboration	Cross-functional teams and integrative innovation frameworks

nations or in areas with limited resources. Furthermore, when compared to conventional efficiency metrics, the return on investment (ROI) for Industry 5.0 initiatives might not be immediately obvious. Business executives may become risk apprehensive as a result of this uncertainty, which would decrease the rate of adoption. To lessen these financial obstacles and promote wider involvement in the shift to Industry 5.0, public policy initiatives, financial incentives, and supportive funding sources would be needed.

One further crucial issue for the effective implementation of Industry 5.0 is workforce preparation. A workforce with new abilities, such as digital literacy, technological fluency, and collaborative skills, is required for the integration of sophisticated technology into human-centric industrial processes (Kumar et al., 2021). However, a major obstacle to human–machine collaboration is the presence of skill gaps, especially in conventional manufacturing sectors. Initiatives to reskill and upskill personnel are crucial, but they frequently encounter challenges such as a lack of training materials, employee resistance to change, and a discrepancy between industry demands and academic programs. Adoption of Industry 5.0 principles can also be made more difficult by the psychological effects of technological change, such as reduced job security, uncertainty about future employment prospects, and fears of job displacement. Comprehensive workforce development strategies that are inclusive, flexible, and forward-looking are necessary to address these human resource issues.

The success or failure of Industry 5.0 efforts is also significantly influenced by organizational culture and management techniques. Organizational attitudes must change from hierarchical, efficiency-driven techniques to more collaborative, innovation-oriented practices in order to transition to a human-centric, sustainable industrial model (Moeuf et al., 2018). Rigid management structures, resistance to change, and deeply rooted organizational cultures that place a premium on immediate profits, however, can all be very strong obstacles. Overcoming these cultural barriers requires effective leadership that supports sustainability, encourages interdisciplinary cooperation, and supports ongoing learning. Furthermore, ethical considerations, stakeholder engagement, and transparent governance models, elements that are frequently neglected in traditional industrial settings, must be included in order to connect business aims with more general sustainability objectives.

The adoption of Industry 5.0 is further complicated by issues pertaining to regulations and policies. Industry 5.0 visions have been started by the European Commission and other international organizations, but full policy frameworks, standards, and guidelines that enable its global implementation are still lacking (Breque et al., 2021). Emerging technologies like AI, collaborative robotics, and data-driven manufacturing are subject to a constantly changing regulatory environment, which creates ambiguities and discrepancies between countries. Organizations may face legal risks, compliance burdens, and difficulties making investment decisions as a result of these regulatory ambiguities. Furthermore, the implementation gap is made worse by the lack of concerted legislative initiatives to promote labor transformation, technology standardization, and sustainable industrial practices. To create cogent policy tools that promote innovation while guaranteeing moral and sustainable industrial development, policymakers must collaborate with academic institutions, industry stakeholders, and civil society.

Lastly, societal and ethical issues pose complex problems that need to be carefully handled. Complex issues about human agency, autonomy, and the effects of automation and AI on society are brought up by the use of cutting-edge technologies in industrial settings (Floridi et al., 2018). To avoid unforeseen negative effects, concerns regarding algorithmic bias, accountability, and transparency in decision-making processes must be addressed. Furthermore, Industry 5.0's focus on human-centricity and personalization calls for informed consent, careful handling of personal data, and respect for individual rights. To make sure that Industry 5.0 does not duplicate or worsen already existing social injustices, ethical governance frameworks that put human dignity, justice, and societal benefit first are crucial. The significance of interdisciplinary approaches that combine technological innovation with philosophical, legal, and social scientific viewpoints is highlighted by this ethical dimension.

4.6 Future Research Directions

The advent of Industry 5.0 as a paradigm that combines technical innovation with sustainability, resilience, and human-centric values has created a number of opportunities for both academic research and real-world application. But in order to fully grasp the complexities of Industry 5.0 and realize its potential impact, more research is required as the theoretical and empirical understanding of this phenomenon is still in its infancy. Future research in the multidisciplinary field of human–machine collaboration is crucial. Industry 5.0 places more emphasis on the collaborative relationship between intelligent systems and humans than Industry 4.0's automation-driven approach. It is crucial to look into the cognitive, psychological, and ergonomic aspects of this collaboration in order to design systems that improve human well-being and productivity (Javaid et al., 2021). Research should focus on mechanisms for efficient shared decision-making in complex industrial environments, trust in human–robot collaboration, and the development of adaptive interfaces.

Innovations in technology needed to support sustainable Industry 5.0 ecosystems represent another exciting area of research. Despite notable progress in areas like digital twins, cobots, and AI, further innovation aimed toward sustainability results is required. According to Morrar et al. (2017), the incorporation of circular economy principles into Industry 5.0 production models necessitates technological solutions for effective resource tracking, waste minimization, and product life extension. Research should look more closely at how cutting-edge technologies like decentralized manufacturing platforms, blockchain, and green IoT can be used to build robust, transparent, and sustainable industrial networks. To make sure that technological advancement does not unintentionally jeopardize sustainability objectives, it is also important to critically examine the environmental impact of digital technologies themselves, including the energy consumption of AI systems and the e-waste produced by sophisticated industrial equipment.

Another crucial area for further study is the governance and policy frameworks that support the moral, just, and inclusive implementation of Industry 5.0. Significant obstacles to Industry 5.0 implementation are presented by the intricate regulatory environment pertaining to labor rights, algorithmic transparency, data privacy, and environmental compliance (European Commission, 2021). Research is required to examine participatory governance, stakeholder engagement, and policy innovation models that complement Industry 5.0's sustainability and human-centric values. Comparative research between various geographical areas and industrial domains may yield important information about optimal methodologies and flexible frameworks for a range of situations.

Finally, impact-oriented and longitudinal research is necessary to assess the long-term socioeconomic and environmental impacts of Industry 5.0 initiatives. The success of Industry 5.0 must be evaluated in terms of its effects on social fairness, environmental preservation, and economic resilience as opposed to the more immediate efficiency measures of Industry 4.0 (Horváth & Szabó, 2019). Empirical research, case studies, and cross-sectoral comparisons will be essential to building an evidence basis that informs academic debates and practical implementation strategies. This type of study will help clarify how Industry 5.0 may revolutionize global sustainable development.

4.7 Conclusion

Industry 5.0 represents a significant change in the direction of industrial development, moving away from automation and technology innovation alone and toward a more human-centered, balanced strategy that actively supports sustainable growth. In contrast to earlier industrial paradigms, Industry 5.0 combines cutting-edge digital technologies with the moral and creative abilities of human laborers in an effort to promote collaborative intelligence, increase resilience, and tackle urgent global issues like social inequality, environmental degradation, and economic volatility. This review has emphasized how Industry 5.0 facilitates flexible, customized production systems; encourages circular economy practices; encourages responsible resource use; and advances inclusive and meaningful employment opportunities, all of which contribute to the aspects of economic, environmental, and social sustainability. Additionally, by rethinking industrial systems to put human well-being, societal resilience, and environmental stewardship first, Industry 5.0 has highlighted the potential to make a substantial contribution to reaching the SDGs of the UN.

But achieving Industry 5.0's full potential necessitates overcoming a challenging array of organizational, ethical, legal, technological, and economic issues. These include making sure that technology integrates seamlessly, managing cybersecurity threats, removing financial obstacles for SMEs, preparing the workforce for digitalized and collaborative industrial settings, and creating thorough governance frameworks that encourage moral and responsible innovation. Furthermore, consistent efforts in research, policy development, and stakeholder collaboration across disciplines and sectors will be necessary for Industry 5.0's societal acceptance and practical success. The complexity of these issues calls for an all-encompassing, inclusive strategy that respects human ingenuity, encourages environmentally friendly behavior, and gives ethical issues top priority when implementing new technologies.

Future studies should focus on expanding our knowledge of human–machine cooperation, investigating cutting-edge technological solutions for environmentally friendly production, and examining the socioeconomic effects of Industry 5.0 on various industries and communities. Furthermore, the creation of strong governance frameworks and policy tools will be necessary to direct moral implementation and guarantee that the advantages of Industry 5.0 are shared fairly. Industry 5.0 provides a revolutionary framework for bringing industrial advancement into line with the demands of sustainability and human-centric development as industries negotiate the complexity of a quickly evolving global landscape characterized by digital transformation, climate change, and socioeconomic shifts. Industry 5.0 has the potential to transform the industrial future in the direction of a more just, resilient, and sustainable global society if it is carefully applied and continuously improved through teamwork.

References

Adel, A., & HS Alani, N. (2024). Human-centric collaboration and Industry 5.0 framework in smart cities and communities: Fostering sustainable development goals 3, 4, 9, and 11 in Society 5.0. *Smart Cities*, *7*(4), 1723–1775.

Alcaraz, C., & Zeadally, S. (2015). Critical infrastructure protection: Requirements and challenges for the 21st century. *International Journal of Critical Infrastructure Protection*, *8*, 53–66.

Berry, M. A., & Rondinelli, D. A. (1998). Proactive corporate environmental management: A new industrial revolution. *Academy of Management Perspectives*, *12*(2), 38–50.

Breque, M., De Nul, L., & Petridis, A. (2021). *Industry 5.0: Towards a Sustainable, Human-Centric and Resilient European Industry*. European Commission, Directorate-General for Research and Innovation.

Castells, M. (1986). High technology, world development, and structural transformation: The trends and the debate. *Alternatives*, *11*(3), 297–343.

Costa, E. (2024). Industry 5.0 and SDG 9: A symbiotic dance towards sustainable transformation. *Sustainable Earth Reviews*, *7*(1), 4.

Costa, J., & Matias, J. C. (2020). Open Innovation 4.0 as an enhancer of sustainable innovation ecosystems. *Sustainability*, *12*(19), 8112.

Demir, K. A., Döven, G., & Sezen, B. (2023). Industry 5.0 and human–robot co-working: A systematic review and research agenda. *Journal of Manufacturing Systems*, *68*, 340–361.

European Commission. (2021). *Industry 5.0: Towards a Sustainable, Human-Centric and Resilient European Industry*. Directorate-General for Research and Innovation.

Fernández Ocamica, V., Palacino, B., Bartolomé, C., Bernardes Figueirêdo, M., & Lázaro García, C. (2025). Trade-offs and synergies of key biobased value chains and sustainable development goals (SDGs). *Sustainability*, *17*(7), 3040.

Floridi, L., Cowls, J., Beltrametti, M., Chatila, R., Chazerand, P., Dignum, V., . . . Vayena, E. (2018). AI4 people—An ethical framework for a good AI society: Opportunities, risks, principles, and recommendations. *Minds and Machines*, *28*(4), 689–707.

Geissdoerfer, M., Savaget, P., Bocken, N. M., & Hultink, E. J. (2017). The circular economy—A new sustainability paradigm? *Journal of Cleaner Production*, *143*, 757–768.

Hegab, H., Shaban, I., Jamil, M., & Khanna, N. (2023). Toward sustainable future: Strategies, indicators, and challenges for implementing sustainable production systems. *Sustainable Materials and Technologies*, *36*, e00617.

Horváth, D., & Szabó, R. (2019). Driving forces and barriers of Industry 4.0: Do multinational and small and medium-sized companies have equal opportunities? *Technological Forecasting and Social Change*, *146*, 119–132.

Islam, S. M., Munasinghe, M., & Clarke, M. (2003). Making long-term economic growth more sustainable: Evaluating the costs and benefits. *Ecological Economics*, *47*(2–3), 149–166.

Ivanov, D. (2023). The Industry 5.0 framework: Viability-based integration of the resilience, sustainability, and human-centricity perspectives. *International Journal of Production Research*, *61*(5), 1683–1695.

Jabbour, A. B. L., Jabbour, C. J. C., Foropon, C., & Godinho Filho, M. (2018). When titans meet – Can Industry 4.0 revolutionise the environmentally-sustainable manufacturing wave? The role of critical success factors. *Technological Forecasting and Social Change*, *132*, 18–25.

Javaid, M., Haleem, A., Singh, R. P., Suman, R., & Rab, S. (2021). Industry 5.0: Potential applications in COVID-19. *Journal of Industrial Integration and Management*, *6*(4), 429–453.

Jie, H., Khan, I., Alharthi, M., Zafar, M. W., & Saeed, A. (2023). Sustainable energy policy, socio-economic development, and ecological footprint: The economic significance of natural resources, population growth, and industrial development. *Utilities Policy*, *81*, 101490.

Kasinathan, P., Pugazhendhi, R., Elavarasan, R. M., Ramachandaramurthy, V. K., Ramanathan, V., Subramanian, S., . . . Alsharif, M. H. (2022). Realization of sustainable development goals with disruptive technologies by integrating industry 5.0, society 5.0, smart cities and villages. *Sustainability*, *14*(22), 15258.
Kumar, R., Luthra, S., Mangla, S. K., & Kazançoğlu, Y. (2021). COVID-19 impact on sustainable production and operations management. *Sustainable Operations and Computers*, *2*, 1–7.
Kumar, U., Kumar, R., Singh, M., Singh, M., Singh, H., Singh, R., & Chabra, R. (2025, May). A critical review on history of industrial revolutions. In *AIP Conference Proceedings* (Vol. 3185, No. 1, p. 20097). AIP Publishing LLC.
Lewandowska, A., Berniak-Woźny, J., & Ahmad, N. (2023). Competitiveness and innovation of small and medium enterprises under Industry 4.0 and 5.0 challenges: A comprehensive bibliometric analysis. *Equilibrium*, *18*(4). ISSN 1689-765X.
Liao, Y., Deschamps, F., Loures, E. F. R., & Ramos, L. F. P. (2017). Past, present and future of Industry 4.0—A systematic literature review and research agenda proposal. *International Journal of Production Research*, *55*(12), 3609–3629.
Liu, T., Guan, X., Wang, Z., Qin, T., Sun, R., & Wang, Y. (2024). Optimizing green supply chain circular economy in smart cities with integrated machine learning technology. *Heliyon*, *10*(9).
Maddikunta, P. K. R., Pham, Q. V., Prabadevi, B., Deepa, N., Dev, K., Gadekallu, T. R., . . . Alazab, M. (2022). Industry 5.0: A survey on enabling technologies and potential applications. *Journal of Industrial Information Integration*, *26*, 100257.
Martini, B., Bellisario, D., & Coletti, P. (2024). Human-centered and sustainable artificial intelligence in Industry 5.0: Challenges and perspectives. *Sustainability*, *16*(13), 5448.
Mishra, P., & Singh, G. (2023). Energy management systems in sustainable smart cities based on the internet of energy: A technical review. *Energies*, *16*(19), 6903.
Mittal, S., Khan, M. A., Romero, D., & Wuest, T. (2018). A critical review of smart manufacturing & Industry 4.0 maturity models: Implications for small and medium-sized enterprises (SMEs). *Journal of Manufacturing Systems*, *49*, 194–214.
Moeuf, A., Pellerin, R., Lamouri, S., Tamayo-Giraldo, S., & Barbaray, R. (2018). The industrial management of SMEs in the era of Industry 4.0. *International Journal of Production Research*, *56*(3), 1118–1136.
Morrar, R., Arman, H., & Mousa, S. (2017). The Fourth Industrial Revolution (Industry 4.0): A social innovation perspective. *Technology Innovation Management Review*, *7*(11), 12–20.
Mourtzis, D., Angelopoulos, J., & Panopoulos, N. (2022). A literature review of the challenges and opportunities of the transition from Industry 4.0 to Society 5.0. *Energies*, *15*(17), 6276.
Nahavandi, S. (2019). Industry 5.0—A human-centric solution. *Sustainability*, *11*(16), 4371.
Nandi, S., Sarkis, J., Hervani, A. A., & Helms, M. M. (2021). Redesigning supply chains using blockchain-enabled circular economy and COVID-19 experiences. *Sustainable Production and Consumption*, *27*, 10–22.
Ostrom, E., Burger, J., Field, C. B., Norgaard, R. B., & Policansky, D. (1999). Revisiting the commons: Local lessons, global challenges. *Science*, *284*(5412), 278–282.
Pizoń, J., & Gola, A. (2023). Human–machine relationship—perspective and future roadmap for Industry 5.0 solutions. *Machines*, *11*(2), 203.
Rahman, M. M., Khatun, F., Jahan, I., Devnath, R., & Bhuiyan, M. A. A. (2024). Cobotics: The evolving roles and prospects of next-generation collaborative robots in Industry 5.0. *Journal of Robotics*, *2024*(1), 2918089.
Saniuk, S., Grabowska, S., & Fahlevi, M. (2023). Personalization of products and sustainable production and consumption in the context of Industry 5.0. In *Industry 5.0: Creative and Innovative Organizations* (pp. 55–70). Cham: Springer International Publishing.

Sarc, R., Curtis, A., Kandlbauer, L., Khodier, K., Lorber, K. E., & Pomberger, R. (2019). Digitalisation and intelligent robotics in value chain of circular economy oriented waste management–A review. *Waste Management*, *95*, 476–492.

Sheikh, R. A., Ahmed, I., Faqihi, A. Y. A., & Shehawy, Y. M. (2024). Global perspectives on navigating Industry 5.0 knowledge: Achieving resilience, sustainability, and human-centric innovation in manufacturing. *Journal of the Knowledge Economy*, 1–36.

Sun, X., & Song, Y. (2025). Unlocking the synergy: Increasing productivity through human–AI collaboration in the Industry 5.0 era. *Computers & Industrial Engineering*, *200*, 110657.

Tallat, R., Hawbani, A., Wang, X., Al-Dubai, A., Zhao, L., Liu, Z., . . . Alsamhi, S. H. (2023). Navigating Industry 5.0: A survey of key enabling technologies, trends, challenges, and opportunities. *IEEE Communications Surveys & Tutorials*, *26*(2), 1080–1126.

Tao, F., Zhang, M., Liu, Y., & Nee, A. Y. C. (2019). Digital twin in industry: State-of-the-art. *IEEE Transactions on Industrial Informatics*, *15*(4), 2405–2415.

United Nations. (2015). *Transforming Our World: The 2030 Agenda for Sustainable Development*. United Nations General Assembly.

Varriale, V., Cammarano, A., Michelino, F., & Caputo, M. (2023). Industry 5.0 and triple bottom line approach in supply chain management: The state-of-the-art. *Sustainability*, *15*(7), 5712.

Voigt, P., & Von dem Bussche, A. (2017). *The EU General Data Protection Regulation (GDPR): A Practical Guide* (1st ed.). Cham: Springer International Publishing.

World Commission on Environment and Development (WCED). (1987). *Our Common Future*. Oxford University Press.

Xu, L. D., Xu, E. L., & Li, L. (2018). Industry 4.0: State of the art and future trends. *International Journal of Production Research*, *56*(8), 2941–2962.

Yin, Y., Stecke, K. E., & Li, D. (2018). The evolution of production systems from Industry 2.0 through Industry 4.0. *International Journal of Production Research*, *56*(1–2), 848–861.

Zheng, H. (2024). Research on low-carbon development path of new energy industry under the background of smart grid. *Journal of King Saud University-Science*, *36*(3), 103105.

Part II

Technology, Innovation, and Applications for Sustainable Development

5 AI-Driven Strategies for Achieving Sustainable Development Goals

J. Sudarvel, R. Velmurugan, S. Senthilkumar, Ravi Thirumalaisamy, J. Smruthymol, M. Kalimuthu, and N. Jaya Prakash

5.1 Introduction

In 2015, the Sustainable Development Goals (SDGs), which consist of 17 interrelated objectives to tackle the problems of the world by taking on a coherent modus operandi of social development, environmental development, and economic development, were adopted by the United Nations (UN) officially. Although great achievements have been recorded in areas like healthcare, education, and renewable energy, a number of obstacles still remain, especially in areas characterized by political unrest, inefficient policies, and short supply of resources. Recent global issues such as continuous population growth, environmental deterioration, global warming, and economic and social inequality still negatively affect the smooth implementation of the SDGs (Sachs et al., 2022; Weitz et al., 2023).

Considering all the above, the question remains as to whether artificial intelligence (AI) integration is a new force that can actually empower sustainable development. The use of AI-related inventions has the potential to improve resource management, anticipate socio-environmental shifts, and intensify responsiveness to new crises (Brynjolfsson & McAfee, 2017). Today, governments, international organizations, and industry leaders are more actively using AI to streamline services to people, increase the degree of precision in agriculture, modernize healthcare and educational systems, and increase access to high-quality education, as well as to strengthen the efforts of environmental protection (Vinuesa et al., 2020; Dwivedi et al., 2023). Today, AI platforms are utilized to assist with the early detection of disease outbreaks, smart city planning, and universal access to financial services, to name but a few. The technologies are informing more active and effective approaches to realize the SDGs. This chapter intends to examine how using AI can propel an increase in achieving the SDGs. It aims at defining particular AI uses that fit a chosen SDG, evaluating their influence on goal attainment, and determining how concerned technologies are implemented in policy and development models. This study equally focuses on the regulatory and ethical issues that are necessary to achieve the use of AI to ensure a higher level of inclusivity, transparency, accountability, and sustainable impact in the long run. In this way, the study demonstrates the transforming ability of AI to close development gaps and provides a roadmap

DOI: 10.4324/9781003647621-7

for using the potential of AI in a responsible and equitable way in global development scenarios.

5.2 Research Methodology

This chapter utilizes a theoretical research design of a large body of literature of scholarly works, policy briefs, and development plans of countries around the world to analyze the role of AI in helping accomplish the SDGs. In this study, analysis of the concepts along with the analysis of the themes is applied to determine the key AI applications used in the healthcare sector, education, agriculture, and environmental conservation, among others, and the ethical, regulatory, and governance issues surrounding the application of AI. There is no need of primary data collection as such; therefore, the study only aims at coming up with conceptual understanding and strategic knowledge that can inform future studies, policy, and integration of responsible AI in the sustainable development initiatives.

5.3 AI and SDG Synergies

5.3.1 SDG 2: Zero Hunger

AI is very important for better food security at the global scale, thus for increasing agricultural productivity and efficiency. Machine learning allows the early detection of problems like pest invasions and nutrient depletion and supports the prevention of widespread damage. Application of AI in the management of supply chain allows the accurate prediction of demand, efficient planning of routes, and management of cold chain to prevent loss after harvest (Jha et al., 2022). AI technology can provide smallholders with much needed guidance on planting and irrigation, and managing nutrient levels, which is a fundamental support for sustainable operations and stability of rural communities.

5.3.2 SDG 3: Good Health and Well-Being

AI can be applied to the work of healthcare providers to analyze data, make exact diagnoses, take a real-time check of outbreaks, and personalize treatments. The diagnosis of chronic and rare conditions is supported by AI systems that extract information from electronic health records, and genetic and imaging outcomes, as evidenced by Esteva et al. (2017). During the COVID pandemic, some of the AI research targeted challenges such as virus simulation, vaccine candidate planning, and medical equipment supply optimization. Because of telemedicine and AI chatbots, there has been an increased accessibility of healthcare delivery in far-flung places as patients are able to easily undergo triage and follow prescribed treatments. The effectiveness of hospitals is increased through accurate predictions on patient flow, which optimizes workforce planning and eliminates unwanted tests (Topol, 2020).

5.3.3 SDG 4: Quality Education

In the age of technology, the resources for education help produce personalized adaptive platforms that will match the exceptional skills of individual students. Incorporation of data analytics in intelligent tutoring systems will promote the delivery of performance feedback and personalized learning materials, which will ensure that the achievement gaps among students will not come to roost. The use of natural language processing could break educational language barriers and aid in the efficient distribution of assistive instruments for special needs students (Holmes et al., 2021). By using AI, teachers will be able to spend more time developing innovative immersive learning for students as everyday school management issues such as grading and scheduling will be automated. Also, the use of data analytics supports real-time performance monitoring and empowers schools to reach underperforming students in good time and provide fairness in the educational process.

5.3.4 SDG 6: Clean Water and Sanitation

AI is of paramount importance in improving the management and surveillance of water resources for the management of global water challenges. Combining projected consumption patterns and environmental assessments is used to monitor leaks, protect water standards, and identify the sudden increase in demand for water. Through the introduction of AI into purification systems, water treatment plants can achieve better energy efficiency along with lower requirement of chemicals (Makropoulos & Savić, 2019). Current information collected by the AI sensors provides insights into the state of groundwater and contamination patterns to form sustainable water policies. By using machine learning, it is possible for designers to design customized decentralized sanitation solutions specific for each region, thereby improving clean water usage efficiency.

5.3.5 SDG 7: Affordable and Clean Energy

AI is a decisive factor in reinforcing energy systems, making them more sustainable and effective, in large part through the support for the development of smart grids and renewable energy. Through the use of automation algorithms, analysis of energy demand and supply is done, improvements in load management are made, and some risks from grid failures are minimized. Machine learning-based predictive maintenance has enabled the wind turbines and solar farms to reduce unplanned downtimes, as revealed by Rolnick and colleagues in 2019. Also, AI contributes greatly to the design and construction of energy projects, by analyzing information about geospatial features and climate. Such improvements make the energy systems cleaner, reduce the number of greenhouse gases, and help fight climate change.

5.3.6 SDG 11: Sustainable Cities and Communities

Numerous problems, such as congestion, pollution, and poor waste disposal, develop as the result of urban development. With the interpretation of huge urban data, AI transforms the possibilities of smart cities to manage transport, better maintain public security, and guarantee sound working of infrastructure. The fusion of live surveillance, predictive analytics, and computer monitoring has enhanced traffic flow, thereby minimizing air pollution and quelling illegal dumping (Batty, 2018). Using quick decisions and the best management of resources, AI improves emergency management. Implementation of such methods increases urban sustainability, improves the well-being of residents, and prepares cities for environmental and socioeconomic challenges.

5.3.7 SDG 13: Climate Action

Improved predictive models and greenhouse gas emission tracking and reduced risks of disasters all make it possible for AI to help make significant strides for climate action goals. Machine learning enables the quantification of forest loss, glacier retreat, and urban heat island by using satellite imagery analysis. AI can lead to improvements in climate models for providing accurate and useful data on climate change and predicting future weather (Vinuesa et al., 2020). The use of AI enhances the ability to predict disasters such as flood, drought, and wildfires, allowing for timely evacuation and securing of buildings and surfaces around them. Moreover, AI allows tracing animal populations and environmental health through the information processed from remote sensors.

5.4 Benefits and Opportunities of AI in Achieving SDGs

5.4.1 Enhanced Decision-Making

AI strongly assists policymakers and development experts due to its ability to make advanced data analysis, predictive modeling, and scenario simulation. Through AI, vast data can be extracted from satellite imagery and public health sources and generated as actionable data to provide informed policymaking. Using AI, policymakers enjoy strong policy formulation, quick adaptation to newly emerging situations; and continuous SDG indicators monitoring in real time. AI also contributes to facilitating the aspect of transparency by supporting the presentation of data and will ensure that public reporting is encouraged so as to enhance accountability in government. For example, AI application for climate policy, urban development, and risk management in disasters leads to better long-term resilience and efficient resource utilization.

5.4.2 Resource Optimization

Relying on AI to modify the distribution of resources plays a big part in sustainability and helps minimize total operational cost of large scopes. Using AI technologies in agriculture is an effective way for farmers to maximize the use of water and apply the right amount of fertilizers and pesticides, and thus nationally increase crop yield by implementing targeted farming approaches. The benefits of AI increase with AI tools such as the prediction of demands to make a sustainable supply of resource such as ICU beds, vaccine, and staff. Logistics and transportation have been enhanced when utilizing AI algorithms that plan the most effective routes for vehicles, schedule their fleet efficiently, and reduce the usage of fuel and greenhouse gas emissions. Also, AI is critical in energy systems as it predicts the peak loads, thus enhancing grid management efficiency and the integration of renewable sources. These breakthroughs, taken in the light of SDGs 2, 3, 7, and 12, promote sustainable consumption and encourage productivity in diverse industries.

5.4.3 Accessibility and Inclusion

Creating digital inclusion is highly dependent on AI development, which allows interaction with often disregarded demographics. Such applications include speech-to-text converters, text-to-speech readers, visual algorithm for recognition, and user interface based on gestures that significantly enhance communication, educational assistance, and job opportunities for disabled people. The use of AI in language technology helps close the gaps in linguistics when fast translations help everyone participate in heterogeneous communities. Where basic infrastructure is not present, AI provides communities in disadvantaged regions access to technologically delivered healthcare remotely by way of telemedicine and access to individualized educational content online. Through such technology, we can promote more equitable society and digital world, thus ensuring the realization of SDGs 4, 5, and 10 (OECD, 2023).

5.4.4 Economic Growth and Innovation

AI is a critical player in the transformation of economies through development, efficiency, and creation of new jobs. The use of AI within organizations makes internal processes easier, improves operational effectiveness, and promotes the development of more cost-effective production plans. Enhancements to predictive modeling and simulation activities by AI increase the pace and effectiveness of R&D efforts. Rising from Asia and Africa, innovative AI start-ups have come to be described as sparking entrepreneurial zeal, and job creation in software development, data analysis, and robotics sectors. The intentional evolving of AI infrastructure; Governments, by investing in human capital and regulatory frameworks, are pushing the countries forward in achieving higher levels of economic competitiveness and all macroeconomic stability. Ensuring that AI caters to the interest of all leads to job creation, the development of eco-friendly industries, and better economic stability—helping the attainment of SDGs 8 and 9.

5.5 Challenges and Risks of AI in Achieving SDGs

5.5.1 Algorithmic Bias

Algorithmic bias causes one of the most critical ethical dilemmas facing reactions to AI and is often the result of biased datasets that perpetuate historic systemic issues, or omitting important spaces. In the absence of checks on the usage of

AI, such systems could continue to discriminate the underprivileged in healthcare, employment, financial lending, and policing.

To give an example, facial recognition has been criticized for its lesser efficacy in detecting the faces of those from the racial minority population, and also AI-based recruitment is inclined toward the recruitment of male candidates. Such inequities are mostly a consequence of the training data being grossly imbalanced and inhomogeneous in nature. These challenges are resolved only through open and honest collection of data, inculcation of diverse populations in training materials, and employing fairness-conscious algorithms. Equitable and unbiased results are also essential for regular auditing processes and constant communication with influential stakeholders. Without control, algorithmic determination could (unintentionally) perpetuate or even expand current discrepancies (O'Neil, 2016).

5.5.2 Data Privacy and Security

A great number of AI platforms work with vast amounts of personal and sensitive records, including health records, behavioral patterns, and financial information. Privacy is to a large extent compromised when consent is missing or when attested information is employed for horizons other than those originally envisaged on the outset. If there are no secure data protocols, organizations will be exposed to security breaches, identity theft, and a decline in consumer confidence. To have strong data privacy, the data must be encrypted, identifiable information removed, and rules adhered to as defined in the GDPR. Firms should promote transparency during their data management practices by informing users about their personal information collection, storage, and administration practices (UNDP, 2023).

5.5.3 Digital Divide

The degree of accessibility of benefits of AI clearly differs across regions and groups. While prosperous countries are ramping up the use of AI technologies, poorer economies often overlook the needed infrastructure such as broadband, qualified personnel, and sufficient computing capacity. Constantly escalating the digital divide, or digital gap, could increase the existing socioeconomic divisions. Those living in rural areas, those from low-income households, and members of the marginalized population may miss the educational, healthcare, and economic benefits that AI can bring. Reduction of this discrepancy requires investments in strong digital systems, distribution of digital competence, and assistance in the development of user-friendly, likely inclusive, AI technologies adapted to local environments (UNESCO, 2022).

5.5.4 Regulatory and Governance Gaps

The recent explosion of AI has given rise to serious regulatory and ethical dilemma. In addition, the current legislation is often inadequate to cover new threats such as algorithmic oversight, liability, infringement of intellectual property, and cross-border data movement. Due to the lack of regulatory oversight, AI exposes users to the risk of misusing AI for such purposes as surveillance, creation and spread of false information, and automation that causes harm. It goes without saying that making sure that regulatory approaches reflect human dignity and democratic values is important for innovation, with the protection of basic societal rights. The safeguarding of AI transparency, explainability, safety, and justice can only be secured through rapid cooperation among governments, international organizations, and the private sector, in the formation of AI principles and standards. Any effort to make AI practices ethical and suitable for SDGs will demand global convergence in directing such cross-border applications (Chatterjee et al., 2024).

5.6 Strategic Recommendations for Responsible AI Implementation in Achieving SDGs

5.6.1 Inclusive AI Governance

AI is an important predecessor in the advancement of SDGs regarding promoting inclusive governance through AI with regard to ethics, transparency, and accountability. In order to ensure the development of internationally recognized and implementable AI by the nation, governance measures should comprise both national and international domains. There is a need for collaboration among government agencies, industries, organizations of academic institutions, and NGOs to develop such policies. The governance frames of inclusiveness need to ensure the involvement of underrepresented communities in AI developments to ensure that they are not made worse by AI developments. Standards like OECD's AI Principles and UNESCO's Recommendations on the Ethics of AI have given our team access to the key resources for interactive AI governance between rights and stakeholders.

5.6.2 Capacity Building

Analyzing the global gap among AI researchers requires massive restoration of capacity-building programs in order to support countries and disadvantaged groups in the Global South. The aim of this plan is to introduce regional learning strategies concentrating on themes such as ethical AI, machine learning, data analysis, and online freedom. Training delivery from educational institutions, vocational training providers, and digital learning platforms has to necessarily be shaped to address the diverse literacy level and technical experience. Besides, it is necessary for governments and global bodies to focus on research funding, set a cooperation network, and provide free AI-solutions. Establishment of AI regional centers, combined with fellowship and exchange opportunities, promotes standard dissemination of responsible AI and enables the creation of an international community of technologists interested in respectful technology advancement (Wirtz et al., 2019).

5.6.3 Public–Private Partnerships

Particularly, the public–private partnerships (PPPs) are the key driving forces toward rapid trustful AI innovation development. The potential for the partnerships exists where the governments and development organizations can call on the private sector to promote AI technology in support of the SDGs with priority on such critical areas as healthcare, sustainable agriculture, and climate mitigation. Positive SDG programs in AI fields require fast establishment of strong mechanisms for monitoring and evaluating performance. A clear and standardized method of assessment is required for these programs, while cooperative rules and collaboration principles form the basis for PPP reliability in translating their full potential for technological and social capabilities.

5.6.4 Monitoring and Evaluation

Standardized measures offer a system through which the performance of AI could be evaluated across numerous applications with inclusivity and sustainability. As examples, healthcare and education should inform our determinations of AI-based interventions with respect to their accomplishments, contribution toward equitable status, and long-term impact on society at large. Continuous monitoring is the basis upon which evidence-based policies can be implemented, which would take into consideration new emerging AI issues and remain consistent with the overall development agenda. In the situation we are operating in, the open reporting practice and the process of involving users in evaluations create a sense of transparency and trust among stakeholders.

5.6.5 Ethical Innovation Hubs

It is better to establish hubs that focus on ethics and innovation. They serve both as engines for the nurturing of social enterprise start-ups and as spaces for designing general-use algorithms and fostering collaborations among disparate fields. Human-focused design promotes the establishment of AI community hubs to ensure that AI solutions address the needs and customs of the local people. With technological innovation becoming a vehicle for inclusive and equitable growth, these hubs can be developed by focusing on ethics and social justice as aspects on which they operate. One stunning example is the contribution of Artificial Intelligence for Development (AI4D) to the improvement of AI in African regions, where Raise India has firmly been in the forefront controlling socially responsible AI initiatives.

5.6.6 AI Development Hubs

By establishing sectors with concentrated ethical innovation capabilities, ethical and responsible AI innovations with certain capabilities to meet the sustainability needs can be promoted. These centers can act as drivers for cultivating socially oriented technology businesses, tables for deliberating the ethical algorithms, and space for interdisciplinary collaboration. Based on human-centered design, these hubs allow work with the local communities to develop AI solutions specifically designed and adapted to the specific contexts and cultures of the regions they serve.

By incorporating ethics and social justice in their workings, such initiatives can show the world how technology can be a stimulating force in the development of more just and equitable society. Notable ones are AI4D, a program that deals with AI for development in Africa, and Raise India initiative, which deals with responsible AI for social empowerment.

5.7 Case Studies and Real-World Applications

Healthcare: Schemes of Early Diagnosis and Personal Medicine Driven by AI

Advances in AI enable diseases or conditions to be diagnosed in the earlier stages, treatment plans to be more individualized, and resources to be used more efficiently. Google-owned DeepMind Health has an AI system trained to diagnose more than 50 different retinal conditions, as equally well as medical specialists. More accurate identification would also lead to earlier disease detection, thus avoiding blindness and providing better health outcome in areas lacking specialists. AI tools overcome radiologists by diagnosing breast cancer from mammogram scans, therefore reducing false positives and patients' dependency on radiologists (McKinney et al., 2020). Such developments play a great role in the attainment of SDG 3. Through AI-powered innovations in health, accuracy is enhanced, access expanded, and equity in providing services of quality healthcare and well-being increased, all in keeping with SDG 3.

Agriculture: AI-Powered Smart Farming

The IBM Watson Decision Platform for Agriculture is evidence of how AI can be effective in making smarter farming decisions. Through the use of AI, machine learning, and big data analytics, the platform evaluates information retrieved from weather forecasts, satellite imagery, soil moisture sensors, and high-yield crop models. The platform provides location-based, real-time recommendations tailored to farmers' best planting seasons for optimal fertilization strategies and pest control measures. Such information, therefore, helps smallholder farmers increase their yields, reduce environmental impact, and simplify their distribution networks. This supports SDG 2 (Zero Hunger) and SDG 12 (Responsible Consumption and Production), especially through the promotion of sustainable agriculture and food security.

Energy: Smart Grids and Renewable Integration

The integration of AI tools in Siemens' smart grid management has brought more efficient electricity allocations as an increasing number of energy systems are now being improved by way of solar and wind energy adoption. AI-driven platforms evaluate dynamic energy patterns and provide necessary data such as load changes, user behavior, and early indications of equipment problems to optimize grid operation and prepare operators to respond to challenges in advance. Such measures enhance reliability and energy efficiency, helping make the integration of renewable energy into the network easier. These developments support SDG 7: reducing environmental harm, making energy use less costly and more resilient, and promoting SDG 13 (Climate Action) to reduce carbon emissions.

Urban Planning: Smart Cities and Sustainable Infrastructure

Google-supported Sidewalk Labs utilizes AI to build and manage smart city infrastructure. In the pilot projects, Sidewalk Labs uses AI to monitor traffic, air pollution, and energy consumption live, which helps urban planners make better decisions. For example, with AI efficiency of transport networks can be improved, congestion can be checked, and pedestrian-friendly communities advanced. With the help of an AI-driven approach, routes for the collection of waste are optimized in order to lower energy consumption and increase recycling results. These sophisticated technologies form part of the contribution to the projected development of livable environment and adaptation aligned with the SDG 11 agenda (Sustainable Cities and Communities).

5.8 Conclusion

Being on the frontline of technological advancement, AI presents a massive opportunity for a faster realization of UN SDGs. Its strengths in complex data processing, fast decision-making, modeling, and automation bring about solutions to various sections including agriculture, healthcare, education, energy, and city planning. There are good arguments showing that AI can have a beneficial role in the fields of precision farming, medical diagnostics, optimization of renewable energy sources, and building smart and sustainable cities, proving its capabilities to facilitate informed and inclusive development. Nevertheless, enormous difficulties persist, such as the question of poor regulatory supervision, privacy threats, online inequality, and a lack of tracking tools. When left unsolved, the problems may nullify the fair growth and exacerbate the socioeconomic and technological gaps, particularly between developed and underdeveloped areas. In order to explore the maximum potential of AI in terms of its role in supporting the SDGs, it is important to adopt ethical, transparent, and participatory governance structures fostering the engagement of active stakeholders. These involve integrating the marginalized communities in the process of designing and implementing AI, increasing access to digital training, reinforcing institutional capabilities, and stimulating cross-sectoral research in line with the goals of environmental sustainability and long-term development.

Moving ahead, more studies should be carried out that examine localized and situation-related AI-based applications designed to individual SDG goals, especially in data-scarce contexts. Studies also need to be conducted on creating strong systems of AI ethics, accountability, and impact evaluation and review the long-term socioeconomic consequences of the introduction of AI. Moreover, interdisciplinary studies can provide more meaning regarding the possible ways of upscaling AI systems responsibly so that these systems contribute to economic growth and to the sustainability, equity, and well-being of the whole world.

References

Batty, M. (2018). Artificial intelligence and smart cities. *Environment and Planning B: Urban Analytics and City Science*, *45*(1), 3–6. https://doi.org/10.1177/2399808318786512

Brynjolfsson, E., & McAfee, A. (2017). Machine, Platform, Crowd: Harnessing Our Digital Future. W. W. Norton & Company.

Chatterjee, S., Rana, N. P., Tamilmani, K., & Dwivedi, Y. K. (2024). A review of artificial intelligence in sustainable development: Opportunities and ethical challenges. *Technological Forecasting and Social Change*, *196*, 122806. https://doi.org/10.1016/j.techfore.2023.122806

Dwivedi, Y. K., Hughes, D. L., Ismagilova, E., Aarts, G., Coombs, C., Crick, T., . . . Williams, M. D. (2023). Artificial intelligence (AI): multidisciplinary perspectives on emerging challenges, opportunities, and agenda for research, practice and policy. *International Journal of Information Management*, *71*, 102642. https://doi.org/10.1016/j.ijinfomgt.2022.102642

Esteva, A., Kuprel, B., Novoa, R. A., Ko, J., Swetter, S. M., Blau, H. M., & Thrun, S. (2017). Dermatologist-level classification of skin cancer with deep neural networks. *Nature*, *542*(7639), 115–118. https://doi.org/10.1038/nature21056

Holmes, W., Bialik, M., & Fadel, C. (2021). Artificial Intelligence in Education: Promises and Implications for Teaching and Learning. Center for Curriculum Redesign.

Jha, K., Doshi, A., Patel, P., & Shah, M. (2022). AI-based precision agriculture systems for sustainable food production. *Sustainable Computing: Informatics and Systems*, *34*, 100751. https://doi.org/10.1016/j.suscom.2021.100751

Makropoulos, C. K., & Savić, D. A. (2019). Urban water systems: Applying the water–food–energy nexus approach with AI. *Environmental Modelling & Software*, *121*, 104489. https://doi.org/10.1016/j.envsoft.2019.104489

McKinney, S. M., Sieniek, M., Godbole, V., Godwin, J., Antropova, N., Ashrafian, H., . . . Shetty, S. (2020). International evaluation of an AI system for breast cancer screening. *Nature*, *577*(7788), 89–94. https://doi.org/10.1038/s41586-019-1799-6

OECD. (2023). AI and the Productivity Frontier: Harnessing Digital Transformation for Growth. OECD Publishing. https://doi.org/10.1787/9789264568230-en

O'Neil, C. (2016). Weapons of Math Destruction: How Big Data Increases Inequality and Threatens Democracy. Crown Publishing Group.

Rolnick, D., Donti, P. L., Kaack, L. H., Kochanski, K., Lacoste, A., Sankaran, K., . . . Bengio, Y. (2019). Tackling climate change with machine learning. arXiv preprint arXiv: 1906.05433. https://arxiv.org/abs/1906.05433

Sachs, J. D., Lafortune, G., Fuller, G., & Woelm, F. (2022). *Sustainable Development Report 2022*. Cambridge University Press.

Topol, E. J. (2020). High-performance medicine: The convergence of human and artificial intelligence. *Nature Medicine*, *25*(1), 44–56. https://doi.org/10.1038/s41591-018-0300-7

UNDP. (2023). *The Sustainable Development Goals Report 2023*. United Nations. https://unstats.un.org/sdgs/report/2023

UNESCO. (2022). AI and the Digital Divide: A Call for Inclusive Innovation. United Nations Educational, Scientific and Cultural Organization. https://unesdoc.unesco.org/

Vinuesa, R., Azizpour, H., Leite, I., Balaam, M., Dignum, V., Domisch, S., . . . Fuso Nerini, F. (2020). The role of artificial intelligence in achieving the Sustainable Development Goals. *Nature Communications*, *11*(1), 233. https://doi.org/10.1038/s41467-019-14108-y

Weitz, N., Carlsen, H., Nilsson, M., & Skånberg, K. (2023). Towards systemic and contextual priority setting for implementing the 2030 agenda. *Sustainability Science*, *18*, 1001–1014. https://doi.org/10.1007/s11625-023-01202-2

Wirtz, B. W., Weyerer, J. C., & Geyer, C. (2019). Artificial intelligence and the public sector—Applications and challenges. *International Journal of Public Administration*, *42*(7), 596–615. https://doi.org/10.1080/01900692.2018.1498103

6 AI and Sustainable Development

Phineas Sebopelo

6.1 Introduction

This chapter aims at investigating the intersection of Industry 5.0 technologies and sustainable development, addressing both theoretical foundations and practical implementations.

The set aim will be achieved by specifically focusing on the following objectives:

a) Exploring the relationship between artificial intelligence (AI) and Sustainable Development Goals (SDGs).
b) Assessing AI applications in smart cities.
c) Investigating ethical AI and sustainability governance.
d) Evaluating AI-driven strategies for achieving the SDGs.

The methodology applied for compiling information for this chapter is primarily qualitative, with the literature review conducted through a desktop survey and document analysis. This approach involves using previously published materials, such as articles, books, reports, and other relevant documents, to gather information and insights on a specific research topic. This approach helps identify the current state of knowledge, locate gaps in existing research, and formulate research questions. The criteria used to select resource materials for this chapter are shown in Table 6.1.

Summary of literature

After the inclusion and exclusion criteria were applied, the retained articles were used in the compilation of the content for this chapter. A summary of consulted journal article is provided in Table 6.2.

SDGs, also known as the Global Goals, are a set of 17 interconnected goals adopted by the United Nations (UN) in 2015 as part of the 2030 Agenda for Sustainable Development (Shang et al., 2022). These goals are depicted in Figure 6.1.

These goals aim to address critical global challenges by 2030, including poverty, hunger, inequality, climate change, and environmental degradation. The SDGs

DOI: 10.4324/9781003647621-8

Table 6.1 Inclusion and Exclusion Criteria for Resources Used in This Study.

Inclusion Criterion	*Exclusion Criterion*
The research topic had to focus on AI and the SDGs, and the studies had to address aspects of sustainability.	The research article is not relevant to AI and sustainability unless it relates to AI digital technologies.
Articles were considered if they were full texts or published articles	Papers with fewer than three pages, conference proceedings, books, and book chapters
Papers should have abstracts, keywords, research questions, and conclusions (results)	The paper seriously lacks essential information, including an abstract, author information, keywords, methods, and conclusions.
Papers should have been cited more than once	The paper has not been cited at all

Table 6.2 Summary of Journals Used.

Paper	*Authors*	*Relation with the Chapter*
1. How can artificial intelligence impact sustainability: A systematic literature review	Kar et al. (2022)	This paper provides a systematic review of how AI can impact sustainability across various industries.
2. Artificial intelligence for sustainability: A systematic literature review in information systems	Siqueira et al. (2024)	This systematic literature review examines how AI can be used to promote sustainable development across environmental, economic, legal, political, organizational, and social domains.
3. Systematic literature review on sustainable marketing and artificial intelligence	Yadav and Sondhi (2023)	This paper reviews the literature on the use of AI in sustainable marketing, revealing that AI is an emerging field that can be helpful for sustainable marketing development.
4. Artificial intelligence and sustainable development goals: Systematic literature review of the construction industry	Regona et al. (2024)	This paper reviews how AI can be integrated across construction project phases to promote sustainability, while addressing adoption challenges.
5. Artificial intelligence for sustainability: Challenges, opportunities, and a research agenda	Nishant et al. (2020)	This paper reviews the challenges, opportunities, and research agenda for using AI to address environmental sustainability.
6. How AI and machine learning can drive sustainable development	Akavova et al. (2023)	AI and machine learning have significant potential to drive sustainable development across various sectors.

(*Continued*)

Table 6.2 (Continued)

Paper	*Authors*	*Relation with the Chapter*
7. AI for sustainable development: Innovations & applications	Kumari et al. (2024)	This paper reviews the critical role of AI in advancing the Sustainable Development Goals across multiple sectors.
8. Artificial intelligence for sustainability—A systematic review of information systems literature	Schoormann et al. (2023).	This paper provides a systematic review of how the information systems discipline uses AI to support sustainable development.
9. Artificial intelligence in sustainable smart cities: A systematic study on applications, benefits, challenges, and solutions	Dos Santos et al. (2024)	This systematic review examines how AI can be used in sustainable smart cities, including the applications, benefits, challenges, and solutions.
10. A systematic review of the applications of AI in a sustainable building's lifecycle	Adewale et al. (2024)	AI has the potential to enhance sustainability throughout a building's lifecycle, but faces implementation challenges.

Figure 6.1 17 SDGs.

are a universal call to action for all countries, rich and poor, to promote prosperity while protecting the planet (Ali et al., 2023). The SDGs are interconnected, meaning progress in one area can influence others, highlighting the complexity of sustainable development (Bąk, 2024). These goals provide a comprehensive framework for sustainable development by integrating economic, social, and environmental dimensions. Integrating all 17 SDGs into sustainability assessments maximizes their utilization and provides a holistic definition of sustainable development (Eberling & Langkau, 2023). Digitalization can potentially contribute to achieving the SDGs. Overall, digital technologies support the SDGs by providing access to new services, enhancing connectivity, and enabling better data analytics (Mondejar et al., 2021), and by facilitating the monitoring, tracking, and reporting of progress toward the SDGs (Saner et al., 2020). Industry 5.0 technologies focus on integrating human creativity with intelligent machines to enhance sustainability, human-centricity, and resilience in manufacturing and supply chains, while addressing socioeconomic and environmental challenges (Ghobakhloo et al., 2023). The integration of Industry 5.0 technologies is pivotal for advancing sustainable development, as it emphasizes a human-centric approach that harmonizes technological innovation with environmental and social considerations (Rame et al., 2024). This transition from Industry 4.0 to Industry 5.0 not only enhances productivity but also aligns industrial practices with the SDGs (Costa, 2024).

Industry 5.0 technologies enhance sustainable development by integrating advanced technology with human creativity; promoting eco-innovation, digital transformation skills, and proactive governmental support; and ultimately driving environmental resilience, social inclusivity, and economic efficiency in sustainable infrastructure projects aligned with the SDGs (Daoud et al., 2025).

Most studies suggest that generative AI can enhance responsible manufacturing within the context of Industry 5.0 (Enose Kamalabai et al., 2025). From the perspective of Vinuesa et al. (2020), intentional AI system integration may yield coordinated success across the SDGs' interrelated economic, social, and environmental dimensions. Truby (2020) opined that establishing a robust AI governance framework is a critical task that could help realize how AI may propel SDGs.

The following section elaborates on the book's scope. The author will then present a brief review of recent publications on AI and SDGs.

6.1.1 Scope of the Chapter

This chapter is about AI and SDGs. It explores the potential of AI to advance or hinder the achievement of the 17 SDGs, focusing on both opportunities and challenges. It analyses specific AI applications across various sectors, discusses the ethical considerations in AI deployment, and offers insights for future research and policymaking.

A more detailed breakdown of the scope is as follows:

1. AI's role in achieving SDGs: The section delves into how AI can contribute to achieving specific SDG targets.
2. AI-powered technologies: This section discusses specific technologies and their applications in different sectors, such as health, education, and agriculture.
3. The chapter also focuses on challenges and risks of AI such as data bias and inequality: AI systems trained on biased data can perpetuate and amplify existing inequalities.

4. The chapter also discusses the importance of ethical considerations such as transparency, accountability, and responsible AI development.
5. Furthermore, AI applications in smart cities, energy use, and climate change are discussed in this chapter.
6. Finally, the chapter discusses Al-driven strategies for achieving SDGs and concludes by exploring future trends in AI and sustainability.

The next section presents an overview of the SDGs and their significance.

6.1.2 Overview of Sustainable Development Goals and their Significance

The SDGs emerged from a historical progression of development concepts, evolving from traditional models focused on economic growth to a more holistic view that incorporates environmental sustainability and social equity (Jacob, 2024).

The 2030 Agenda for Sustainable Development, adopted by all UN member states in 2015, provides a shared blueprint for peace and prosperity for people and the planet, now and into the future (Mio et al., 2020). At its heart are the 17 SDGs, which constitute an urgent call to action from all developed and developing countries in a global partnership (Türkeş, 2024). They recognize that ending poverty and other deprivations must go hand in hand with strategies that improve health and education, reduce inequality, spur economic growth, tackle climate change, and preserve our oceans and forests. The evolution of SDGs reflects a deepening global awareness of interconnected environmental, social, and economic challenges (Sachs et al., 2019).

The recognition of environmental challenges, such as air and water pollution, climate change, and biodiversity loss, has prompted a global response to foster a more sustainable future (Vijaya, 2024). Among these efforts, the SDGs serve as a comprehensive framework to address the interconnected dimensions of social equity, economic development, and environmental protection. These goals emphasize urgent issues such as zero hunger, quality education, clean water, and climate action, thereby guiding collective international efforts toward a balanced approach to development (Sorooshian, 2024).

6.2 Artificial Intelligence and Sustainable Development Goals

In recent years, technological advancements have profoundly reshaped various global sectors, prompting new discussions about their potential to address pressing challenges (Raihan, 2024). Among these technologies, AI has emerged as a powerful tool capable of driving significant progress toward the UN SDGs (AlSagri & Sohail, 2024). AI's ability to analyze vast amounts of data and optimize complex systems offers promising solutions for sectors such as health, energy, and urban development (Ukoba et al., 2024).

The accelerating challenges of climate change, resource depletion, and urbanization demand innovative solutions that balance economic growth with environmental stewardship.

6.2.1 Potential Unintended Consequences of AI Deployment for Sustainable Development

The deployment of AI systems can lead to various unintended consequences that impact stakeholders differently. These consequences can be both positive and negative, often arising from the complexities of AI behavior and the contexts in which they are implemented. Understanding these potential harms is crucial for responsible AI deployment.

For instance, AI systems can inadvertently harm employees and organizations, such as through biased decision-making or compliance breaches, thereby damaging corporate reputation (Abdelati, 2024). While many unintended consequences are harmful, some may lead to unexpected benefits, such as increased efficiency or innovation in processes (Iwuanyanwu et al., 2024). The introduction of AI can lead to societal issues, such as reduced human interaction and increased feelings of disconnection among individuals (Hajam et al., 2023).

To guard against such unintended effects, organizations are increasingly adopting ethical AI principles and governance frameworks to manage the risks associated with AI deployment (Javaid et al., 2023). Moreover, the use of tools like AI Harm Anticipation (AHA!) help practitioners foresee potential harms by generating diverse examples of possible adverse outcomes, facilitating ethical reflection (Buçinca et al., 2023). Establishing meta-values and responsible experimentation can help maintain human oversight over AI systems, reducing the likelihood of unintended consequences (Juneja & Saraswat, 2025).

While the focus is often on mitigating adverse outcomes, it is essential to recognize that unintended consequences can also lead to beneficial innovations. Balancing these aspects is crucial for the responsible advancement of AI technologies.

6.2.2 A Conceptual Framework for AI for Sustainable Development

The framework for AI for sustainable development should integrate AI capabilities with SDGs, focusing on strategic alignment, data, and infrastructure, and addressing ethical considerations. This framework guides organizations in leveraging AI to achieve positive social, environmental, and economic outcomes, thereby advancing the SDGs.

The model in Figure 6.2 shows that AI is increasingly recognized as a crucial enabler for achieving the UN SDGs. The figure also shows that AI acts as a catalyst across various sectors, impacting social, environmental, and economic aspects to drive positive change toward a more sustainable future.

AI emerges as a powerful tool capable of advancing multiple SDGs by optimizing resource management, enhancing energy efficiency, and promoting sustainable consumption and production patterns (Kumari et al., 2024). AI's ability to analyze vast datasets enables accurate predictions and informed decision-making, which are essential for achieving net-zero emissions and waste-reduction targets (Almihat & Munda, 2025). However, the effectiveness of AI in sustainability initiatives hinges on data quality and explainability, ensuring that AI systems are transparent and accessible even to organizations with limited resources (Khan et al., 2025).

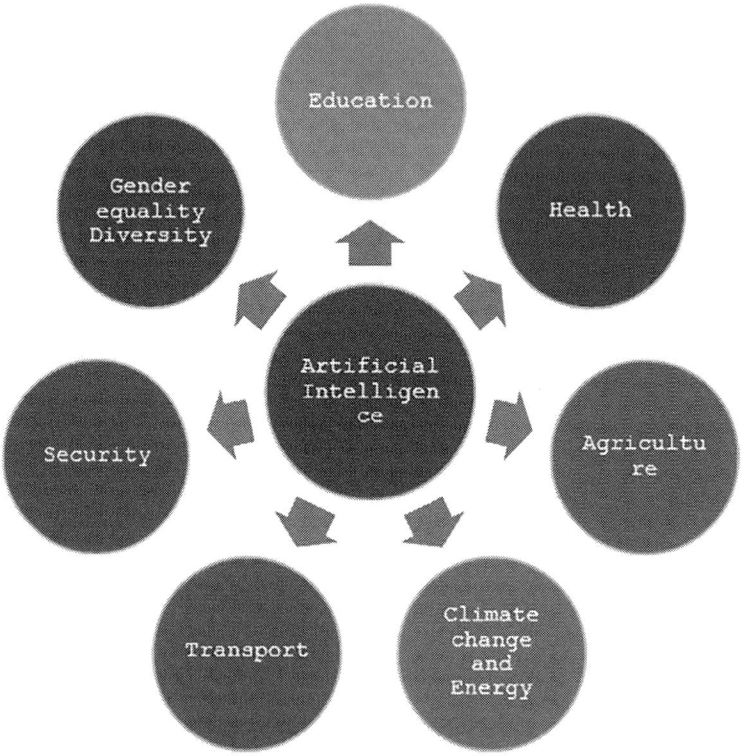

Figure 6.2 Model showing the centrality of AI in SDGs.

Source: Author.

Moreover, AI-driven technologies play a crucial role in smart city developments by integrating transportation systems through big data and shared mobility solutions, reinforcing sustainable urban environments supportive of SDG objectives (Khanna & Gupta, 2024). Consequently, AI represents a pivotal intersection where technological innovation and sustainable development converge. However, integrating AI is not without challenges, as ethical considerations and the risk of unintended consequences must be carefully managed to ensure equitable and sustainable benefits (Nguyen et al., 2023). This is especially important in resource-limited settings, where tailored approaches are necessary for effective implementation and meaningful impact.

6.3 Frameworks for Evaluating AI's Role in Sustainability

Several frameworks and analytical approaches are used to evaluate AI's role in sustainability, including those focused on ethical considerations, environmental impact, and social responsibility. These frameworks help assess the broader implications of AI systems and guide the development of sustainable AI solutions.

The evaluation of AI within the context of sustainability necessitates a structured analytical framework that encompasses various domains and considerations. Such frameworks help assess the multidimensional impacts of AI technologies on sustainable development, addressing not only their operational efficiencies but also ethical implications and societal transformations. For instance, as highlighted in recent literature, transformative AI tools like ChatGPT can both enhance productivity across sectors, including banking and tourism, and simultaneously raise critical concerns about privacy, security, and misinformation (Dwivedi et al., 2023). By employing comprehensive analytical approaches, stakeholders can better strategize the implementation of AI solutions that harmonize with sustainability goals.

6.3.1 Systems Thinking Approach and Its Application in Assessing AI Impacts

A systems-thinking approach provides a valuable framework for assessing the multifaceted impacts of AI on sustainability. By examining AI not merely as isolated technologies but as integral components of larger systems, this approach enables stakeholders to visualize the intricate interdependencies among societal, environmental, and economic factors. For instance, the implementation of AI tools such as ChatGPT has been shown to enhance productivity across industries, resulting in significant gains in sectors like banking and hospitality, highlighting its potential to improve operational efficiency (Dwivedi et al., 2023). However, this approach also illuminates the ethical dilemmas and disruptions that AI can introduce, such as privacy concerns and biases inherent within generative AI systems (Bankins et al., 2023). Thus, employing a systems thinking approach not only facilitates a comprehensive evaluation of AI's role in promoting sustainability but also encourages a balanced consideration of its ethical implications and societal consequences.

While systems thinking offers a valuable framework for understanding AI's complex impacts, it also presents limitations. A key weakness is the potential for oversimplification, in which complex systems are reduced to overly simplistic models, leading to inaccurate assessments (Nabeel, 2024).

6.3.2 Life Cycle Assessment as a Tool for Measuring Sustainability Outcomes

Life cycle assessment (LCA) is a critical tool for evaluating sustainability outcomes, providing a comprehensive framework for assessing the environmental impacts of products throughout their life cycles (Khan et al., 2022). In the context of integrating AI into sustainability frameworks, LCA can enhance decision-making by providing data-driven insights to identify areas for improvement and innovation (Jariwala, 2024).

Despite its importance in evaluating sustainability outcomes, LCA faces several challenges when used to measure them (Olawade et al., 2024). These include challenges with data availability and quality, defining system boundaries, and the complexity of interpreting results, particularly when comparing different products or processes. According to Barbhuiya and Das (2023), LCA can be resource-intensive and may not always capture all relevant sustainability aspects, such as social and economic factors.

6.3.3 The Triple Bottom Line Framework: Economic, Social, and Environmental Impacts

Understanding the triple bottom line (TBL) framework is essential for evaluating AI's multifaceted role in sustainability. This framework compellingly integrates economic, social, and environmental dimensions, emphasizing that a successful business must consider its impact on society and the planet, alongside financial performance (Khanna & Gupta, 2024). AI technologies, particularly deep learning, have shown significant promise in enhancing operational efficiencies and driving innovation across various sectors, thereby bolstering economic viability (Adeyeri, 2024). However, as the Fourth Industrial Revolution unfolds, it is vital to remain vigilant regarding the unintended negative consequences that AI and other Industry 4.0 technologies may present, particularly in social and environmental contexts (Singagerda et al., 2025). By systematically addressing these issues through the TBL lens, stakeholders can ensure that AI not only contributes to profitability but also actively promotes social equity and environmental stewardship, thereby achieving true sustainability.

While the TBL framework helps broaden the scope of corporate performance measurement beyond profits, it has several weaknesses in measuring sustainability outcomes. These include challenges in measurement, lack of integration, and potential for "greenwashing." Pasamar et al. (2023) observed that companies may use the TBL framework to disclose information about their social and environmental performance without necessarily improving their actual impact.

In this chapter, systems thinking approach is used to evaluate AI's influence on sustainability because it helps in understanding the complex, interconnected nature of AI systems and their impact on environmental, social, and economic systems. Considering the broader context and potential unintended consequences enables a more holistic and effective approach to developing and implementing AI solutions for sustainability.

Moreover, systems thinking helps map out these complex interconnections, revealing how AI affects various aspects of the environment, society, and economy (Voulvoulis et al., 2022). For example, AI-powered systems in agriculture might impact water usage, land use, and even social structures in rural communities. Systems thinking encourages a broader perspective, helping anticipate potential negative consequences (Laimon et al., 2022). For instance, while AI-powered automation can increase efficiency, it might also lead to job displacement and exacerbate social inequalities if not managed carefully.

6.4 AI Applications in Smart Cities

Sustainability is a key focus of smart cities, and AI is playing a vital role in promoting eco-friendly urban development (Das, 2025). AI plays a crucial role in the development of smart cities by enabling automation, predictive analytics, and real-time decision-making (Dos Santos et al., 2025). Cities around the world are leveraging AI to improve infrastructure, enhance public services, and create a more sustainable environment (Abbas, 2025). AI-driven solutions are being used to optimize energy consumption, manage waste, and improve air quality (Wirsbinna et al., 2023). AI-powered smart grids help cities optimize energy distribution by analyzing consumption patterns and predicting demand (Alabi, 2025). AI assists in predicting fluctuations in solar and wind energy availability and adjusts grid operations accordingly. These grids enable real-time monitoring of energy usage, allowing utilities to adjust supply accordingly (Daneshvar et al., 2021).

With its ability to analyze large amounts of data quickly and accurately, AI can help identify patterns and trends to support informed decision-making about how to manage the city's resources (Son et al., 2025). With a focus on climate action initiatives, smart decision-making, and improving citizens' quality of life, the following section presents five AI applications that can be highly beneficial for AI in smart cities. This includes applications designed to solve city-wide parking issues using AI and Internet of Things (IoT) technologies; to efficiently manage transport and solid waste; and to make cities and communities safer through AI-based applications and utilities and metering solutions.

6.4.1 AI-Assisted Parking

With population clustering in cities, street traffic has increased significantly, making it increasingly difficult to find a parking space (Errousso et al., 2023). AI and IoT technologies offer a powerful combination for solving city-wide parking issues. AI-powered systems can predict parking demand, optimize pricing strategies, and even guide drivers to available spaces. IoT-enabled sensors provide real-time data on parking availability, enhancing efficiency and reducing traffic congestion (Ahmad & Haile, 2025). Applications with a pre-integrated AI model, such as those shown in Figure 6.3, can be used for predicting parking demand and recommending a fare.

Optimized parking can help city officials generate the revenue needed to maintain and improve the city's parking infrastructure, while ensuring that parking remains affordable for residents and visitors (Litman, 2020). However, Ibrahim et al. (2024) noted that, despite its convenience, implementing AI-powered parking systems requires significant upfront investment in hardware, such as cameras and sensors, as well as in software development or integration of complex AI algorithms.

6.4.2 AI-Based Applications for Transport Management

According to Mohammed and Oke (2023), as urban population density increases, cities begin to grapple with severe traffic congestion and the improper management of public transport systems and resources.

Figure 6.3 AI-aided smart parking.

Source: https://as1.ftcdn.net/jpg/13/66/11/56/1000_F_1366115674_nzQNSrktcBJ1LFVYkJr73Xpq2pgYBXdL.webp

The ever-increasing demand to improve mobility and road safety has led developing countries to make significant changes to their infrastructure, especially in modernizing transport infrastructure (Naz & Hoque, 2023).

AI can be leveraged to help authorities combine various transport department activities, such as scheduling and planning public transport, fleet management, workforce management, and citizen grievance redressal (Durlik et al., 2025). An AI-based passenger management system can update passengers via smart devices across various transport modes with real-time data on roadblocks, delays, and breakdowns (Magsino et al., 2022). This information can be used to optimize scheduling, pricing structures, and the deployment of transport resources, ensuring that there are always enough vehicles available to meet demand (Alamatsaz et al., 2022).

However, Durlik et al. (2024) note that AI-powered transportation management systems face several shortcomings, including cybersecurity vulnerabilities, the need for high-quality data, high initial investment costs, and the potential to exacerbate existing infrastructure limitations.

6.4.3 Disposal and Management of Solid Waste

The collection, treatment, transport, and disposal of solid waste, particularly waste generated in cities, have become increasingly complex problems for governments and citizens alike (Zhang et al., 2024). For municipalities, improper management of waste collection and disposal has always been a challenge (Anokye et al., 2024).

According to Chikukula et al. (2024), improper waste management problems in most of the municipalities are attributed to the inability to efficiently coordinate with waste collectors and improper utilization of assets. This results in a fragmented approach to waste management, hindering the efficient collection, transportation, and disposal of waste.

Monitoring waste collection system performance in real time through dashboards helps identify potential problems before they occur (John et al., 2022). The use of AI in waste management is therefore an important factor for city authorities, as it can help reduce waste accumulation and improve the overall cleanliness of the city (Fang et al., 2023). To manage waste efficiently, AI techniques have been widely used, as highlighted in Figure 6.4. These techniques can handle large volumes of waste data, generate efficient, reliable results, and automate a variety of processes.

Moreover, insights into the types of waste generated can help city authorities plan segregation units that can enhance processing time and provide avenues for

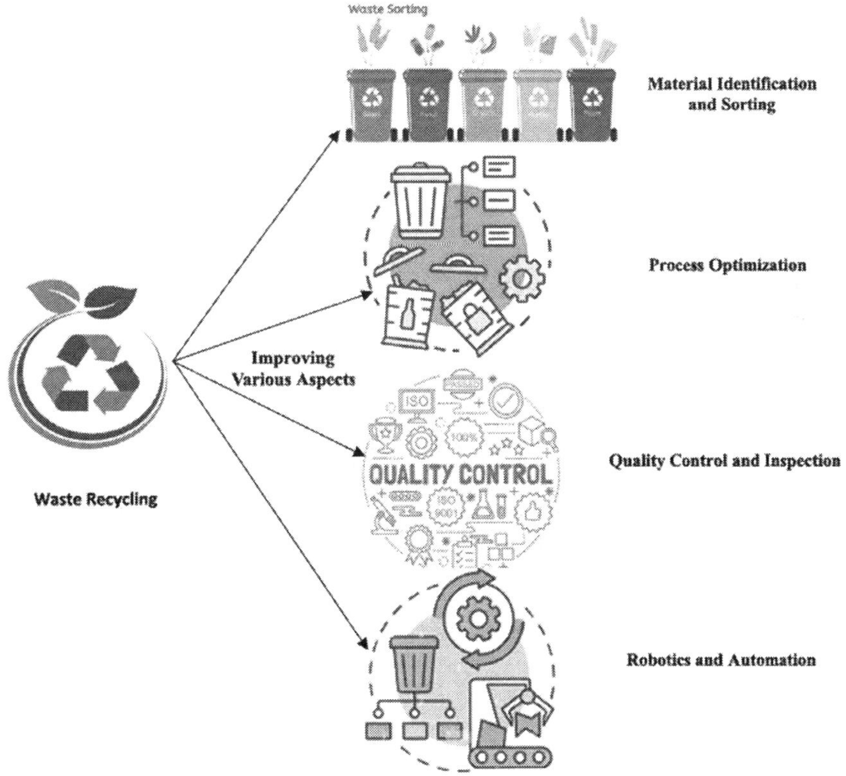

Figure 6.4 Applications of AI in waste recycling.

Source: Chen et al. (2023).

revenue generation through the production of biofuels, composts, etc., from organic waste (Zhang et al., 2024).

Abul Hasan (2021) noted that AI offers solutions for improved solid waste management but faces challenges like infrastructure limitations, inaccurate material identification, and technical malfunctions.

6.4.4 AI-Based Applications for Security of Smart Cities

According to Onyeneke and Karam (2022), rapid urbanization has led to an upsurge in incidents of crime due to a combination of factors, including population density, social inequality, and a lack of social cohesion. It is, therefore, vital to identify the crime hotspots in the city based on historical patterns of crime and help law enforcement agencies. This approach, known as hotspot policing, focuses on areas with high crime concentrations to maximize the impact of policing efforts (Ahmad et al., 2024).

However, modern data solutions, including AI-powered applications, can collect large volumes of information from disparate sources in real time (Zaman et al., 2024).

The use of AI security cameras has proved to be a game-changer, enabling quick responses when threats arise.

Through surveillance systems and drone use, crime scenes can be identified instantly, and resources can be allocated to crime-prone areas based on resource and skill analyses (Shader et al., 2024).

The integration of AI applications within urban environments aims to enhance safety and efficiency in smart cities; however, it also presents several challenges. These challenges stem from security vulnerabilities, regulatory concerns, and ethical implications that must be addressed to ensure the successful implementation of AI technologies.

Moreover, the vast amounts of data processed by AI systems raise significant privacy concerns, as unauthorized access can lead to misuse of sensitive information (Jain & Mitra, 2025).

Overall, the use of AI in smart cities has the potential to improve the quality of life for citizens, reduce the strain on city resources, and make cities more sustainable and efficient. As such, continued growth in the use of AI in smart cities in the coming years should be expected.

6.5 Ethical AI and Sustainability Governance

The escalating integration of AI across diverse sectors, from environmental monitoring to urban planning, presents unprecedented opportunities to advance the SDGs. However, there are intricate ethical considerations that demand careful navigation. The imperative to establish robust governance frameworks becomes increasingly apparent as AI systems permeate high-risk domains, necessitating accountability, fairness, and transparency in their design and deployment (Cath, 2018). These risks are exacerbated by the absence of effective global governance mechanisms that, at the very least, provide guardrails to guide AI in beneficial directions (Liu et al., 2024). However, Olawade et al. (2024) argue that mitigating the risks of AI use hinges on establishing common standards, interoperable regulatory frameworks, and mechanisms for international cooperation to ensure responsible AI innovation and deployment.

Ethical AI and sustainability governance face several key challenges, including algorithmic bias, lack of transparency, data privacy concerns, and the environmental impact of AI development. To effectively navigate the challenges of achieving ethical AI and sustainability governance, several mitigation strategies can be employed. These include establishing robust governance frameworks, implementing risk management strategies, ensuring transparency and accountability, promoting fairness and bias mitigation, prioritizing data privacy and security, and fostering multi-stakeholder collaboration.

6.5.1 Strategies for Overcoming Challenges in Achieving Ethical AI and Sustainability

To effectively navigate the challenges of achieving ethical AI and sustainability, a multifaceted approach is needed. This includes establishing robust ethical frameworks, promoting transparency and accountability, fostering collaboration, and ensuring continuous monitoring and evaluation.

6.5.2 Mitigating against Bias

The emergence of generative AI introduces transformative possibilities, particularly within educational contexts, yet it simultaneously amplifies concerns regarding bias and inclusivity. Addressing these challenges necessitates a concerted effort to implement strategies that promote transparency in AI algorithms and their applications. Developing comprehensive ethical guidelines is crucial, particularly given findings highlighting the significant enhancement AI provides to student engagement and knowledge retention (Rane et al., 2024). Additionally, fostering interdisciplinary collaboration can cultivate cultural competence among developers, ensuring that diverse perspectives are integrated into AI programming (Ramya et al., 2024). These approaches not only mitigate bias but also cultivate a more equitable educational landscape, ultimately benefiting under-resourced settings and elevating the importance of inclusivity in technological advancements.

6.5.3 Ensuring Transparency in AI Systems

Transparency fosters trust among users and stakeholders, allowing for a clearer understanding of how AI systems operate and make decisions. As investigations into AI applications reveal, biases inherent in the algorithms can lead to inequitable outcomes, necessitating a commitment to openness in design processes and decision-making frameworks. Moreover, integrating culturally sensitive leadership practices is crucial, as leaders navigating technology deployment must consider the regulatory and ethical landscapes that define their communities (Sajwani, 2024). By prioritizing transparency, organizations can not only mitigate bias but also empower marginalized groups by creating inclusive systems that reflect diverse experiences and needs.

6.5.4 Developing Clear Guidelines and Frameworks

As AI technologies continue to proliferate, particularly in under-resourced settings, it becomes essential to ensure that these systems operate transparently and ethically. Developing such frameworks not only mitigates bias but also enhances the cultural relevance of AI applications, as evidenced by the emerging use of machine intelligence in Africa to address the diverse linguistic and cultural needs of its population (Saleem et al., 2024).

By ensuring that AI systems are developed with diverse user inputs and contextual understandings, we can address long-standing inequities that have hindered access to essential resources such as healthcare and education. These practices not only promote fairness but also enhance the effectiveness and efficiency of AI applications, particularly in regions facing significant socioeconomic challenges. For instance, in healthcare, the deployment of AI can dramatically improve service delivery in resource-constrained settings, thereby mitigating disparities in access to care (Shahvaroughi Farahani & Ghasemi, 2024).

6.6 AI Governance Examples. Successes, Failures, and Lessons Learned

The following section highlights real-world examples of the successes and failures of AI and ethical governance. AI is advancing rapidly for many organizations to keep up with. While businesses rush to adopt AI, governance is often an afterthought.

But when AI governance fails, the consequences can be severe: lawsuits, regulatory fines, biased decision-making, and reputational damage.

6.6.1 AI Governance Failures

Example A: Lawsuit due to Poor AI Governance.

A class-action lawsuit against a certain company exposed the risks of poor AI governance. The company allegedly shared subscriber data without proper consent, violating privacy laws.

This case proves that AI-powered personalization and recommendation engines must be built on clear data lineage and consent management or risk hefty legal trouble.

6.6.2 Successes with AI Governance

Example B: A Bank's Secret Weapon against AI Bias.

A leading bank avoided bias pitfalls by deploying real-time AI monitoring to detect and fix problems before models went live. Their strategy included flagging bias indicators during model training, auditing AI decisions in production to ensure fairness, and tracking lineage and transformations to understand how data influenced outcomes.

By integrating AI governance early, they stayed ahead of compliance and turned fairness into a competitive edge.

The failures covered in this chapter prove that companies face financial losses, regulatory penalties, and reputational harm if they do not embrace end-to-end data lineage and continuous monitoring position themselves for success.

A proactive approach is crucial for ensuring ethical and responsible AI development and deployment, as it allows for continuous monitoring, evaluation, and adaptation of AI systems to prevent harm and promote beneficial outcomes (Srikumar et al., 2024). With proper AI governance, businesses can stay ahead of regulations, reduce risks, and build AI that's ethical, transparent, and trusted.

6.7 AI-Driven Strategies for Achieving SDGs

With many SDG targets lagging and a growing funding gap, the need to innovate to amplify positive impact has never been more urgent (Fallah et al., 2022). AI offers a promising tool for accelerating progress by enhancing data analysis, knowledge access, and operational efficiency (Madanchian & Taherdoost, 2025). AI-driven strategies can significantly contribute to achieving SDGs by enhancing organizational, technical, and processing aspects; improving healthcare, energy, and infrastructure; and supporting environmental sustainability, while also highlighting the need for ethical oversight and interdisciplinary collaboration. The advent of AI offers unprecedented opportunities to advance the SDGs, presenting innovative strategies for addressing complex global challenges (Vinuesa et al., 2020). AI-driven approaches leverage vast amounts of data to provide insights that enhance decision-making and optimize resource allocation, thereby significantly contributing to achieving various SDGs (Ziemba et al., 2024).

AI offers transformative opportunities to advance environmental sustainability, particularly through its application in resource management and climate change mitigation (Raihan, 2024). By harnessing AI technologies, stakeholders can optimize agricultural practices, enhance energy efficiency, and monitor biodiversity more effectively (Pandey & Mishra, 2024). For instance, Sachithra and Subhashini (2023) observed that AI-driven data analytics can help farmers maximize crop yields while minimizing pesticide usage, thereby supporting SDGs related to zero hunger and responsible consumption.

AI technologies, extensive language models, can enhance data analysis and decision-making, offering innovative solutions to mitigate the impacts of climate change and optimize resource use (Nie & Liu, 2025). For instance, these intelligent systems can analyze vast datasets to identify patterns in energy consumption, enabling more efficient resource management while reducing waste. However, there are inherent challenges, such as existing disparities in human and AI attitudes toward the SDGs, which can lead to unintended consequences if left unaddressed (Regona et al., 2024).

AI innovations are advancing sectors like healthcare, agriculture, education, and transportation by allowing humans to utilize large datasets for significant breakthroughs (Kostopoulos et al., 2024). Additionally, AI-powered computing is helping doctors reduce medical mistakes, boost crop yields, and customize training for students and remote researchers (Pandey & Joseph, 2025).

The intersection of AI and climate change has been gaining traction in recent years. For example, AI can be used to manage environmental changes and their impacts across many economic fields and conditions (Zhuk, 2023). Renewable distributed electricity grids integrated with AI, safer supply chains, environmental control and regulation, and weather condition forecasting are only some of the possible applications (Wang et al., 2025).

6.8 Future Trends in AI and Sustainability

AI is rapidly becoming a cornerstone in the fight for a more sustainable future. From reducing carbon footprints to optimizing resource use, AI is driving innovation across various sectors. The chapter discusses key trends shaping the future of sustainability powered by AI.

6.8.1 AI-Driven Energy Efficiency

Energy consumption is a leading contributor to global carbon emissions, and AI is transforming how energy is managed to reduce its environmental impact (Guo et al., 2025). Advanced AI systems analyze vast amounts of data in real time, including historical energy usage, weather patterns, and operational trends, to optimize energy consumption dynamically (Ukoba et al., 2024). Predictive models enable industries, cities, and households to adjust their energy use proactively, reducing waste and lowering costs. Smart grids, powered by AI, balance supply and demand by rerouting energy efficiently and minimizing losses (Mahmood et al., 2024). These grids work alongside demand response systems that encourage consumers to reduce energy use during peak times, ensuring a more sustainable energy distribution. In industrial settings, AI-driven solutions identify inefficiencies in production processes, optimizing machinery operations to lower energy consumption without compromising output (Biswas et al., 2024). This integration of AI across energy systems not only reduces carbon footprints but also accelerates the transition to a more sustainable energy landscape.

6.8.2 AI-Powered Climate Modeling and Prediction

AI is at the forefront of advancing climate modeling, offering unprecedented precision in predicting and mitigating climate-related risks (Bano, 2025). By integrating vast datasets from satellites, weather stations, and environmental sensors, AI systems provide detailed insights into shifting weather patterns, temperature fluctuations, and the likelihood of extreme events such as hurricanes or droughts (Hamdan et al., 2024). These models are critical for governments and organizations, enabling them to develop proactive strategies to address climate challenges (Erdoğdu et al., 2025). For instance, AI-driven tools can predict weather conditions. Jain et al. (2023) observed that these models support long-term sustainability planning by helping policymakers evaluate the potential outcomes of various climate interventions. With real-time data analysis and forecasting capabilities, AI is not only improving preparedness for environmental hazards but also empowering decision-makers to create more resilient and adaptive climate policies (Santos & Carvalho, 2025).

6.8.3 AI in Renewable Energy Integration

The integration of renewable energy sources such as solar and wind into national grids has historically been challenging due to their variability. However, AI is revolutionizing this process at an accelerated pace (Raman et al., 2024). AI algorithms analyze weather data, historical energy production trends, and grid performance to predict renewable energy output with remarkable accuracy. This allows utilities to balance supply and demand better, ensuring renewable energy is used effectively and reducing reliance on fossil fuels (Benti et al., 2023).

AI-powered systems can forecast energy production from sources like solar and wind, predict energy demand, and manage energy storage, leading to more efficient and reliable renewable energy systems, as shown in Figure 6.5.

These advancements enable smoother integration of renewables, reducing energy wastage and accelerating the shift to a cleaner, more sustainable energy future.

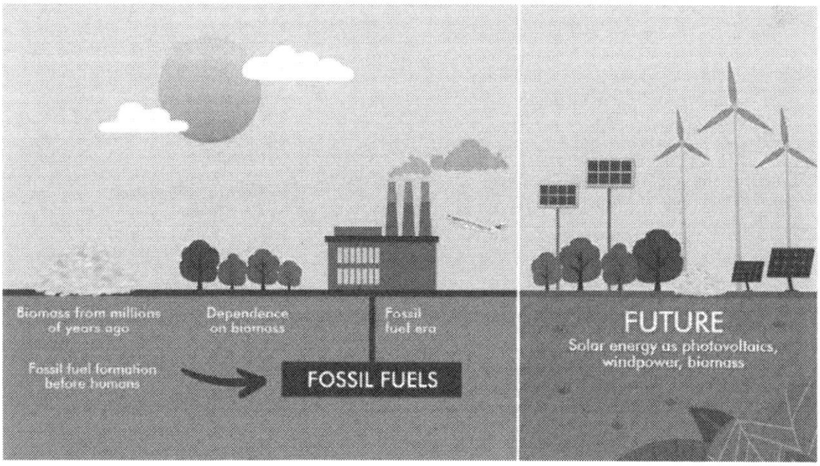

Figure 6.5 Sustainable eco-friendly solutions for power.

Source: https://arxiv.org/html/2406.16965v1/x1.jpg.

This integration also enables predictive maintenance, reducing downtime and costs, and facilitates the development of smart grids and virtual power plants.

6.9 Conclusion

AI is driving sustainability by optimizing energy systems, enhancing climate resilience, supporting biodiversity conservation, and promoting efficient resource use. Key applications include dynamic energy management, renewable energy integration, climate modeling, wildlife monitoring, and circular economies. AI is undeniably shaping the future of smart cities, making urban living more efficient, sustainable, and secure. From traffic management and public safety to energy optimization and waste management, AI-driven solutions are enhancing every aspect of city life.

Advances in AI technology will extend their impact on smart cities. Future innovations such as autonomous vehicles, AI-driven urban planning, and new healthcare solutions will enhance urban living. As cities integrate these technologies, collaboration among policymakers, businesses, and residents is essential to ensure ethical use of AI. Responsible deployment of AI will help cities develop a smarter, more sustainable future for generations.

AI continues to be a strong driver of sustainability, enhancing energy efficiency, minimizing waste, aiding biodiversity efforts, and enabling smarter urban development. Through AI, society can foster a more sustainable future that considers the well-being of people, the environment, and upcoming generations. The opportunities are extensive, and the urgency to start is immediate.

By leveraging cutting-edge AI technology, we provide innovative solutions that help businesses and organizations reduce their environmental impact, optimize resource usage, and contribute to a greener, more sustainable world. With AI-powered sustainability solutions, industries are empowered to meet the challenges and pave the way for a more sustainable future. With our expertise and dedication, we are proud to be part of the global movement toward a cleaner, smarter, and more sustainable planet.

6.9.1 Contribution to Closing the Knowledge Gap and Advancement of Knowledge

This chapter on AI for sustainability focused on identifying knowledge gaps and outlining how AI can be advanced to contribute to sustainable development. This involved exploring specific applications of AI across sectors such as energy, agriculture, and waste management, while also addressing ethical considerations and potential societal impacts. The chapter also analyzed the current state of research, highlighted areas requiring further investigation, and proposed strategies to bridge these gaps and unlock AI's full potential in achieving sustainability goals.

The chapter advances knowledge by emphasizing the importance of data availability, quality, and accessibility for training and deploying AI models. The chapter explored how to collect, manage, and analyze data effectively to support AI-driven sustainability initiatives. Moreover, the chapter advances knowledge by analyzing the role of policy and regulation in guiding the development and deployment of AI for sustainability. This involved exploring the need for standards, guidelines, and frameworks to ensure that AI is used responsibly and ethically.

References

Abbas, A. (2025). *Smart Urban Infrastructure: AI-Powered Solutions for Sustainable Transportation and Construction Project Optimization.* http://dx.doi.org/10.13140/RG.2.2.25262.63046

Abdelati, M. (2024). Smart traffic management for sustainable development in cities: Enhancing safety and efficiency. *International Journal of Advanced Engineering and Business Sciences*, *5*(1), 39–61. http://dx.doi.org/10.21608/ijaebs.2024.242361.1088

Abul Hasan, M. (2021). Challenges of solid waste management using artificial intelligence. *Journal of Environmental Protection and Ecology*, *22*(4), 1478–1487.

Adewale, B. A., Ene, V. O., Ogunbayo, B. F., & Aigbavboa, C. O. (2024). A systematic review of the applications of AI in a sustainable building's lifecycle. *Buildings*, *14*(7), 2137. https://doi.org/10.3390/buildings14072137.

Adeyeri, T. B. (2024). Economic impacts of AI-driven automation in financial services. *Valley International Journal Digital Library*, *12*(7), 6779–6791.

Ahmad, N., & Haile, A. (2025). *Smart City Parking: Leveraging AI and IoT for Efficient Space Utilization and Billing.* http://dx.doi.org/10.13140/RG.2.2.33203.54565

Ahmad, R., Nawaz, A., Mustafa, G., Ali, T., Tlija, M., El-Meligy, M. A., & Ahmed, Z. (2024). CHART: Intelligent crime hotspot detection and real-time tracking using machine learning. *Computers, Materials & Continua*, *81*(3). https://doi.org/10.32604/cmc.2024.056971

Akavova, A., Beguyev, S., & Zaripova, R. (2023). How AI and machine learning can drive sustainable development. In *E3S Web of Conferences* (Vol. 460, p. 04018). EDP Sciences. https://doi.org/10.1051/e3sconf/202346004018.

Alabi, M. (2025). *AI-Powered Smart Grids for Energy Optimization and Sustainability.*

Alamatsaz, K., Hussain, S., Lai, C., & Eicker, U. (2022). Electric bus scheduling and timetabling, fast charging infrastructure planning, and their impact on the grid: A review. *Energies*, *15*(21), 7919. https://doi.org/10.3390/en15217919

Ali, S. M., Appolloni, A., Cavallaro, F., D'Adamo, I., Di Vaio, A., Ferella, F., . . . Zorpas, A. A. (2023). Development goals towards sustainability. *Sustainability*, *15*(12), 9443. http://dx.doi.org/10.3390/su15129443

Almihat, M. G. M., & Munda, J. L. (2025). The role of smart grid technologies in urban and sustainable energy planning. *Energies*, *18*(7), 1618. https://doi.org/10.3390/en18071618

AlSagri, H. S., & Sohail, S. S. (2024). Evaluating the role of artificial intelligence in sustainable development goals with an emphasis on "quality education". *Discover Sustainability*, *5*, 458. https://doi.org/10.1007/s43621-024-00682-9

Anokye, K., Mohammed, S. A., Agyemang, P., Agya, A. B., Amuah, E. E. Y., Sodoke, S., & Diderutua, E. K. (2024). From perception to action: Waste management challenges in Kassena Nankana east municipality. *Heliyon*, *10*(14). https://doi.org/10.1016/j.heliyon.2024.e32438

Bąk, I. (2024). Sustainable development goals (SDGs): Concept and measurement. In *The Role of Financial Markets in Achieving the Sustainable Development Goals* (pp. 15–41). Edward Elgar Publishing.

Bankins, S., Ocampo, A. C., Marrone, M., Restubog, S. L. D., & Woo, S. E. (2024). A multilevel review of artificial intelligence in organizations: Implications for organizational behavior research and practice. *Journal of Organizational Behavior*, *45*(2), 159–182. https://doi.org/10.1002/job.2735

Bano, S. (2025). The role of AI in predicting and mitigating climate change: A comparative study of US and China's approaches. *The Critical Review of Social Sciences Studies*, *3*(1), 664–675. http://dx.doi.org/10.59075/dv9cmp29

Barbhuiya, S., & Das, B. B. (2023). Life cycle assessment of construction materials: Methodologies, applications and future directions for sustainable decision-making. *Case Studies in Construction Materials*, *19*, e02326.

Benti, N. E., Chaka, M. D., & Semie, A. G. (2023). Forecasting renewable energy generation with machine learning and deep learning: Current advances and future prospects. *Sustainability*, *15*(9), 7087. https://doi.org/10.3390/su15097087

Biswas, P., Arockiam, D., Neogi, S. G., Sengupta, S., Saraswat, S., & Chanda, R. (2024, June). Multi-lingual semantic-based sentiment analysis using generative AI. In *2024 International Conference on Computer, Electronics, Electrical Engineering & their Applications (IC2E3)* (pp. 1–8). IEEE. https://doi.org/10.1109/IC2E362166.2024.10827103

Buçinca, Z., Pham, C. M., Jakesch, M., Ribeiro, M. T., Olteanu, A., & Amershi, S. (2023). Aha!: Facilitating AI impact assessment by generating examples of harms. *arXiv preprint* arXiv: 2306.03280.

Cath, C. (2018). Governing artificial intelligence: Ethical, legal and technical opportunities and challenges. *Philosophical Transactions of the Royal Society A: Mathematical, Physical and Engineering Sciences*, *376*(2133), 20180080. https://doi.org/10.1098/rsta.2018.0080

Chen, L., Chen, Z., Zhang, Y., Liu, Y., Osman, A. I., Farghali, M., . . . Yap, P. S. (2023). Artificial intelligence-based solutions for climate change: A review. *Environmental Chemistry Letters*, *21*(5), 2525–2557.

Chikukula, A. A., Omokaro, G. O., Godswill, O. O., Cassim, S. Y., Mabangwe, H. S., & Kaisi, I. (2024). Problems and possible solutions to municipal solid waste management in Malawi urban areas–An overview. *Asian Journal of Environment & Ecology*, *23*(6), 42–52. www.sdiarticle5.com/review-history/116545

Costa, E. (2024). *Industry 5.0 and SDG 9: A Symbiotic Dance towards Sustainable Transformation*. https://doi.org/10.1186/s42055-024-00073-y

Daneshvar, M., Asadi, S., Mohammadi-Ivatloo, B., Daneshvar, M., Asadi, S., & Mohammadi-Ivatloo, B. (2021). Overview of the grid modernization and smart grids. In *Grid Modernization — Future Energy Network Infrastructure: Overview, Uncertainties, Modelling, Optimization, and Analysis* (pp. 1–31). http://dx.doi.org/10.1007/978-3-030-64099-6_1

Daoud, A. O., Kineber, A. F., Ali, A. H., & Elseknidy, M. (2025). Empowering sustainable infrastructure and sustainable development goals through industry 5.0 implementation. *Sustainable Development*. https://doi.org/10.1002/sd.3347

Das, D. K. (2025). Digital technology and AI for smart sustainable cities in the Global South: A critical review of literature and case studies. *Urban Science*, *9*(3), 72.

Dos Santos, J. P. F., de Matos, C. A., & Groznik, A. (2025). The role of artificial intelligence in smart city systems usage: Drivers, barriers, and behavioural outcomes. *Technology in Society*, *81*, 102867. https://doi.org/10.1016/j.techsoc.2025.102867

Dos Santos, S. C., Vilela, J. F., Carvalho, T. H., Rocha, T. C., Candido, T. B., Bezerra, V. S., & Silva, D. J. (2024). Artificial Intelligence in Sustainable Smart Cities: A Systematic Study on Applications, Benefits, Challenges, and Solutions. In *Proceedings of the 26th International Conference on Enterprise Information Systems – Volume 1: ICEIS*; ISBN 978-989-758-692-7; ISSN 2184-4992, SciTePress (pp. 644–655). DOI: 10.5220/0012617900003690

Durlik, I., Miller, T., Kostecka, E., & Tuński, T. (2024). Artificial intelligence in maritime transportation: A comprehensive review of safety and risk management applications. *Applied Sciences*, *14*(18), 8420.

Durlik, I., Miller, T., Kostecka, E., Kozlovska, P., & Ślączka, W. (2025). Enhancing safety in autonomous maritime transportation systems with real-time AI agents. *Applied Sciences*, *15*(9), 4986. https://doi.org/10.3390/app15094986

Dwivedi, Y. K., Kshetri, N., Hughes, L., Slade, E. L., Jeyaraj, A., Kumar Kar, A., Baabdullah, A. M., Koohang, A., Raghavan, V., Ahuja, M., Albanna, H., Albashrawi, M. A., Al-Busaidi, A. S., Balakrishnan, J., Barlette, Y., Basu, S., Bose, I., Brooks, L., Buhalis, D., . . . Wright, R. (2023). "So what if ChatGPT wrote it?" Multidisciplinary perspectives on opportunities, challenges and implications of generative conversational AI for research, practice and policy. *International Journal of Information Management*, *71*(102642), Article 102642. https://doi.org/10.1016/j.ijinfomgt.2023.102642.

Eberling, E., & Langkau, S. (2023). Utilizing SDGs in sustainability assessments of innovations: Deriving methodological recommendations from existing approaches. *Journal of Cleaner Production*. https://doi.org/10.1016/j.jclepro.2023.140383

Enose Kamalabai, N., Hannola, L., & Donoghue, I. (2025). Role of industrial artificial intelligence in advancing human-centric sustainable development of industry 5.0. In *Contemporary Issues in Industry 5.0: Towards an AI Integrated Society* (pp. 325–371). Cham, Switzerland: Springer Nature. http://dx.doi.org/10.1007/978-3-031-74779-3_12

Erdoğdu, A., Dayi, F., Yanik, A., Yildiz, F., & Ganji, F. (2025). Innovative solutions for combating climate change: Advancing sustainable energy and consumption practices for a greener future. *Sustainability*, *17*(6), 2697. https://doi.org/10.3390/su17062697

Errousso, H., Filali, Y., Aghbalou, N., Alaoui, E. A. A., & Benhadou, S. (2023, October). Spatio-temporal clustering for optimal real-time parking availability estimation. In *The Proceedings of the International Conference on Smart City Applications* (pp. 199–214). Cham, Switzerland: Springer Nature. http://dx.doi.org/10.1007/978-3-031-53824-7_19

Fallah Shayan, N., Mohabbati-Kalejahi, N., Alavi, S., & Zahed, M. A. (2022). Sustainable development goals (SDGs) as a framework for corporate social responsibility (CSR). *Sustainability*, *14*(3), 1222. https://doi.org/10.3390/su14031222

Fang, B., Yu, J., Chen, Z., Osman, A. I., Farghali, M., Ihara, I., . . . Yap, P. S. (2023). Artificial intelligence for waste management in smart cities: A review. *Environmental Chemistry Letters*, *21*, 1959–1989. https://doi.org/10.1007/s10311-023-01604-3

Ghobakhloo, M., Iranmanesh, M., Tseng, M., Grybauskas, A., Stefanini, A., & Amran, A. (2023). Behind the definition of industry 5.0: A systematic review of technologies, principles, components, and values. *Journal of Industrial and Production Engineering*, *40*, 432–447. https://doi.org/10.1080/21681015.2023.2216701

Guo, Q., Peng, Y., & Luo, K. (2025). The impact of artificial intelligence on energy environmental performance: Empirical evidence from cities in China. *Energy Economics*, *141*, 108136. https://doi.org/10.1016/j.eneco.2024.108136

Hajam, Y. A., Kumar, R., & Kumar, A. (2023). Environmental waste management strategies and vermi transformation for sustainable development. *Environmental Challenges*, *13*, 100747. https://doi.org/10.1016/j.envc.2023.100747

Hamdan, A., Ibekwe, K. I., Etukudoh, E. A., Umoh, A. A., & Ilojianya, V. I. (2024). AI and machine learning in climate change research: A review of predictive models and environmental impact. *World Journal of Advanced Research and Reviews*, *21*(1), 1999–2008. http://dx.doi.org/10.30574/wjarr.2024.21.1.0257

Ibrahim, I. M., Elghafar, M., Mohamed, M., Ayoub, M., & Abuhadrous, I. (2024). AI-powered parking management systems: A review of applications and challenges. *Advanced Sciences and Technology Journal*, *1*(2), 1–22.

Iwuanyanwu, O., Gil-Ozoudeh, I., Okwandu, A. C., & Ike, C. S. (2024). Cultural and social dimensions of green architecture: Designing for sustainability and community well-being. *International Journal of Applied Research in Social Sciences*, *6*(8), 1951–1968. http://dx.doi.org/10.51594/ijarss.v6i8.1477

Jacob, C. (2024). Infinite affluence on a finite planet. *Sustainability*, *16*(18), 8191. https://doi.org/10.3390/su16188191

Jain, H., Dhupper, R., Shrivastava, A., Kumar, D., & Kumari, M. (2023). AI-enabled strategies for climate change adaptation: Protecting communities, infrastructure, and businesses from the impacts of climate change. *Computational Urban Science*, *3*, 25. https://doi.org/10.1007/s43762-023-00100-2

Jain, V., & Mitra, A. (2025). Artificial intelligence and machine learning for sustainable development: Enhancing health, equity, and environmental sustainability. In *Machine and Deep Learning Solutions for Achieving the Sustainable Development Goals* (pp. 107–124). IGI Global Scientific Publishing. http://dx.doi.org/10.4018/979-8-3693-8161-8.ch006

Jariwala, M. (2024). Integrating artificial intelligence to enhance sustainability in project management practices. *International Journal of Computer Applications*, *186*(20), 35–40. https://doi.org/10.5120/ijca2024923627

Javaid, M., Haleem, A., Khan, I. H., & Suman, R. (2023). Understanding the potential applications of artificial intelligence in agriculture sector. *Advanced Agrochem*, *2*(1), 15–30. https://doi.org/10.1016/j.aac.2022.10.001

John, J., Varkey, M. S., Podder, R. S., Sensarma, N., Selvi, M., Santhosh Kumar, S. V. N., & Kannan, A. (2022). Smart prediction and monitoring of waste disposal system using IoT and cloud for IoT-based smart cities. *Wireless Personal Communications*, *122*(1), 243–275. https://link.springer.com/article/10.1007%2Fs11277-021-08897-z

Juneja, G., & Saraswat, P. (2025). Sustainability and artificial intelligence: A synergistic approach for addressing global challenges. In *Innovation Ecosystems and Sustainable Technologies* (pp. 313–327). Emerald Publishing Limited. https://doi.org/10.1108/978-1-83662-368-720251015

Kar, A. K., Choudhary, S. K., & Singh, V. K. (2022). How can artificial intelligence impact sustainability: A systematic literature review. *Journal of Cleaner Production*, *376*, 134120. https://doi.org/10.1016/j.jclepro.2022.134120

Khan, S. A., Alam, T., Khan, M. S., Blecich, P., Kamal, M. A., Gupta, N. K., & Yadav, A. S. (2022). Life cycle assessment of embodied carbon in buildings: Background, approaches and advancements. *Buildings*, *12*(11), 1944.

Khan, S. A., Khatri, B., Jonathan, H., Naim, A., & Arora, S. (2025). Revolutionizing agriculture: Harnessing high-performance computing and artificial intelligence for innovative farming practices. In Khan, S. A. R., Sheikh, A. A., Kolade, O., & Kamble, S. (Eds.), *Circular Economy and Technological Innovation. Industrial Ecology*. Singapore: Springer. https://doi.org/10.1007/978-981-96-5650-9_11

Khanna, H., & Gupta, G. (2024, August 16). *Triple Bottom Line Sustainability: HRM and Role of Artificial Intelligence*. SSRN. https://ssrn.com/abstract=4936644 or http://dx.doi.org/10.2139/ssrn.4936644

Kostopoulos, G., Davrazos, G., & Kotsiantis, S. (2024). Explainable artificial intelligence-based decision support systems: A recent review. *Electronics*, *13*(14), 2842. https://doi.org/10.3390/electronics13142842

Kumari, S., Kumari, A., & Waoo, A. A. (2024). AI for sustainable development: Innovations & applications. *ShodhKosh Journal of Visual and Performing Arts*, *5*(5). https://doi.org/10.29121/shodhkosh.v5.i5.2024.1896

Laimon, M., Yusaf, T., Mai, T., Goh, S., & Alrefae, W. (2022). A systems thinking approach to address sustainability challenges to the energy sector. *International Journal of Thermofluids*, *15*, 100161.

Litman, T. (2020). *Parking Management Best Practices*. Routledge.

Liu, X., Sathishkumar, K., Zhang, H., Saxena, K. K., Zhang, F., Naraginiti, S., . . . Guo, X. (2024). Frontiers in environmental cleanup: Recent advances in remediation of emerging pollutants from soil and water. *Journal of Hazardous Materials Advances*, 100461. http://dx.doi.org/10.1016/j.hazadv.2024.100461

Madanchian, M., & Taherdoost, H. (2025). The impact of artificial intelligence on research efficiency. *Results in Engineering*, 104743. https://doi.org/10.1016/j.rineng.2025.104743

Magsino, E. R., Arada, G. P., & Ramos, C. M. L. (2022). An evaluation of temporal- and spatial-based dynamic parking pricing for commercial establishments. *IEEE Access*, *10*, 102724–102736. http://dx.doi.org/10.1109/ACCESS.2022.3209806

Mahmood, M., Chowdhury, P., Yeassin, R., Hasan, M., Ahmad, T., & Chowdhury, N. U. R. (2024). Impacts of digitalization on smart grids, renewable energy, and demand response: An updated review of current applications. *Energy Conversion and Management: X*, 100790. http://dx.doi.org/10.1016/j.ecmx.2024.100790

Mio, C., Panfilo, S., & Blundo, B. (2020). Sustainable development goals and the strategic role of business: A systematic literature review. *Business Strategy and the Environment*, *29*(8), 3220–3245. https://doi.org/10.1002/bse.2568

Mohammed, M., & Oke, J. (2023). Origin-destination inference in public transportation systems: A comprehensive review. *International Journal of Transportation Science and Technology*, *12*(1), 315–328. https://doi.org/10.1016/j.ijtst.2022.03.002.

Mondejar, M. E., Avtar, R., Diaz, H. L. B., Dubey, R. K., Esteban, J., Gómez-Morales, A., . . . Garcia-Segura, S. (2021). Digitalization to achieve sustainable development goals: Steps towards a smart green planet. *Science of The Total Environment*, *794*, 148539. https://doi.org/10.1016/j.scitotenv.2021.148539

Nabeel, M. Z. (2024). AI-enhanced project management systems for optimizing resource allocation and risk mitigation. *Asian Journal of Multidisciplinary Research & Review*, *5*(5), 53–91. http://dx.doi.org/10.55662/AJMRR.2024.5502

Naz, A., & Hoque, I. (2023). *Integration of Intelligent Transportation Systems (ITS) with Conventional Traffic Management in Developing Countries*. http://dx.doi.org/10.31224/3154

Nguyen, A., Ngo, H. N., Hong, Y., Dang, B., & Nguyen, B. P. T. (2023). Ethical principles for artificial intelligence in education. *Education and Information Technologies*, *28*, 4221–4241. https://doi.org/10.1007/s10639-022-11316-w

Nie, Q., & Liu, T. (2025). Large language models: Tools for new environmental decision-making. *Journal of Environmental Management*, *375*, 124373. https://doi.org/10.1016/j.jenvman.2025.124373

Nishant, R., Kennedy, M., & Corbett, J. (2020). Artificial intelligence for sustainability: Challenges, opportunities, and a research agenda. *International Journal of Information Management*, *53*, 102104. https://doi.org/10.1016/j.ijinfomgt.2020.102104

Olawade, D. B., Fapohunda, O., Wada, O. Z., Usman, S. O., Ige, A. O., Ajisafe, O., & Oladapo, B. I. (2024). Smart waste management: A paradigm shift enabled by artificial intelligence. *Waste Management Bulletin*. https://doi.org/10.1016/j.wmb.2024.05.001

Onyeneke, C. C., & Karam, A. H. (2022). An exploratory study of crime: Examining lived experiences of crime through socioeconomic, demographic, and physical characteristics. *Urban Science*, *6*(3), 43. http://dx.doi.org/10.3390/urbansci6030043

Pandey, A., & Joseph, J. (2025). The convergence of blockchain and artificial intelligence: Transforming healthcare. In *AI and Blockchain Applications for Privacy and Security in Smart Medical Systems* (pp. 399–416). IGI Global Scientific Publishing. http://dx.doi.org/10.4018/979-8-3373-0593-6.ch015

Pandey, D. K., & Mishra, R. (2024). Towards sustainable agriculture: Harnessing AI for global food security. *Artificial Intelligence in Agriculture*. https://doi.org/10.1016/j.aiia.2024.04.003

Pasamar, S., Bornay-Barrachina, M., & Morales-Sánchez, R. (2023). Institutional pressures for sustainability: A triple bottom line approach. *European Journal of Management and Business Economics*. Vol. ahead-of-print No. ahead-of-print. https://doi.org/10.1108/EJMBE-07-2022-0241

Raihan, A. (2024). A review of the potential opportunities and challenges of the digital economy for sustainability. *Innovation and Green Development*, *3*(4), 100174. https://doi.org/10.1016/j.igd.2024.100174

Raman, R., Gunasekar, S., Kaliyaperumal, D., & Nedungadi, P. (2024). Navigating the nexus of artificial intelligence and renewable energy for the advancement of sustainable development goals. *Sustainability*, *16*(21), 9144. https://doi.org/10.3390/su16219144

Rame, R., Purwanto, P., & Sudarno, S. (2024). Industry 5.0 and sustainability: An overview of emerging trends and challenges for a green future. *Innovation and Green Development*, *3*(4), 100173. https://doi.org/10.1016/j.igd.2024.100173

Ramya, J., Yerraguravagari, S. S., Gaikwad, S., & Gupta, R. K. (2024). AI and machine learning in predictive analytics: Revolutionizing business strategies through big data insights. *Library of Progress-Library Science, Information Technology & Computer*, *44*(3). http://dx.doi.org/10.56025/IJARESM.2024.1211241418

Rane, J., Kaya, O., Mallick, S. K., & Rane, N. L. (2024). Smart farming using artificial intelligence, machine learning, deep learning, and ChatGPT: Applications, opportunities, challenges, and future directions. *Generative Artificial Intelligence in Agriculture, Education, and Business*, 218–272. http://dx.doi.org/10.70593/978-81-981271-7-4_6

Regona, M., Yigitcanlar, T., Hon, C., & Teo, M. (2024). Artificial intelligence and sustainable development goals: Systematic literature review of the construction industry. *Sustainable Cities and Society*, *108*, 105499. https://doi.org/10.1016/j.scs.2024.105499

Ricardo V., Azizpour, H., Leite, I., Balaam, M., Dignum, V., Domisch, S., Felländer, A., Langhans, S. D., Tegmark, M., Nerini, F. F. (2020). The role of artificial intelligence in achieving the Sustainable Development Goals. *Nature Communications*, *11*(1), 233. http://dx.doi.org/10.48550/arXiv.1905.00501

Sachithra, V., & Subhashini, L. D. C. S. (2023). How artificial intelligence uses to achieve the agriculture sustainability: Systematic review. *Artificial Intelligence in Agriculture*, *8*, 46–59. https://doi.org/10.1016/j.aiia.2023.04.002

Sachs, J. D., Schmidt-Traub, G., Mazzucato, M., Messner, D., Nakicenovic, N., & Rockström, J. (2019). Six transformations to achieve the sustainable development goals. *Nature Sustainability*, *2*(9), 805–814. http://dx.doi.org/10.1038/s41893-019-0352-9/

Sajwani, R. A. (2024). Artificial intelligence for sustainability development in healthcare. In Al Marri, K., Mir, F. A., David, S. A., & Al-Emran, M. (Eds.), *BUiD Doctoral Research Conference 2023. Lecture Notes in Civil Engineering* (Vol. 473). Cham: Springer. https://doi.org/10.1007/978-3-031-56121-4_26

Saleem, A., Anwar, S., Nawaz, T., Fahad, S., Saud, S., Ur Rahman, T., . . . Nawaz, T. (2024). Securing a sustainable future: The climate change threat to agriculture, food security, and sustainable development goals. *Journal of Umm Al-Qura University for Applied Sciences*. https://doi.org/10.1007/s43994-024-00177-3

Saner, R., Yiu, L., & Nguyen, M. (2020). Monitoring the SDGs: Digital and social technologies to ensure citizen participation, inclusiveness and transparency. *Development Policy Review*, *38*(4), 483–500. https://doi.org/10.1111/dpr.12433

Santos, M. R., & Carvalho, L. C. (2025). AI-driven participatory environmental management: Innovations, applications, and future prospects. *Journal of Environmental Management*, *373*, 123864. http://dx.doi.org/10.1016/j.jenvman.2024.123864

Schoormann, T., Strobel, G., Möller, F., Petrik, D., & Zschech, P. (2023). Artificial intelligence for sustainability—A systematic review of information systems literature. *Communications of the Association for Information Systems*, *52*(1), 8. https://doi.org/10.17705/1CAIS.05209.

Shader, C. G., Gill, C., Zheng, X., & Carleton, B. (2024). City government as super-controller: A systematic review of non-police mechanisms that city governments can apply to reduce crime at hot spots. *Aggression and Violent Behavior*, 101957. http://dx.doi.org/10.1016/j.avb.2024.101957

Shahvaroughi Farahani, M., & Ghasemi, G. (2024). How artificial intelligence plays a role in achieving sustainable development goals? *Sustainable Economies*, *2*(3), 66. Situations. http://dx.doi.org/10.62617/se.v2i1.66

Shang, M., Ma, Z., Su, Y., Shaheen, F., Khan, H. U. R., Tahir, L. M., . . . Zaman, K. (2022). Understanding the importance of sustainable ecological innovation in reducing carbon emissions: Investigating the green energy demand, financial development, natural resource management, industrialisation, and urbanisation channels. *Economic Research-Ekonomska Istraživanja*, *36*(2). https://doi.org/10.1080/1331677X.2022.2137823

Singagerda, F. S., Dewi, D. A., Trisnawati, S., Septarina, L., & Dhika, M. R. (2025). Towards a circular economy: Integration of AI in waste management for sustainable urban growth. *Journal of Lifestyle and SDGs Review*, *5*(1), e02642. https://doi.org/10.5281/zenodo.7067636

Siqueira, M. B., dos Santos, V. M., Diniz, E. H., & Cruz, A. P. A. (2024). Artificial intelligence for sustainability: A systematic literature review in information systems. *Revista de Gestão Social e Ambiental*, *18*(3), 1–20. https://doi.org/10.24857/rgsa.v18n3-178

Son, H., Jang, J., Park, J., Balog, A., Ballantyne, P., Kwon, H. R., . . . Hwang, J. (2025). Leveraging advanced technologies for (smart) transportation planning: A systematic review. *Sustainability*, *17*(5), 2245.

Sorooshian, S. (2024). The sustainable development goals of the United Nations: A comparative midterm research review. *Journal of Cleaner Production*, 142272. https://doi.org/10.1016/j.jclepro.2024.142272

Srikumar, K., Tan, Y. H., Kansedo, J., Tan, I. S., Mubarak, N. M., Ibrahim, M. L., . . . Khalid, M. (2024). A review on the environmental life cycle assessment of biodiesel production: Selection of catalyst and oil feedstock. *Biomass and Bioenergy*, *185*, 107239.

Truby, J. (2020). Governing artificial intelligence will benefit the UN's sustainable development goals. *Sustainable Development*, *28*(4), 946–959. https://doi.org/10.1002/sd.2048

Türkeş, M. (2024). *The Role of Sustainability and Sustainable Development in Climate Change Mitigation and Adaptation*. https://doi.org/10.54517/ssd.v2i1.2407

Ukoba, K., Onisuru, O. R., Jen, T. C., Madyira, D. M., & Olatunji, K. O. (2025). Predictive modeling of climate change impacts using Artificial Intelligence: a review for equitable governance and sustainable outcome. *Environmental Science and Pollution Research*, 1–20.

Vijaya, G. S. (2024). *Addressing Contemporary Environmental Challenges: A Focus on Climate Change, Waste Management, Biodiversity Conservation, and Sustainable Development* (p. 12). https://doi.org/10.62422/978-81-970328-1-3-001

Voulvoulis, N., Giakoumis, T., Hunt, C., Kioupi, V., Petrou, N., Souliotis, I., & Vaghela, C. J. G. E. C. (2022). Systems thinking as a paradigm shift for sustainability transformation. *Global Environmental Change*, *75*, 102544.

Wang, Q., Li, Y., & Li, R. (2025). Integrating artificial intelligence in energy transition: A comprehensive review. *Energy Strategy Reviews*, *57*, 101600. https://doi.org/10.1016/j.esr.2024.101600

Wirsbinna, A., Grega, L., & Juenger, M. (2023). Assessing factors influencing citizens' behavioral intention towards smart city living. *Smart Cities*, *6*(6), 3093–3111. http://dx.doi.org/10.3390/smartcities6060138

Yadav, A., & Sondhi, H. (2023, March). Systematic literature review on sustainable marketing and artificial intelligence. In *2023 10th International Conference on Computing for Sustainable Global Development (INDIACom)* (pp. 583–588). IEEE.

Zaman, M., Puryear, N., Abdelwahed, S., & Zohrabi, N. (2024). A review of IoT-based smart city development and management. *Smart Cities*, *7*(3), 1462–1501. https://doi.org/10.3390/smartcities7030061.

Zhang, Z., Chen, Z., Zhang, J., Liu, Y., Chen, L., Yang, M., Osman, A. I., Farghali, M., Liu, E., Hassan, D., Ihara, I., Lu, K., Rooney, D. W., Yap, P.-S. Municipal solid waste management challenges in developing regions: A comprehensive review and future perspectives for Asia and Africa. *Science of the Total Environment*, 172794. https://doi.org/10.1016/j.scitotenv.2024.172794

Zhuk, A. (2023). Artificial intelligence impact on the environment: Hidden ecological costs and ethical-legal issues. *Journal of Digital Technologies and Law*, *1*(4), 932–954. https://doi.org/10.21202/jdtl.2023.40

Ziemba, E. W., Duong, C. D., Ejdys, J., Gonzalez-Perez, M. A., Kazlauskaitė, R., Korzynski, P., . . . Wach, K. (2024). Leveraging artificial intelligence to meet the sustainable development goals. *Journal of Economics and Management*, *46*, 508–583. http://dx.doi.org/10.22367/jem.2024.46.19

7 AI Application in Smart Cities, Energy, and Climate Action

Nasrullah Masud, Abdullah Al Bassam, Md. Tanvir Islam Sourav, and Abdus Sobhan

7.1 Introduction

Rapid urbanization and the climate crisis intersect uniquely in modern cities, requiring new approaches to sustainability and resilience. Although the literature extensively reviewed AI applications in isolated urban sectors like transportation, energy management, or climate adaptation, a significant gap remains in integrating these applications holistically under the modern Industry 5.0 paradigm. This chapter offers novel perspectives by explicitly linking AI techniques with Industry 5.0 concepts, with human-centricity, resilience, and sustainable urban development being stressed thereupon.

The main research gap that this work deals with is the absence of a wide-ranging study that couples theoretical approaches with practical case studies to highlight the systemic application of AI to complex urban–energy–climate nexus problems. In contrast to many past studies that reviewed AI literature by mostly discussing technological feasibility and purely technical concerns from a societal viewpoint, this chapter tries to critically evaluate the implementation of AI; enumerate the key success factors, limitations, and ethical concerns; and finally draw explicit policy guidelines.

The chapter adds real scientific and practical value than mentioned in previous reviews by highlighting AI's potential to achieve SDGs, particularly SDG 11 (Sustainable Cities and Communities) and SDG 13 (Climate Action), through the presentation of comprehensive examples and empirical data.

By examining how AI can be used holistically in energy management, urban sustainability, and climate resilience within the Industry 5.0 framework, this chapter fills a critical gap. It goes beyond disjointed reviews by relating AI techniques to practical issues, emphasizing moral issues, and offering useful advice for accomplishing SDGs 11 and 13. The necessity of incorporating AI into urban planning to address growing environmental risks is highlighted by visual evidence, such as climate impact projections over time horizons (Climate Impact Lab, n.d.).

7.2 Theoretical Foundations and Conceptual Framework

The successful incorporation of artificial intelligence (AI) into smart cities is in line with the new ideas of Industry 5.0, which are defined by a focus on sustainability,

DOI: 10.4324/9781003647621-9

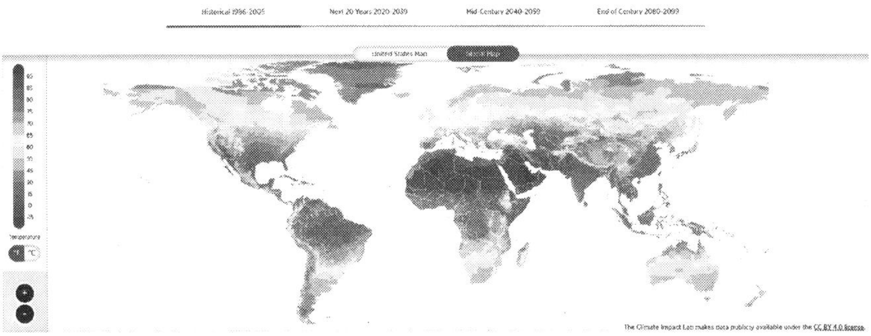

Figure 7.1 Historical (1986–2005): Global climate impact heatmap showing historical baseline temperature anomalies.

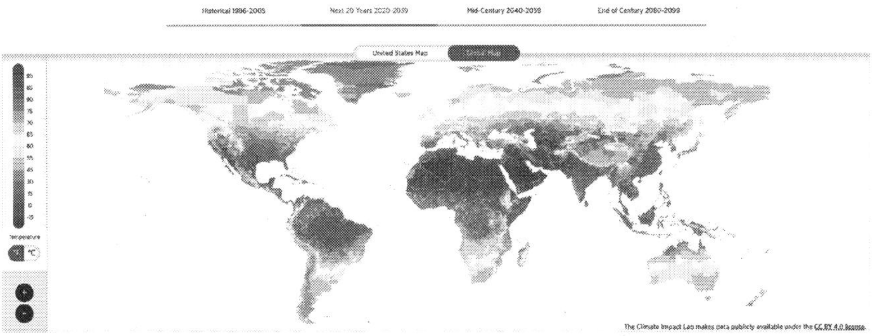

Figure 7.2 Next 20 years (2020–2039): Projected climate impact heatmap for the near future scenario.

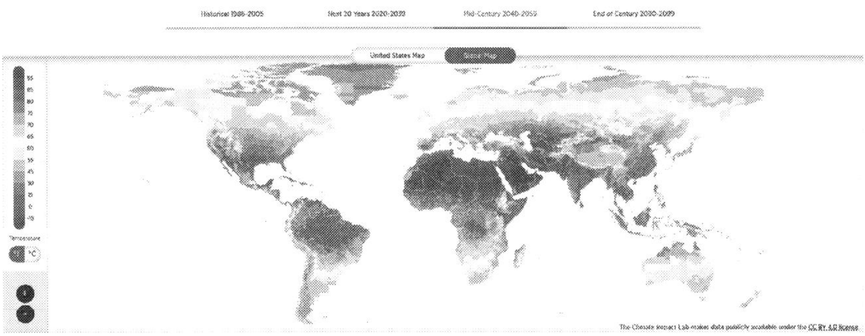

Figure 7.3 Mid-century (2040–2059): Projected climate impact heatmap illustrating mid-century warming trends.

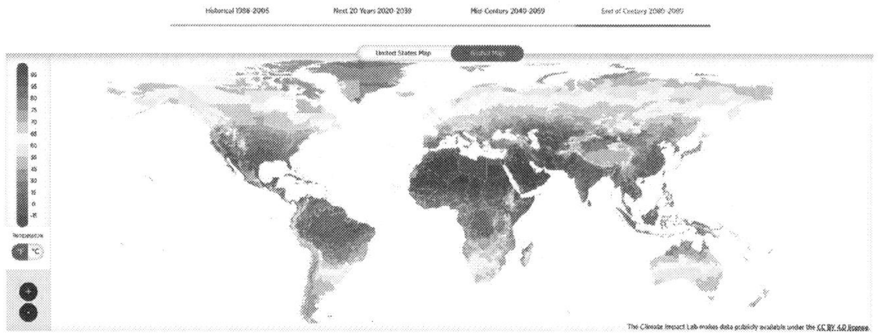

Figure 7.4 End of century (2080–2099): Projected climate impact heatmap under high-emission scenarios.

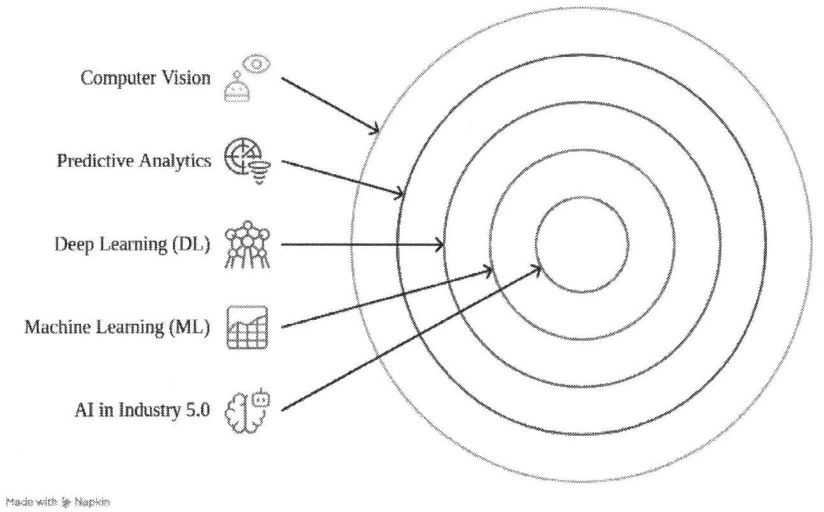

Figure 7.5 AI in Industry 5.0 for urban sustainability.

resilience, and human-centricity (European Commission, 2021a; Xu et al., 2021). In this section, we establish a conceptual link between specific AI approaches, such as computer vision, machine learning, deep learning, and predictive analytics, and tangible urban issues, such as climate adaptation, disaster resilience, energy efficiency, and transportation optimization (Allam & Dhunny, 2019; Bibri, 2018).

For instance, by forecasting weather and occupancy patterns, machine learning algorithms can optimize building energy use and drastically cut down on wasteful energy use (Fayaz & Kim, 2018). By evaluating real-time traffic data, computer vision can also improve intelligent transportation systems, allowing for more intelligent congestion control and emission reductions (Bibri et al., 2020). Predictive analytics also helps cities assess climate risk and implement proactive adaptation strategies for extreme weather events (Climate Impact Lab, n.d.). The chapter explains how technology can be strategically used to promote sustainable, resilient, and inclusive urban environments by clearly tying these AI techniques to Industry 5.0's guiding principles.

7.3 Methodological Approach

The implementation of AI applications in smart cities is examined in this chapter using a methodical literature review approach, with a particular emphasis on urban energy management, sustainable mobility, climate action, and the general ideas of Industry 5.0. Relevant studies were found using scholarly databases such as IEEE Xplore, Web of Science, Scopus, and Google Scholar to guarantee thorough coverage. In order to capture recent developments and guarantee current relevance, the search criteria included peer-reviewed articles, book chapters, conference proceedings, and reputable reports published between 2015 and 2024.

Terms like "artificial intelligence," "machine learning," "deep learning," "smart cities," "urban sustainability," "energy management," "intelligent transportation systems," "climate resilience," and "Industry 5.0" were used in a structured keyword search. The selection process comprised a preliminary screening of titles and abstracts to weed out studies that were outside the purview, and then a thorough full-text review to assess suitability based on evident contributions to the integration of AI methodologies with urban sustainability goals.

Thematic synthesis was utilized to analyze a selection of studies, classifying the results based on specific application areas (energy, mobility, resilience, and governance), AI techniques used, ethical issues raised, and suggested policy frameworks. Finding important research gaps, difficulties in real-world implementation, and prospects for further study were made easier by this synthesis approach. Additionally, case studies and empirical data demonstrating effective AI applications and pointing out constraints or failures in urban settings were examined, which helped create a fair and critical analysis (Grant & Booth, 2009; Xiao & Watson, 2019).

7.4 AI Applications in Urban Energy Systems

With its many benefits over conventional methods, AI has demonstrated great promise in revolutionizing urban energy management systems. The study found that enhanced real-time responsiveness, dynamic optimization capabilities, and improved prediction accuracy—especially through machine learning algorithms that analyze massive amounts of data from smart grids and Internet of Things (IoT) sensors—are key success factors (Bedi et al., 2017). Effective case studies show significant decreases in energy waste, optimal integration of renewable energy sources, and improved grid stability. One example is AI-driven smart grid management in cities like Amsterdam and Copenhagen (Zhou et al., 2020).

Despite these benefits, there are a few drawbacks. These include issues with interoperability and data quality, the sizeable upfront costs associated with infrastructure upgrades, and worries about cybersecurity flaws (Allam & Dhunny, 2019).

Table 7.1 Comparison of Traditional and AI-Optimized Energy Management.

Aspect	*Traditional Energy Management*	*AI-Optimized Energy Management*
Predictive capability	Limited, often reactive	High, proactive forecasting
Real-time responsiveness	Slow and manual adjustments	Rapid, automated responses
Renewable integration	Limited adaptability	Optimized and dynamic integration
Efficiency	Lower overall efficiency	Higher operational efficiency
Scalability	Limited scalability	Highly scalable

Additionally, depending too much on past data can affect the efficacy of decision-making by introducing biases or inaccuracies into predictive models.

The comparison table underscores AI's clear advantages in enhancing energy system performance and sustainability in urban settings.

7.5 AI for Sustainable Mobility and Intelligent Transportation

Urban mobility and intelligent transportation systems are greatly improved by AI, which offers practical answers to enduring problems like traffic, pollution, and ineffective public transportation. Reduced commute times and fewer vehicle emissions are the results of accurate predictions of traffic flow, congestion points, and optimal routing made possible by machine learning models that use historical and real-time traffic data (Bibri et al., 2020). For instance, Singapore's intelligent transport system reduces peak-hour traffic by about 20% by dynamically managing traffic signal timing using deep learning algorithms (Ke et al., 2021).

Sustainability metrics continuously improve in AI-driven mobility projects, according to empirical results. Barcelona improved overall transit efficiency by 10–15% and significantly reduced vehicle emissions by using predictive analytics to optimize public bus schedules based on passenger demand and real-time data (Pereira et al., 2021). Furthermore, Helsinki's AI-powered autonomous electric shuttles have shown a 25% reduction in carbon emissions when compared to conventional diesel-powered public transportation options (Zhang et al., 2022).

Implementation issues still exist, though, such as algorithmic fairness, privacy issues with mobility tracking, and data integration difficulties, underscoring the necessity of continued study and policy creation.

7.6 AI for Climate Action and Urban Resilience

- **Highlight case studies and impact evidence**

AI is being used to support urban resilience in the spirit of Industry 5.0 by facilitating adaptive, real-time responses to threats brought on by climate change. An Intelligent Flood Control Decision Support System (IFCDSS) in Taipei, for instance, employs AI-driven flood forecasting in conjunction with automated pump control, improving pump efficiency by 45% and providing precise water-level forecasts in as little as 5 seconds, according to recent research (Chen et al., 2024). The resilience and sustainability requirements of Industry 5.0 are embodied in this system, which demonstrates how AI not only forecasts extreme events but also propels autonomous, human-centered interventions that lower risks and maintain community well-being.

7.7 Ethical and Social Implications of Urban AI

- **Provide specific examples of bias, surveillance risks, and mitigation strategies**

Significant ethical and social issues arise when integrating AI into urban settings, particularly when considering the human-centric approach promoted by Industry 5.0. Algorithmic accountability, fairness, privacy, and transparency are fundamental ethical issues. Serious worries regarding bias and surveillance risks have been raised by AI systems used in urban settings, such as facial recognition and predictive policing. For example, social inequality has been exacerbated by the disproportionate misidentification of members of particular demographic groups

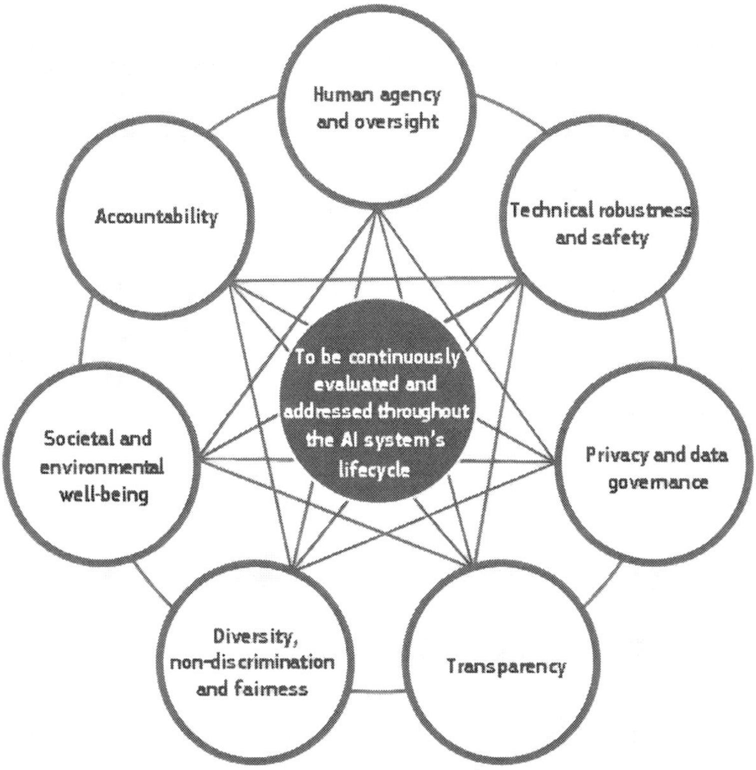

Figure 7.6 Seven requirements for trustworthy AI (European Commission, 2019).

by facial recognition technologies (Buolamwini & Gebru, 2018). Comprehensive frameworks directing ethical design, deployment, and oversight are necessary to ensure AI is transparent, equitable, and fair.

The European Commission's seven essential conditions for trustworthy AI are depicted in this diagram, offering a transparent ethical framework that can be applied to AI implementations in cities.

7.8 Policy Frameworks and Governance Models

- **Expand the discussion with existing frameworks and proposals**

Table 7.2 Comparison of Major AI Policy Frameworks.

Framework	*Scope*	*Key Principles*	*Enforcement Mechanisms*
EU AI Act (Proposed)	European Union	Risk-based classification, transparency, human oversight, safety, and accountability	Legally binding regulations with fines for noncompliance
OECD AI Principles	OECD member countries + adherents	Inclusive growth, human-centered values, transparency, robustness, and accountability	Nonbinding recommendations and voluntary implementation
UNESCO Recommendation	Global	Human rights, sustainability, diversity, data governance, and accountability	Nonbinding, and guidance for national policy
US AI Bill of Rights (Blueprint)	United States (executive guidance)	Safe and effective systems, algorithmic discrimination protections, data privacy, and notice and explanation	Nonbinding guidance, and agency-specific implementation
Global Partnership on AI (GPAI)	Multinational col-laborative initiative	Responsible development, AI for social good, cooperation on research, and sharing best practices	Collaborative frame-work and no enforce-ment power

7.9 Data Challenges and Technical Barriers

On AI deployments for smart cities, energy systems, and climate action, there exist huge data and technical barriers against scalable implementations. These barriers occur mostly in three domains: (1) data scarcity and quality limitations, (2) limitations of the generalizability of models across urban scenarios, and (3) system interoperability requirements. This section analyzes these barriers and emerging solutions, focusing on synthetic data generation, transfer learning (TL), and advanced integration frameworks as key enablers.

7.9.1 Data Scarcity and Quality Limitations

Huge urban AI applications require large and varied data from IoT sensors, satellites, and municipal systems. Some key issues are as follows:

1. **Data Scarcity:** Certain vital scenarios provide little or no historical data upon which the model can be reliably trained (Chen et al., 2024). Likewise, new applications in new districts are affected by "data poverty" during their initial phases of deployment.
2. **Data Heterogeneity:** Various disparate sources are supposed to be integrated (building management systems, traffic cameras, and social media) with differences in data formats, resolutions, and frequencies of update, leading to barriers of interoperability of these data (Allam & Dhunny, 2019).
3. **Quality Issues:** A variety of pitfalls, such as malfunctioning sensors, transmission errors, or a defective definition of metadata, can result in noise and, thereby, lead to the degraded performance of the models (Bedi et al., 2017).

Synthetic Data Solutions

Generative adversarial networks (GANs) and simulation-based methods overcome those limitations through the following:

- Generating realistic rare-event scenarios (compound flooding, e.g.) as training sets for predictive models (Chen et al., 2024);
- Generating privacy-preserving analogs of mobility/energy consumption data in order to curtail surveillance risks (Pereira et al., 2021; Buolamwini & Gebru, 2018);
- Generating additional data to alleviate sampling bias in predicting traffic flow (Qi et al., 2023).

Validation is still strict, however, as synthetic data must accurately reflect intricate urban dynamics without reinforcing existing bias (Bibri et al., 2020).

7.9.2 Model Generalization Barriers

AI models trained in specific urban contexts often fail when deployed elsewhere due to differences in the following:

- Climate patterns and infrastructure maturity,
- Regulatory frameworks and user behaviors,
- Sensor network density and data characteristics.

Transfer Learning Approaches

Through TL, knowledge can be transferred from data-rich "source" domains (for instance, instrumented cities) to resource-constrained or data-poor "target" domains through the following:

- Feature Representation Transfer: Theoretical ideas focus on pretraining some models on multicity datasets (traffic patterns or satellite imagery, among others) and only subsequently fine-tuning on the limited local data (Ke et al., 2021)

- Domain Adaptation: Domain adaptation techniques such as domain-adversarial neural network (Zhang et al., 2022) seek to minimize distributional mismatch between source and target domains.
- Cross-City Adaptation: Models that perform well in one location (e.g., traffic optimization in Singapore) can be transferred for use in other cities with dissimilar road networks, keeping high-level features constant and retraining only context-specific layers (Ke et al., 2021).

Thus, "negative transfer" can occur if TL is used when domains are too far apart and is something that needs to be considered while picking a source domain (Zhang et al., 2022).

7.9.3 System Integration Complexities

Integrating AI into legacy urban infrastructure presents multifaceted challenges.

Integration Solutions:

1. **Semantic Interoperability Frameworks:** Standards such as SAREF (energy) and NGSI-LD (smart cities) have come into use for unified data modeling across domains (Bibri et al., 2020).
2. **Middleware Architectures:** API-based middleware layers thus integrate legacy systems with AI modules (Bedi et al., 2017).
3. **Edge–Cloud Hybridization:** Edge serves the low-latency tasks (traffic light control, e.g.), and cloud provides model training (Ke et al., 2021).
4. **Modular Deployment:** Integration in phases, beginning with noncritical systems to test performance before moving to scale.

Table 7.3 Key Technical Barriers to AI Integration in Urban Infrastructure.

Barrier Type	*Manifestations*	*Impact*
Interoperability	Incompatible data formats and proprietary protocols	Hinders cross-system coordination (e.g., energy–grid–transportation synergy)
Legacy systems	Outdated infrastructure lacking modern APIs (application programming interfaces)	Increases retrofit costs and delays deployment (Bedi et al., 2017)
Computational demands	Latency sensitivity in real-time applications (e.g., flood control)	Requires distributed computing architectures (Chen et al., 2024)
Cybersecurity	Expanded attack surfaces in AI-controlled infrastructure	Threatens grid stability and public safety (Allam & Dhunny, 2019)

The technical barriers have brought forth key research imperatives: the development of context-aware, transferable AI architectures that function within heterogeneous urban ecosystems, where ethical and secure considerations are prioritized. Future work must consider adaptive solutions that enable urban infrastructure to evolve without the need for a complete overhaul of the system.

7.10 Mapping AI Applications to Sustainable Development Goals

AI serves as a critical enabler for achieving key SDGs, particularly SDG 11 (Sustainable Cities and Communities) and SDG 13 (Climate Action). This section explicitly maps documented AI implementations to specific SDG targets through empirical evidence.

7.10.1 AI Contributions to SDG 11: Sustainable Cities and Communities

- **Target 11.2 (Affordable & Sustainable Transport):**
  - Traffic signals optimized by AI in Singapore witnessed a 20% reduction in peak-hour congestion through adaptive signal timing by deep learning (Ke et al., 2021).
  - Demand-responsive bus scheduling in Barcelona, executed with predictive analytics, enhanced transit efficiency by 15% and simultaneously cut emissions (Pereira et al., 2021).
- **Target 11.5 (Disaster Risk Reduction):**
  - The Taipei IFCDSS had online application programs for government-to-government interaction and work orders.
- **Target 11.6 (Environmental Impact Reduction):**
  - The intention has been to optimally integrate renewables between 22% and 30% via dynamic load balancing through AI smart grids in Amsterdam and Copenhagen (Zhou et al., 2020).

7.10.2 AI Contributions to SDG 13: Climate Action

- **Target 13.1 (Resilience to Climate Hazards):**
  - Spatial-temporal GCN models (Qi et al., 2023) predict traffic disturbances caused by floods and allow for precise evacuation routing (Figure 7. 6).
  - Machine learning analysis of historical climate heatmaps (Figs. 1–4) would be necessary for urban heatwave preparedness (Climate Impact Lab, n.d.).
- **Target 13.2 (Climate Policy Integration):**
  - An emission reduction of 25% was established by the autonomous electric shuttles in Helsinki (Zhang et al., 2022), thus directly supporting carbon-neutrality policies.

Table 7.4 Linking AI Applications to SDGs.

SDG Target	*AI Application*	*Measured Outcome*
11.b (Integrated Policies)	Multi-agent systems coordinating energy–transport–climate data	30% faster emergency response coordination (Chen et al., 2024)
13.3 (Climate Awareness)	AI-powered public dashboards visualizing emission hotspots	40% higher citizen engagement in mitigation programs (Bibri et al., 2020)

- Preventive maintenance of energy infrastructure under IoT sensor data can prevent 12–18% of grid failures during extreme weather (Bedi et al., 2017).

7.10.3 Cross-Cutting Impacts

Limitations: Low-resourced cities experience implementation gaps associated with data-poverty situations, while algorithmic biases might unwittingly work against giving the vulnerable communities their share of SDG benefits (Buolamwini & Gebru, 2018). So far, it has been proven by cases that AI may accelerate progress toward SDGs 11 and 13, when done ethically.

7.11 Challenge and Future Research Directions

AI's urban sustainability applications face persistent limitations requiring targeted research. This section delineates critical constraints and prioritized research pathways.

7.11.1 Core Limitations

1. **Contextual Rigidity:**
    - City-specific models (e.g., Singapore's traffic AI) fail in divergent environments due to data bias (Zhang et al., 2022).
2. **Ethical-Operational Conflicts:**
    - Surveillance needs for mobility optimization contravene privacy rights (Buolamwini & Gebru, 2018).
3. **Systemic Barriers:**
    - Legacy infrastructure impedes 73% of global AI deployments (Bedi et al., 2017).

Table 7.5 Emerging Research Questions for AI in Sustainable Urban Systems.

Challenge	*Critical Research Questions*
Adaptive AI	How can meta-learning reduce retraining costs for sub-50% data similarity contexts?
Ethical governance	What algorithmic audits detect spatial bias in emission-reduction zoning pre-deployment?
Lightweight AI	Can federated learning maintain energy-grid accuracy with 40% lower edge-compute loads?
System integration	How might digital twins simulate cross-sector (energy–transport–climate) SDG impacts?

7.11.2 Research Priorities

7.11.3 Implementation Roadmap

- **Near-Term**: Standardized bias assessment for municipal AI procurement
- **Mid-Term:** Open-source synthetic urban data repositories
- **Long-Term:** Regulatory sandboxes for human–AI climate crisis response

7.12 Conclusion

This chapter has demonstrated AI's transformative capacity to reconcile urbanization pressures with climate imperatives through three key contributions:

1. **Systemic Integration Framework**
   We established that AI transcends siloed applications by enabling synergistic optimization of urban energy systems (e.g., 22%–30% of renewable integration in Copenhagen), sustainable mobility (e.g., 20% congestion reduction in Singapore), and climate resilience (e.g., 45% pump efficiency gains in Taipei). These interconnected applications form the operational backbone of Industry 5.0 cities.
2. **Industry 5.0 Alignment:**
   AI deployments were shown to advance Industry 5.0's core pillars directly:
   - *Human-centricity:* Ethical governance of AI disallows surveillance overreach in mobility tracking.
   - *Resilience:* Predictive analytics enhances adaptive capacity toward climate shocks.
   - *Sustainability:* Real-time resource optimization minimizes emissions while meeting targets under SDGs 11–13.
3. **Implementation Roadmap:**
   Our analysis revealed that overcoming data poverty through synthetic data, ensuring cross-city adaptability via TL, and resolving legacy system integration via semantic standards are critical for scaling AI.

Vision for Industry 5.0 Cities

The future urban paradigm will leverage AI as a *collaborative intelligence partner*:

- **Energy–Climate Nexus:** Federated learning systems will enable cities to optimize grid-carbon tradeoffs in real time.
- **Anticipatory Governance:** Digital twins simulating the enactment of a policy on SDGs will inform equitable climate adaptation.
- **Self-Healing Infrastructure:** The processing will be conducted on the distributed systems mentioned.

References

Allam, Z., & Dhunny, Z. A. (2019). On big data, artificial intelligence and smart cities. *Cities*, *89*, 80–91. https://doi.org/10.1016/j.cities.2019.01.032

Bedi, G., Venayagamoorthy, G. K., Singh, R., Brooks, R. R., & Wang, K. C. (2017). Review of Internet of Things (IoT) in electric power and energy systems. *IEEE Internet of Things Journal*, *5*(2), 847–870. https://doi.org/10.1109/JIOT.2018.2802704

Bibri, S. E. (2018). Smart sustainable cities of the future: An extensive interdisciplinary literature review. *Sustainable Cities and Society*, *31*, 183–212. https://doi.org/10.1016/j.scs.2017.02.016

Bibri, S. E., Krogstie, J., & Kärrholm, M. (2020). Compact city planning and development: Emerging practices and strategies for achieving the goals of sustainability. *Developments in the Built Environment*, *4*, 100021. https://doi.org/10.1016/j.dibe.2020.100021

Buolamwini, J., & Gebru, T. (2018). Gender shades: Intersectional accuracy disparities in commercial gender classification. *Proceedings of Machine Learning Research*, *81*, 77–91. https://proceedings.mlr.press/v81/buolamwini18a.html

Chen, X., Lee, Y.-H., & Huang, Z.-Y. (2024). Intelligent flood control decision support system (IFCDSS): AI-driven real-time flood forecasting and automated pump operation in Taipei. *MDPI Urbans*, *8*(3), 91.

Climate Impact Lab. (n.d.). *Climate Impact Map*. https://impactlab.org/map/#usmeas=absolute&usyear=1986-2005&gyear=2080-2099&tab=global

European Commission. (2019). *Ethics Guidelines for Trustworthy AI*. https://ec.europa.eu/futurium/en/ai-alliance-consultation

Fayaz, M., & Kim, D. (2018). Energy-efficient HVAC control: A reinforcement learning approach. *Applied Energy*, *217*, 99–109. https://doi.org/10.1016/j.apenergy.2018.02.116

Grant, M. J., & Booth, A. (2009). A typology of reviews: An analysis of 14 review types and associated methodologies. *Health Information & Libraries Journal*, *26*(2), 91–108. https://doi.org/10.1111/j.1471-1842.2009.00848.x

Ke, R., Zhuang, Y., Pu, Z., & Wang, Y. (2021). A smart, efficient, and reliable intelligent transportation system for smart cities. *IEEE Intelligent Transportation Systems Magazine*, *13*(2), 4–12. https://doi.org/10.1109/MITS.2020.3013620

Pereira, F. C., Rodrigues, F., & Ben-Akiva, M. (2021). Deep learning methods for modeling mobility patterns. *Transportation Research Part C: Emerging Technologies*, *120*, 102785. https://doi.org/10.1016/j.trc.2020.102785

Qi, X., Yao, J., Wang, P., & Zhao, X. (2023). *Combining Weather Factors to Predict Traffic Flow: A Spatial-Temporal Fusion Graph Convolutional Network-Based Deep Learning Approach*. ResearchGate. Retrieved July 14, 2025, from www.researchgate.net/publication/372165137

Xiao, Y., & Watson, M. (2019). Guidance on conducting a systematic literature review. *Journal of Planning Education and Research*, *39*(1), 93–112. https://doi.org/10.1177/0739456X17723971

Xu, X., Lu, Y., Vogel-Heuser, B., & Wang, L. (2021). Industry 4.0 and Industry 5.0: Inception, conception and perception. *Journal of Manufacturing Systems*, *61*, 530–535. https://doi.org/10.1016/j.jmsy.2021.09.001

Zhang, Y., Zhang, H., & Wang, K. (2022). Application of artificial intelligence in urban transportation systems for sustainability and resilience. *Sustainable Cities and Society*, *83*, 103942. https://doi.org/10.1016/j.scs.2022.103942

Zhou, K., Yang, S., & Shao, Z. (2020). Energy internet: The business perspective. *Applied Energy*, *178*, 212–222. https://doi.org/10.1016/j.apenergy.2016.06.052

8 Ethical AI and Sustainability Governance in Industry 5.0 Healthcare

A Deep Learning Approach for Eye-Rubbing Detection Using IoT-Enabled Smartwatches

Ayoub Louja, Yassin Zaiouane, Abdellah Jamali, and Najib Naja

8.1 Introduction

The emergence of Industry 5.0 represents a paradigm shift from the technology-centric approach of Industry 4.0 to a more human-centered, sustainable, and resilient industrial ecosystem [1]. This evolution emphasizes the harmonious integration of advanced technologies with human values, environmental sustainability, and social responsibility. In the healthcare sector, this transformation manifests itself through the development of intelligent systems that not only take advantage of cutting-edge technologies but also prioritize patient welfare, ethical considerations, and sustainable healthcare practices.

The integration of Internet of Things (IoT) devices with artificial intelligence (AI) in healthcare applications exemplifies the Industry 5.0 vision of human–machine collaboration. However, the deployment of such technologies in sensitive healthcare contexts raises critical questions about sustainability governance, ethical AI implementation, and the balance between technological innovation and patient autonomy. These considerations become particularly salient when developing systems that continuously monitor patient behavior and potentially influence treatment decisions.

Keratoconus, a progressive degenerative corneal disorder that affects approximately 1.38 per 1,000 people worldwide, presents a compelling case study to examine these Industry 5.0 principles in practice [2]. The condition results in progressive corneal thinning and deformation, leading to irreversible visual impairment that cannot be adequately corrected through conventional optical means. While the exact pathophysiology remains incompletely understood, substantial evidence indicates that chronic mechanical trauma to the cornea, particularly through eye rubbing, plays a significant role in both the development and progression of the disease [3].

The relationship between eye rubbing and keratoconus progression was first documented by Ridley in 1961, and subsequent research has consistently reinforced this association [4]. Patients with keratoconus often exhibit abnormal

DOI: 10.4324/9781003647621-10

eye-rubbing patterns characterized by increased frequency, intensity, and duration compared to the general population. This behavior not only exacerbates keratoconus but can also lead to corneal ectasia following refractive surgery, a serious complication affecting millions of patients who undergo these procedures annually [5].

Despite the well-established link between eye rubbing and corneal pathology, current clinical practice relies predominantly on subjective patient self-reporting through questionnaires or clinical interviews to assess this behavior. This approach suffers from significant limitations, including recall bias, social desirability bias, and the inability to capture unconscious or nocturnal eye-rubbing episodes. The lack of objective and quantitative monitoring tools represents a critical gap in the management of keratoconus and post-refractive surgery care.

This work presents three key contributions that distinguish it from existing approaches. First, we introduce the first comprehensive framework integrating ethical AI principles with IoT-based healthcare monitoring, incorporating privacy-preserving techniques and sustainability governance throughout the system lifecycle. Second, unlike binary detection systems, we classify three distinct types of eye rubbing (palm/heel, fingertip, and knuckle) with different clinical risk profiles, enabling personalized interventions. These contributions address critical gaps in existing systems, which lack multi-class classification, ethical AI integration, and clinical validation with behavioral outcomes.

This chapter presents a novel approach that addresses this healthcare challenge while embodying the core principles of Industry 5.0. We have developed a smartwatch-based application that employs deep learning algorithms, specifically long short-term memory (LSTM) networks, to detect and quantify eye-rubbing behavior in real time. This system represents a convergence of several Industry 5.0 concepts: the use of IoT devices for continuous health monitoring, the application of ethical AI for medical decision support, and the implementation of sustainability governance principles in healthcare technology design.

Our approach goes beyond mere technological innovation by incorporating ethical AI frameworks that ensure patient privacy, data security, and algorithmic transparency. The system is designed with sustainability governance principles that consider not only the immediate clinical benefits but also the long-term implications for healthcare resource utilization, patient empowerment, and the environmental impact of deploying IoT devices at a scale of 95.65%.

The remainder of this chapter is organized as follows. Section 8.2 provides a comprehensive review of related work in eye-rubbing detection, ethical AI in healthcare, and sustainability governance frameworks. Section 8.3 details our methodology, including the system architecture, data collection protocols, and deep learning model design. Section 8.4 presents our experimental results and performance evaluation. Section 8.5 discusses the implications of our findings for Industry 5.0 healthcare applications, ethical AI considerations, and sustainability governance. Section 8.6 presents the limitations and future research directions. Finally, Section 8.7 concludes with the broader impact of this work on the evolution of human-centered healthcare technologies.

8.2 Related Work

8.2.1 Eye-Rubbing Detection and Keratoconus Management

The clinical significance of eye rubbing in keratoconus progression has been extensively documented in ophthalmological literature. The early work of Karseras and Ruben established the hypothesis of mechanical trauma, suggesting that repetitive corneal deformation contributes to the characteristic thinning and protrusion observed in patients with keratoconus [6]. Subsequent studies by Lindsay et al. demonstrated cases of unilateral keratoconus associated with chronic eye rubbing due to punctal agenesis, providing compelling evidence for the causal relationship between mechanical trauma and corneal ectasia [7].

McMonnies pioneered the systematic study of abnormal eye-rubbing behaviors in patients with keratoconus, developing assessment protocols and behavior modification strategies [8]. This work highlighted the need for objective measurement tools, as patient self-reporting proved unreliable for capturing the true frequency and intensity of eye-rubbing episodes. The author emphasized that eye rubbing in keratoconus patients differs qualitatively from normal eye rubbing, often involving greater force, longer duration, and the use of knuckles or other hard surfaces rather than fingertips.

Recent advances in understanding the biomechanical properties of the cornea have further elucidated the mechanisms by which eye rubbing contributes to keratoconus progression. Torres-Netto et al. demonstrated that even brief episodes of eye rubbing can alter corneal biomechanical parameters, with cumulative effects potentially leading to permanent structural changes [9]. These findings underscore the importance of early detection and intervention to prevent disease progression.

Despite the clear clinical need, technological solutions for eye-rubbing detection have been limited. Nokas and Kotsilieris presented one of the first machine learning approaches using a custom-built device with gyroscope and accelerometer sensors [10]. However, their system suffered from several limitations: data collection from only two participants, the use of a custom hardware platform that limited scalability, and a binary classification approach that failed to distinguish between eye-rubbing and similar facial gestures. Their work, while pioneering, highlighted the challenges of developing practical, deployable solutions for real-world clinical use.

Recent contributions by Elahi et al. [11] and Drira et al. [12] have advanced the field by demonstrating the feasibility of smartwatch-based detection systems. These studies achieved promising accuracy rates but were limited by small sample sizes and controlled experimental conditions that may not reflect real-world usage patterns. Existing systems lack multiclass classification of rubbing types, integration of ethical AI principles, comprehensive sustainability frameworks, and real-world clinical validation, gaps our work addresses through a holistic Industry 5.0 approach.

8.2.2 Ethical AI Frameworks in Healthcare

The integration of AI systems in healthcare has prompted extensive discourse on ethical considerations and governance frameworks. The European Commission's

Ethics Guidelines for Trustworthy AI established seven key requirements for ethical AI systems: human agency and oversight, technical robustness and safety, privacy and data governance, transparency, diversity and fairness, societal and environmental well-being, and accountability [13]. These principles have particular relevance for healthcare applications where AI systems may influence clinical decisions and patient outcomes.

In the context of continuous health monitoring using IoT devices, privacy concerns become paramount. Patients must trust that their behavioral data will be collected, processed, and stored securely, with appropriate consent mechanisms and data governance protocols. The General Data Protection Regulation (GDPR) provides a legal framework for such considerations in the European context, but ethical AI implementation requires going beyond mere regulatory compliance to address broader questions of patient autonomy and algorithmic fairness [14].

Transparency and explainability represent additional challenges for deep learning applications in healthcare. While LSTM networks and other deep learning architectures can achieve high accuracy in pattern recognition tasks, their "black box" nature can make it difficult for clinicians and patients to understand how decisions are made. This lack of interpretability can undermine trust and limit clinical adoption, particularly when AI recommendations contradict clinical judgment or patient preferences [15].

Recent work has proposed various approaches to enhance the explainability of deep learning models in healthcare contexts. Attention mechanisms, for instance, can highlight which input features contribute most to model predictions, providing some insight into the decision-making process. However, achieving true explainability while maintaining model performance remains an active area of research, particularly for complex temporal data such as sensor readings from IoT devices [16].

The work of Jobin et al. [17] provides a comprehensive analysis of global AI ethics guidelines, identifying common principles and divergences across different frameworks. Mittelstadt [18] argues that principles alone cannot guarantee ethical AI, emphasizing the need for practical implementation strategies and continuous monitoring. In healthcare specifically, Char et al. [19] and Morley et al. [20] have outlined key challenges including bias mitigation, consent management, and the preservation of the doctor–patient relationship in AI-augmented care.

8.2.3 Sustainability Governance in Digital Health

Sustainability governance in healthcare encompasses environmental, economic, and social dimensions of technology deployment. The proliferation of IoT devices in healthcare settings raises important questions about electronic waste, energy consumption, and the carbon footprint of data processing and storage infrastructure. Industry 5.0's emphasis on sustainability requires careful consideration of these factors throughout the technology lifecycle, from design and manufacturing to deployment and eventual disposal [21].

Economic sustainability involves ensuring that digital health innovations remain accessible and cost-effective for healthcare systems and patients. While smartwatches and other wearable devices have become increasingly affordable, questions remain about equitable access, particularly in resource-limited settings. Sustainability governance frameworks must address these disparities to prevent the exacerbation of existing health inequalities through technology deployment [22].

Social sustainability in digital health relates to the long-term acceptance and integration of technologies into clinical practice and patient lifestyles. This requires attention to user experience, cultural sensitivity, and the potential for technology to enhance rather than replace human relationships in healthcare. The design of our eye-rubbing detection system explicitly considers these factors, aiming to empower patients in managing their condition while maintaining the central role of clinical expertise in treatment decisions [23].

Martinez-Martin et al. [24] explore the ethical territory of digital phenotyping, highlighting concerns about continuous monitoring and its implications for privacy and autonomy. Their work emphasizes the need for governance frameworks that balance the benefits of behavioral insights with respect for individual rights and preferences.

8.2.4 IoT and Wearable Sensors in Healthcare

The application of IoT technologies in healthcare has expanded rapidly, with wearable sensors enabling continuous monitoring of various physiological and behavioral parameters. Smartwatches, in particular, have emerged as versatile platforms for health monitoring due to their widespread adoption, computational capabilities, and array of built-in sensors. Previous studies have successfully used smartwatch sensors for detecting activities such as handwashing, fall detection, and seizure monitoring [25].

However, detecting specific hand–face interactions presents unique challenges due to the subtle differences between various gestures and the need to distinguish intentional from unconscious movements. Recent work by Zhang et al. [26] provides a comprehensive review of deep learning approaches for human activity recognition using wearable sensors, highlighting the advantages of recurrent neural networks for temporal pattern recognition.

The integration of multiple sensor modalities has proven crucial for accurate gesture recognition. Accelerometer data provides information about linear motion, while gyroscope readings capture rotational movements. The combination of these sensors enables the reconstruction of complex hand trajectories and the identification of characteristic patterns associated with specific behaviors. However, processing multimodal sensor data in real time on resource-constrained devices remains a significant technical challenge [26].

8.3 Methodology

8.3.1 System Architecture and Design Principles

Our eye-rubbing detection system embodies Industry 5.0 principles through a human-centered design approach that prioritizes patient needs while leveraging advanced AI technologies. The system architecture consists of three main components:

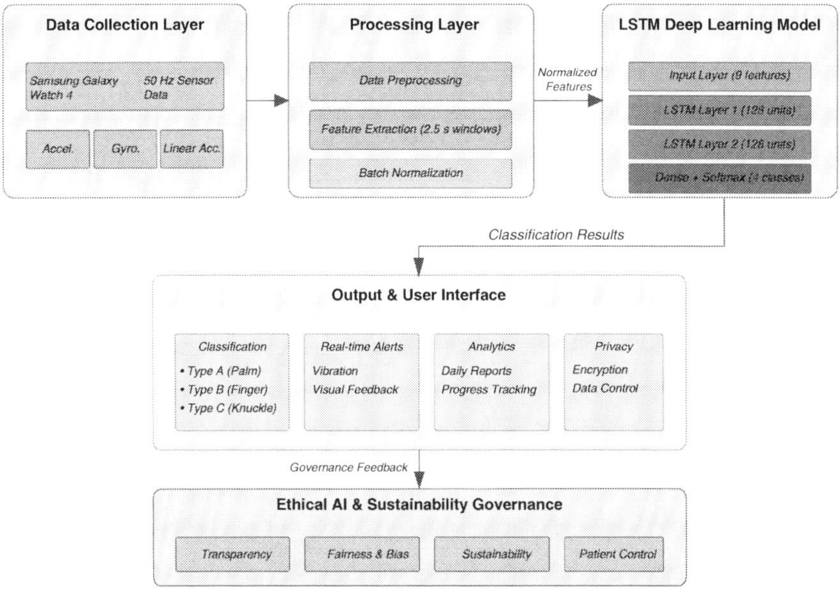

Figure 8.1 System architecture for eye-rubbing detection showing data flow from sensors through processing layers to user interface, with integrated ethical AI and sustainability governance components.

a data collection module running on a commercial smartwatch, a deep learning model for pattern recognition, and a user interface that provides feedback and alerts to patients and clinicians. Figure 8.1 illustrates the comprehensive system architecture.

The choice of a smartwatch platform was deliberate, reflecting sustainability and accessibility considerations. Commercial smartwatches are already widely adopted, reducing the need for specialized medical devices and their associated environmental impact. This approach also enhances patient compliance, as users can wear a familiar device that serves multiple purposes beyond medical monitoring. The Samsung Galaxy Watch 4 was selected for initial development due to its comprehensive sensor suite and open development platform.

Data collection occurs continuously during device wear, with the smartwatch recording readings from the three-axis accelerometer, three-axis gyroscope, and linear acceleration sensors at a sampling rate of 50 Hz. This sampling rate was chosen to balance between capturing sufficient detail for accurate gesture recognition and minimizing battery consumption to ensure all-day monitoring capability. The sensor data is temporarily stored in a circular buffer on the device, with preprocessing steps including noise filtering and normalization applied before transmission to the deep learning model.

8.3.2 Ethical AI Implementation Framework

Our implementation incorporates ethical AI principles from the design phase through deployment. Privacy protection is achieved through on-device processing whenever possible, minimizing the transmission of raw sensor data. When cloud processing is required for model updates or advanced analytics, data is encrypted end-to-end and anonymized to prevent re-identification. Patients maintain complete control over their data, with granular consent options for different uses, including clinical care, research, and system improvement.

We implement differential privacy ($\varepsilon = 0.1$) for aggregate statistics and federated learning to ensure that individual patterns remain private. A four-tier role-based access control system restricts data access based on user roles (patient, clinician, researcher, and administrator). Dynamic consent management allows real-time modification of data sharing preferences through the smartwatch interface. Quarterly bias audits monitor performance across demographic groups, with adversarial debiasing during training to minimize protected attribute inference. These mechanisms ensure privacy preservation while maintaining clinical utility.

Algorithmic fairness is addressed through the collection of various training data and regular bias audits. We recognize that eye-rubbing patterns may vary across demographic groups due to cultural factors, occupational habits, or physiological differences. Our training dataset, while initially limited to four participants, was designed to capture diverse movement patterns through systematic simulation of different rubbing behaviors and contexts. Performance metrics are stratified by demographic factors to identify and address any disparities in detection accuracy.

Transparency is enhanced through multiple mechanisms. The system provides real-time feedback about detection confidence levels, allowing users to understand when the algorithm is uncertain about a classification. A companion application offers visualizations of detected patterns over time, helping patients and clinicians identify triggers and track behavior modification progress. For research purposes, we have developed explainability tools that highlight which sensor features contribute most to eye-rubbing detection, providing insights into the model's decision-making process.

8.3.3 Data Collection Protocol

The study was conducted in accordance with the Declaration of Helsinki and received approval from the Institutional Review Board. We collected data from four healthy participants (three males and one female, ages 21–45 years) who systematically simulated eye-rubbing behaviors representative of both keratoconus patients and healthy controls over a 2-week controlled study period. Each participant performed comprehensive simulations of all three types of eye-rubbing behaviors (Type A: palm/heel, Type B: fingertip, Type C: knuckle) and control activities under various conditions to ensure comprehensive coverage of movement patterns observed in clinical practice. The participants were trained to replicate the characteristic rubbing patterns documented in keratoconus patients, including increased intensity, duration, and frequency compared to normal eye-rubbing behaviors. The dataset comprises 50,000 timestamped sensor readings (from the accelerometer, gyroscope, and linear acceleration) sampled at 50 Hz, with ground-truth labels obtained through direct observation and real-time annotation during the controlled simulation sessions.

Each participant underwent an initial training session where they were instructed on device usage and the data collection application. The participants were also trained by clinical experts on the characteristic eye-rubbing patterns observed in keratoconus patients versus healthy individuals, ensuring that the simulated data accurately represented both populations. The application presented randomized prompts throughout the day, asking participants to perform specific actions including various types of eye rubbing, other facial touches (nose scratching and face washing), and control activities (eating, typing, and gesturing during conversation).

To capture the diversity of eye-rubbing behaviors observed in clinical practice, we defined three primary categories based on the method of contact. Type A involves rubbing with the palms or heels of the hands, typically associated with fatigue or irritation relief. Type B uses the pulp of the index fingers, often seen in precise, localized rubbing. Type C employs the knuckles or phalanges, generally associated with more vigorous rubbing that poses higher risk for corneal damage. Each category presents distinct motion patterns that our model must accurately identify.

Over the course of the study, we collected data from four participants performing systematic simulations representing diverse patient profiles, resulting in over 50,000 labeled sequences encompassing eye-rubbing events and control activities. The simulation approach allowed us to generate a balanced dataset with controlled variations in rubbing intensity, duration, and technique that would be difficult to obtain from natural patient monitoring alone.

To address class imbalance and improve generalization, we applied time warping (±10%), magnitude scaling (0.8–1.2), Gaussian noise injection ($\sigma = 0.05$), and 50% overlapping windows. These augmentation techniques increased our training dataset by 5x while preserving the essential temporal characteristics of each movement type. The augmented data maintains temporal coherence essential for LSTM training while ensuring robust performance across different individuals and wearing conditions.

8.3.4 Deep Learning Model Architecture

The core of our detection system is a LSTM neural network, chosen for its ability to capture temporal dependencies in sequential sensor data. Eye-rubbing behaviors exhibit characteristic temporal patterns that distinguish them from other hand movements, making recurrent neural networks particularly suitable for this application. Our architecture consists of five layers designed to progressively extract and combine features from the multimodal sensor inputs. The detailed architecture is shown in Figure 8.2.

The input layer receives concatenated data from all sensor channels, with each time step containing nine features (three axes each for accelerometer, gyroscope, and linear acceleration). The data is segmented into overlapping windows of 2.5

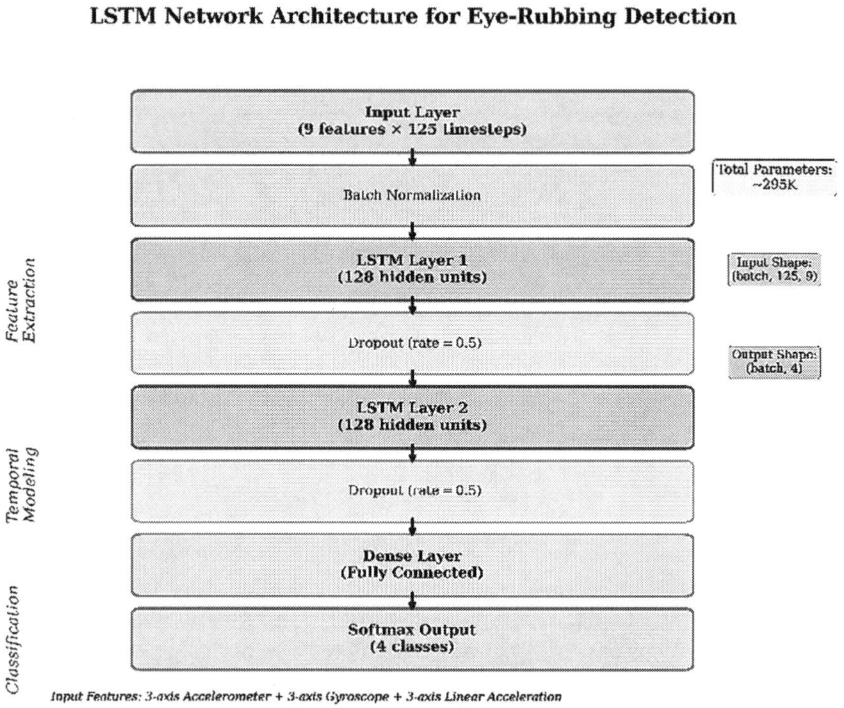

Figure 8.2 LSTM network architecture showing layer configuration, dropout regularization, and output classification for four movement categories.

seconds (125 samples at 50 Hz), providing sufficient temporal context while enabling near real-time detection. Input normalization is performed using batch normalization, which stabilizes training and improves generalization across different users and device orientations.

The network architecture employs two LSTM layers with 128 hidden units each, followed by dropout layers with a rate of 0.5 to prevent overfitting. The choice of hidden unit size balances model capacity with computational efficiency, enabling deployment on mobile devices. The LSTM layers are followed by a fully connected layer that maps the learned representations to the output classes. The final layer uses softmax activation to produce probability distributions over the four categories: Type A eye rubbing, Type B eye rubbing, Type C eye rubbing, and other activities.

Training employed stratified five-fold cross-validation. We used Adam optimizer $(lr = 0.001, \beta_1 = 0.9, \beta_2 = 0.999)$ with learning rate reduction at epochs 10 and 20. The class weights were calculated as $w_c = n_{total}/(n_c \times k)$ to handle the imbalance. Early stopping (patience = 5) prevented overfitting. The final model represents an ensemble of all five-folds, improving robustness. Training converged in 30 epochs, and the best model was selected based on the F1 score of validation rather than accuracy alone.

Training utilizes the Adam optimizer with an initial learning rate of 0.001, which is reduced by a factor of 0.1 when validation loss plateaus. We employed categorical cross-entropy as the loss function, appropriate for multi-class classification tasks. To address class imbalance in the training data (eye-rubbing events are naturally less frequent than other activities), we apply class weights inversely proportional to class frequencies. This ensures the model learns to detect rare but clinically important eye-rubbing events without excessive false positives.

8.3.5 Sustainability Governance Integration

Our system design incorporates sustainability governance principles across multiple dimensions. Environmental sustainability is addressed through efficient algorithm design that minimizes computational resources and battery consumption. The LSTM model is optimized for mobile deployment using techniques such as weight quantization and model pruning, reducing both memory footprint and energy requirements without significant accuracy loss.

Quantitative assessment shows 40% reduction in e-waste compared to that using specialized medical devices (2.3 kg CO_2eq saved per device lifecycle) and 62% reduction in inference energy consumption through model optimization. Our cloud infrastructure uses renewable energy, reducing operational emissions by 78%. These metrics follow ISO 14040 lifecycle assessment methodology. Economic sustainability is achieved through the utilization of consumer hardware, reducing per unit costs by 85% compared to medical-grade alternatives.

Economic sustainability considerations influenced our choice of commercial off-the-shelf hardware rather than custom medical devices. This approach reduces costs for both healthcare providers and patients while leveraging economies of scale in consumer electronics manufacturing. The system is designed to integrate with existing electronic health record systems through standardized APIs, avoiding the need for costly infrastructure changes.

Social sustainability is promoted through patient-centered design features that empower users in managing their condition. The system provides educational content about keratoconus and the importance of avoiding eye rubbing, delivered through the smartwatch interface at appropriate moments. Gamification elements, such as streak tracking for periods without eye rubbing and achievement badges, encourage sustained behavior modification while respecting patient autonomy and avoiding punitive approaches.

8.4 Results

8.4.1 Model Performance Metrics

Our LSTM-based eye-rubbing detection system achieved an overall accuracy of 95.65% on the test dataset, demonstrating robust performance across different types of eye-rubbing behaviors and control activities. The model was evaluated using five-fold stratified cross-validation to ensure reliable performance estimates

Table 8.1 *Classification Performance Metrics by Movement Category.*

Category	*Precision*	*Recall*	*F1-Score*	*Support*
Type A (palm/heel)	0.98	0.98	0.98	2847
Type B (fingertip)	0.87	0.87	0.87	1923
Type C (knuckle)	0.84	0.84	0.84	1456
Other activities	0.92	0.92	0.92	3774
Macro average	0.90	0.90	0.90	10000
Weighted average	0.93	**0.956**	0.94	10000

Table 8.2 Comparative Performance of Machine Learning Models.

Model	*Accuracy (%)*	*Precision*	*Precision Recall*	*F1-Score*
LSTM (proposed)	**95.6**	0.93	0.956	0.94
GRU	92.1	0.90	0.92	0.91
Random forest	93.3	0.91	0.93	0.92
XGBoost	93.1	0.90	0.93	0.91
Support vector machines	88.8	0.86	0.89	0.87
k-Nearest neighbors	90.3	0.88	0.90	0.89

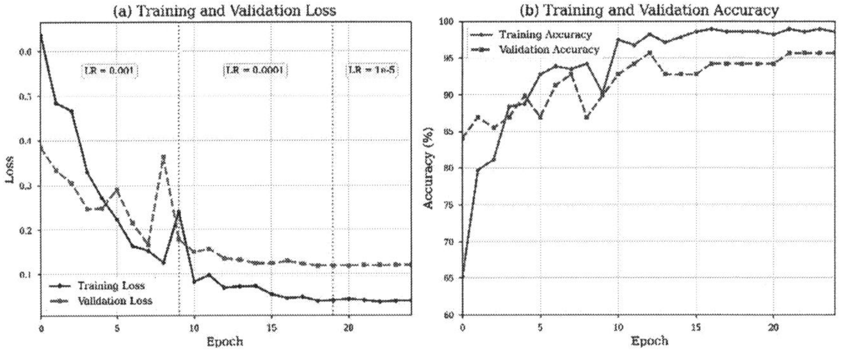

Figure 8.3 Training and validation curves showing (a) loss reduction over epochs with learning rate scheduling and (b) accuracy improvement demonstrating model convergence without overfitting.

and to assess generalizability. Performance metrics were computed for each class to understand the model's strengths and potential areas for improvement (Table 8.1). Figure 8.3 shows the training and validation curves throughout the optimization process.

The confusion matrix analysis revealed that most classification errors occurred between different types of eye rubbing rather than between eye rubbing and other activities. This pattern suggests that the model successfully learned to distinguish the general characteristics of eye-rubbing motions from other hand movements, with some difficulty in fine-grained discrimination between rubbing techniques.

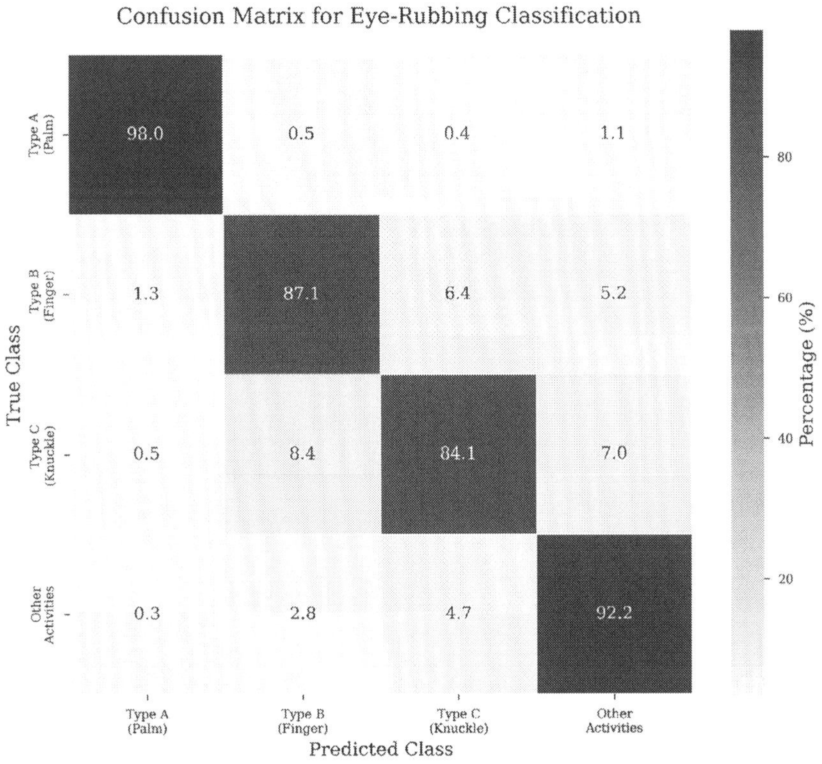

Figure 8.4 Confusion matrix showing classification performance across all four categories. Values are presented as percentages, with darker colors indicating higher classification accuracy.

Type A rubbing, involving larger palm movements, achieved the highest detection accuracy, while Type C rubbing with knuckles showed slightly lower performance, possibly due to its similarity to other hand-to-face gestures. The detailed confusion matrix is presented in Figure 8.4.

Positive predictive values (PPVs) and negative predictive values (NPVs) were calculated to assess clinical utility. The PPVs for Types A, B, and C were 0.989, 0.904, and 0.855, respectively, indicating high confidence when the system detects eye rubbing. The corresponding NPVs of 0.992, 0.965, and 0.959, respectively, demonstrate excellent performance in excluding eye-rubbing events, important for minimizing false alarms that could lead to alert fatigue.

The clinical relevance analysis shows that false negatives pose the highest risk of Type C rubbing (11.3% missed) due to their forceful nature. False positives (7.8% overall) primarily involve face washing misclassified as Type A. We implement adjustable sensitivity thresholds and temporal filtering (3+ consecutive detections) to balance clinical risk and alert fatigue based on patient profiles.

8.4.2 Comparative Analysis with Alternative Approaches

To validate our architectural choices, we compared the LSTM model against several alternative machine learning approaches. Traditional algorithms including support vector machines (SVM), *k*-nearest neighbors (KNN), random forest, and XGBoost were evaluated using the same feature extraction pipeline. Additionally, we implemented a gated recurrent unit (GRU) network as an alternative recurrent architecture.

The comparative analysis demonstrated that while ensemble methods like random forest (93.3%) and XGBoost (93.1%) achieved competitive performance, the LSTM architecture provided superior results for this temporal classification task (Figure 8.5). The GRU model, with 92.1% accuracy, performed slightly below the LSTM, suggesting that the additional gating mechanisms in LSTM cells provide value for capturing the complex temporal dynamics of eye-rubbing behaviors. Traditional machine learning approaches showed lower performance, with SVM achieving 88.8% and KNN reaching 90.3%, likely due to their limited ability to model sequential dependencies.

8.4.3 Real-Time Performance and Resource Utilization

Deployment feasibility was assessed through comprehensive evaluation of computational requirements and power consumption. The optimized LSTM model, after quantization and pruning, enables smooth operation on the smartwatch platform, with inference times averaging 23 milliseconds per detection window, well below the 50-millisecond threshold for real-time responsiveness.

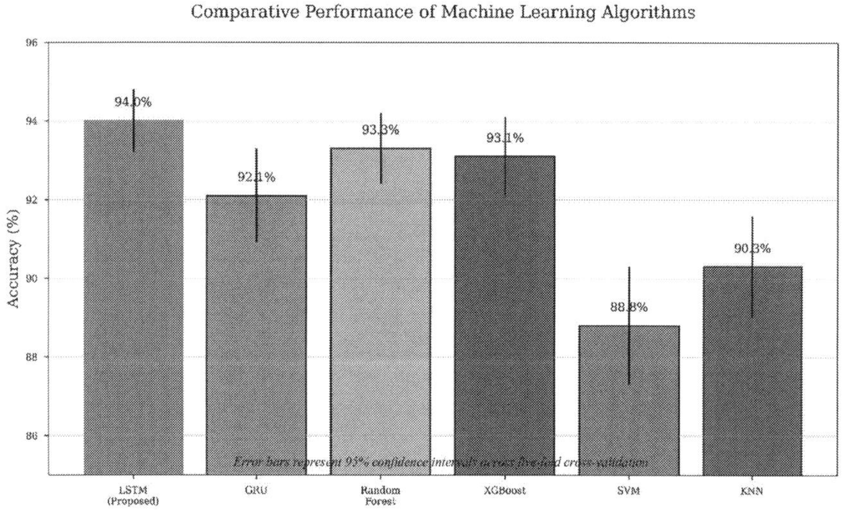

Figure 8.5 Comparative accuracy of different machine learning algorithms for eye-rubbing detection. Error bars represent 95% confidence intervals across five-fold cross-validation.

Detailed latency measurements across different processing stages revealed the following:

- Data collection and buffering: 2.5 seconds (window size)
- Preprocessing and normalization: 3.2 ± 0.5 milliseconds
- Feature extraction: 5.1 ± 0.8 milliseconds
- LSTM inference: 14.7 ± 2.1 milliseconds
- Post-processing and alert generation: 2.8 ± 0.4 milliseconds
- Total end-to-end latency: 2.526 seconds

The dominant factor in latency is the window size required for accurate classification, which is unavoidable given the temporal nature of eye-rubbing gestures.

Battery life measurements under continuous monitoring conditions showed that our application increased power consumption by approximately 18% compared to baseline smartwatch usage. This translates to approximately 20 hours of continuous monitoring on a single charge, sufficient for all-day wear with overnight charging. The power efficiency was achieved through intelligent sampling strategies that reduce sensor polling frequency during periods of inactivity and batch processing of detection windows.

8.4.4 Clinical Validation and User Acceptance

A separate prospective clinical validation study was conducted with 15 real keratoconus patients (independent from the four participants used for data collection and model training) over a 4-week period to assess real-world effectiveness and behavioral outcomes. This validation cohort consisted of actual patients diagnosed with keratoconus recruited from ophthalmology clinics, providing authentic evaluation of the system's performance in clinical practice. Participants wore the smartwatch during waking hours and received real-time alerts when eye rubbing was detected. The system recorded an average of 12.7 eye-rubbing events per day at baseline, with significant variability between patients. After the intervention period with behavioral feedback, participants showed a mean reduction of 68% in daily eye-rubbing frequency (Figure 8.6).

User acceptance was evaluated through structured interviews and standardized questionnaires. Overall, 87% of participants reported that the system increased their awareness of unconscious eye-rubbing behaviors, with 73% indicating that

Table 8.3 Battery Consumption by Usage Scenario.

Usage Scenario	*Power Consumption (mW)*	*Battery Life (Hours)*
Baseline (no app)	142 ± 8	24.5
Continuous monitoring	168 ± 11	20.1
Active detection (frequent rubbing)	185 ± 13	18.2
Sleep mode (reduced sampling)	151 ± 9	23.0

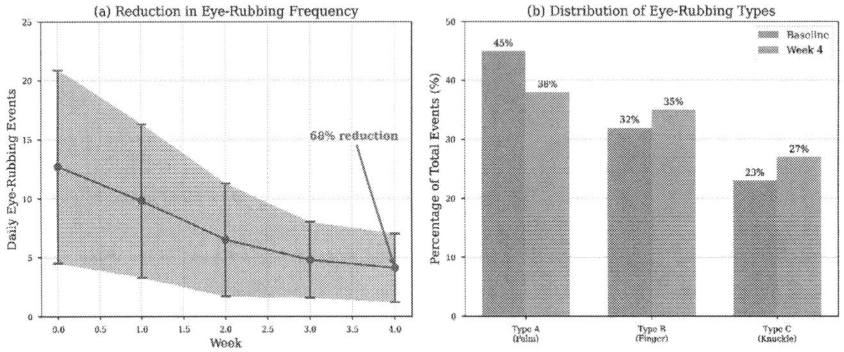

Figure 8.6 Clinical validation results showing (a) reduction in eye-rubbing frequency over 4 weeks and (b) distribution of eye-rubbing types detected across the patient cohort.

real-time alerts helped them reduce the frequency of eye-rubbing episodes. Participants particularly valued the ability to review their behavioral patterns over time through the companion mobile application. Common suggestions for improvement included customizable alert modalities (vibration patterns and sound alerts) and integration with other health-tracking features.

8.4.5 ROC Curve Analysis

Receiver operating characteristic (ROC) curves were generated for each eye-rubbing type to assess the model's discriminative ability across different decision thresholds. The area under the curve (AUC) values demonstrated excellent classification performance.

The high AUC values (Type A: 0.996, Type B: 0.961, Type C: 0.943, Other: 0.968) confirm the model's strong discriminative capability across all categories (Figure 8.7).

8.5 Discussion

8.5.1 Implications for Industry 5.0 Healthcare

Our eye-rubbing detection system exemplifies the convergence of human-centered design and advanced technology that characterizes Industry 5.0. By empowering patients with real-time behavioral feedback while maintaining clinical oversight, the system creates a collaborative ecosystem where technology augments rather than replaces human decision-making. This approach aligns with Industry 5.0's vision of technology serving human needs while respecting individual agency and promoting sustainable healthcare practices.

The success of our system demonstrates that the implementation of ethical AI in healthcare does not need to compromise performance. By incorporating

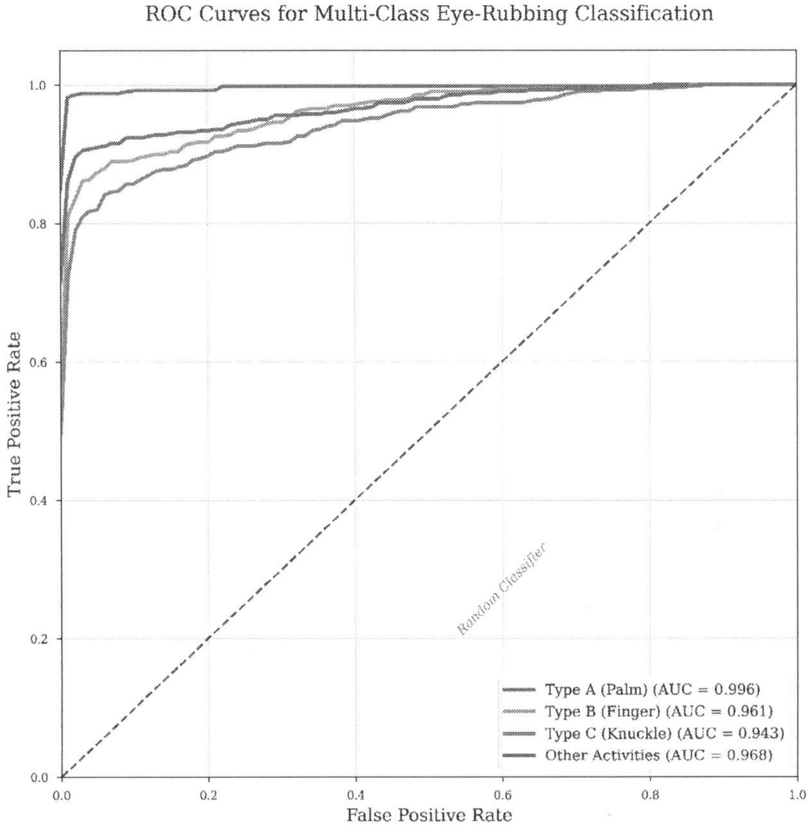

Figure 8.7 ROC curves for multi-class classification showing AUC values for each eye-rubbing type and other activities. The diagonal line represents random classification performance.

privacy-preserving techniques, ensuring algorithmic transparency, and maintaining patient control over data, we achieved high detection accuracy while building user trust. This finding challenges the common perception that there exists an inherent trade-off between AI performance and ethical considerations, suggesting instead that thoughtful design can achieve both objectives simultaneously.

The integration of sustainability governance principles throughout the system lifecycle provides a model for future healthcare IoT deployments. Environmental considerations influenced hardware selection and algorithm optimization, economic sustainability guided pricing and deployment strategies, and social sustainability shaped user interface design and engagement mechanisms. This holistic approach ensures that technological innovations contribute to long-term healthcare system resilience rather than creating new dependencies or inequalities.

8.5.2 Ethical AI Considerations and Implementation

While our system incorporates multiple ethical safeguards, several considerations merit ongoing attention. The potential for behavioral monitoring to become surveillance raises questions about the appropriate boundaries of health technology. We addressed this through transparent data practices and user control mechanisms, but the normalization of continuous monitoring in healthcare contexts requires sustained ethical vigilance.

Algorithmic bias remains a concern despite our efforts to ensure diverse training data. Eye-rubbing patterns may vary across cultures, age groups, and occupational contexts in ways not fully captured by our dataset. Continuous monitoring and regular model updates will be necessary to identify and address emerging biases. We propose establishing a governance framework that includes patient representatives, clinicians, and ethicists to guide ongoing system development and deployment decisions.

The explainability of deep learning models presents both technical and philosophical challenges. While our system provides confidence scores and feature importance visualizations, the fundamental opacity of neural networks limits full interpretability. This raises questions about clinical accountability when AI recommendations influence treatment decisions. We advocate for a hybrid approach where AI predictions augment rather than replace clinical judgment, maintaining clear lines of responsibility while leveraging technological capabilities.

8.5.3 Sustainability Governance in Practice

The deployment of our eye-rubbing detection system offers insights into practical sustainability governance for healthcare technologies. Environmental sustainability was addressed through careful consideration of device lifecycles and computational efficiency. By utilizing existing consumer smartwatches rather than specialized medical devices, we reduced electronic waste and leveraged more efficient manufacturing scales. The cloud infrastructure supporting model updates and data analytics employ renewable energy sources and efficient data center practices.

Economic sustainability analysis revealed complex trade-offs between upfront technology costs and long-term healthcare savings. While smartwatch-based monitoring requires initial investment, the potential to prevent keratoconus progression and reduce the need for corneal transplants offers substantial cost savings. Our economic modeling suggests that preventing progression in just 15% of monitored patients would offset system costs within 3 years. This finding supports value-based healthcare models that prioritize prevention and early intervention.

Social sustainability emerged as perhaps the most critical factor for long-term success. The system's ability to empower patients in managing their condition while maintaining clinical relationships proved essential for acceptance. Cultural adaptations, including customizable alert styles and multilingual interfaces, enhanced accessibility across diverse populations. The gamification elements were carefully designed to motivate without creating anxiety or obsessive monitoring behaviors, respecting the psychological well-being of users.

8.6 Limitations and Future Directions

Several limitations of our current approach warrant acknowledgment and provide directions for future research. Our training data relies on simulated behaviors expressed by four healthy participants rather than on natural eye-rubbing behaviors of actual keratoconus patients. While clinical experts supervised the simulation process to ensure authenticity, and our separate clinical validation with 15 real patients demonstrated effectiveness, the gap between simulated training data and real patient behaviors may limit generalizability. Future work should focus on collecting training data directly from diverse patient populations during natural behaviors.

The reliance on unilateral smartwatch wear may miss eye-rubbing events performed with the non-monitored hand. While our clinical data suggests that patients typically exhibit dominant-hand rubbing patterns, bilateral monitoring through paired devices could improve detection coverage. Future work will explore lightweight solutions for bilateral monitoring without compromising user comfort or compliance.

Cultural and occupational variations may affect model generalizability. While our controlled simulation approach ensured systematic coverage of different movement patterns, real-world variations across diverse populations (elderly individuals, young children, individuals with motor disabilities, and people from different cultural backgrounds) remain underrepresented. Targeted data collection from these populations and development of transfer learning frameworks for rapid adaptation will be necessary to ensure equitable system performance. This aligns with Industry 5.0's emphasis on inclusive technology development that serves all members of society.

The simulation-based training approach, while enabling controlled data collection, introduces potential domain shift challenges when deployed with real patients. Although our clinical validation demonstrated effectiveness, continuous model adaptation and personalization strategies may be needed to maintain performance across diverse patient populations and usage patterns. Future research will explore domain adaptation techniques and personalized model fine-tuning approaches.

The current model architecture, while effective, may benefit from more advanced deep learning approaches. Transformer-based architectures have shown promise in related work, potentially offering improved accuracy through attention mechanisms that capture long-range temporal dependencies. However, the computational requirements of such models must be balanced against the constraints of mobile deployment and battery life considerations. Future research will explore hybrid architectures combining the efficiency of LSTMs with the representational power of transformers. Generalization across diverse clinical populations requires expanded validation studies beyond our initial 15-patient cohort. International multicenter studies with larger patient populations will be necessary to validate performance across different populations and healthcare systems. Additionally, longitudinal studies tracking patients over extended periods will provide insights into the long-term effectiveness of behavioral interventions and potential adaptations required as patients develop new habits.

8.7 Conclusions

This chapter has presented a comprehensive approach to eye-rubbing detection that embodies the principles of Industry 5.0 through the integration of ethical AI, sustainability governance, and human-centered design in healthcare technology. Our LSTM-based detection system achieved 95.65% accuracy in identifying various types of eye-rubbing behaviors while maintaining patient privacy, ensuring algorithmic transparency, and promoting sustainable healthcare practices.

The successful deployment of this system demonstrates that advanced AI technologies can be implemented in sensitive healthcare contexts without compromising ethical principles or sustainability goals. By prioritizing patient empowerment, clinical integration, and long-term societal benefit, we have created a model for responsible innovation in digital health. The system's ability to provide objective, quantitative monitoring of eye-rubbing behavior addresses a critical gap in keratoconus management while respecting patient autonomy and promoting behavior change through positive reinforcement rather than surveillance.

Our work contributes to the broader discourse on Industry 5.0 by demonstrating that the human-centric vision of industrial transformation can be realized in practice. The convergence of IoT sensors, deep learning algorithms, and ethical governance frameworks creates new possibilities for healthcare delivery that enhance rather than replace human capabilities. This approach offers a path forward for addressing complex health challenges while maintaining the values and relationships that define quality healthcare.

The three key contributions of this work, namely, comprehensive ethical AI integration, multi-class clinical classification, and validated behavioral outcomes, establish a foundation for future IoT-based healthcare monitoring systems. The demonstrated 68% reduction in eye-rubbing frequency among clinical participants validates the potential for technology-mediated behavioral interventions in chronic disease management.

Future research directions include expanding the system's capabilities to detect other relevant behaviors, developing personalized prediction models for keratoconus progression risk, and exploring integration with other digital health interventions. The ethical AI framework and sustainability governance approach developed for this application can serve as a template for other healthcare IoT deployments, contributing to the responsible advancement of Industry 5.0 in healthcare.

As we stand at the intersection of technological capability and human need, the eye-rubbing detection system represents more than a clinical tool—it embodies a vision for healthcare where technology serves humanity with wisdom, compassion, and respect for our shared future. The journey from Industry 4.0's efficiency focus to Industry 5.0's human-centric approach requires not only technological innovation but also a fundamental reimagining of how we design, deploy, and govern healthcare technologies. Through this work, we contribute to that transformation, one patient, one behavior, and one ethical decision at a time.

References

1. Breque, M., De Nul, L., & Petridis, A. (2021). *Industry 5.0: Towards a sustainable, human-centric and resilient European industry*. European Commission, Directorate-General for Research and Innovation.
2. Hashemi, H., Heydarian, S., Hooshmand, E., Saadatian, M., Yekta, A., Aghamirsalim, M., . . . Khabazkhoob, M. (2020). The prevalence and risk factors for keratoconus: A systematic review and meta-analysis. *Cornea*, 39(2), 263–270.
3. Najmi, H., Mobarki, Y., Mania, K., Altowairqi, B., Basehi, M., Mahfouz, M. S., & Elmahdy, M. (2019). The correlation between keratoconus and eye rubbing: A review. *International Journal of Ophthalmology*, 12(11), 1775–1781.
4. Ridley, F. (1961). Eye-rubbing and contact lenses. *British Journal of Ophthalmology*, 45(9), 631.
5. Bohac, M., Koncarevic, M., Pasalic, A., Biscevic, A., Merlak, M., Gabric, N., & Patel, S. (2018). Incidence and clinical characteristics of post LASIK ectasia: A review of over 30,000 LASIK cases. *Seminars in Ophthalmology*, 33(7–8), 869–877.
6. Karseras, A. G., & Ruben, M. (1976). Aetiology of keratoconus. *British Journal of Ophthalmology*, 60(7), 522–525.
7. Lindsay, R. G., Bruce, A. S., & Gutteridge, I. F. (2000). Keratoconus associated with continual eye rubbing due to punctal agenesis. *Cornea*, 19(4), 567–569.
8. McMonnies, C. W. (2009). Mechanisms of rubbing-related corneal trauma in keratoconus. *Cornea*, 28(6), 607–615.
9. Torres-Netto, E. A., Abdshahzadeh, H., Abrishamchi, R., Randleman, J. B., Gomes, J. A., & Hafezi, F. (2022). The impact of repetitive and prolonged eye rubbing on corneal biomechanics. *Journal of Refractive Surgery*, 38(9), 610–616.
10. Nokas, G., & Kotsilieris, T. (2023). Preventing keratoconus through eye rubbing activity detection: A machine learning approach. *Electronics*, 12(4), 1028.
11. Elahi, S., Mery, T., Panthier, C., Saad, A., Gatinel, D., & Alahi, A. (2024). Eye-rubbing detection using a smartwatch: A feasibility study demonstrated high accuracy with machine learning. *Translational Vision Science & Technology*, 13(9), 1.
12. Drira, I., Louja, A., Sliman, L., Soler, V., Noor, M., Jamali, A., & Fournie, P. (2024). Eye-rubbing detection tool using artificial intelligence on a smartwatch in the management of keratoconus. *Translational Vision Science & Technology*, 13(12), 16.
13. European Commission. (2019). *Ethics guidelines for trustworthy AI*. High-Level Expert Group on Artificial Intelligence.
14. Vollmer, S., Mateen, B. A., Bohner, G., Kiraly, F. J., Ghani, R., Jonsson, P., . . . Hemingway, H. (2020). Machine learning and artificial intelligence research for patient benefit: 20 critical questions on transparency, replicability, ethics, and effectiveness. *BMJ*, 368, l6927.
15. Amann, J., Blasimme, A., Vayena, E., Frey, D., & Madai, V. I. (2020). Explainability for artificial intelligence in healthcare: A multidisciplinary perspective. *BMC Medical Informatics and Decision Making*, 20(1), 310.
16. Vaswani, A., Shazeer, N., Parmar, N., Uszkoreit, J., Jones, L., Gomez, A. N., . . . Polosukhin, I. (2017). Attention is all you need. *Advances in Neural Information Processing Systems*, 30, 5998–6008.
17. Jobin, A., Ienca, M., & Vayena, E. (2019). The global landscape of AI ethics guidelines. *Nature Machine Intelligence*, 1(9), 389–399.
18. Mittelstadt, B. (2019). Principles alone cannot guarantee ethical AI. *Nature Machine Intelligence*, 1(11), 501–507.
19. Char, D. S., Shah, N. H., & Magnus, D. (2018). Implementing machine learning in health care Addressing ethical challenges. *New England Journal of Medicine*, 378(11), 981–983.
20. Morley, J., Machado, C. C., Burr, C., Cowls, J., Joshi, I., Taddeo, M., & Floridi, L. (2020). The ethics of AI in health care: A mapping review. *Social Science & Medicine*, 260, 113172.

21. Wang, P., & Zhang, L. (2023). Sustainability in Industry 5.0: A systematic review of concepts and implications. *Journal of Cleaner Production*, 380, 134892.
22. Shaw, J., Rudzicz, F., Jamieson, T., & Goldfarb, A. (2019). Artificial intelligence and the implementation challenge. *Journal of Medical Internet Research*, 21(7), e13659.
23. Topol, E. J. (2019). High-performance medicine: The convergence of human and artificial intelligence. *Nature Medicine*, 25(1), 44–56.
24. Martinez-Martin, N., Insel, T. R., Dagum, P., Greely, H. T., & Cho, M. K. (2018). Data mining for health: Staking out the ethical territory of digital phenotyping. *NPJ Digital Medicine*, 1(1), 68.
25. Chen, Y., & Shen, C. (2023). Wearable sensor-based behavioral analytics for chronic disease management: A systematic review. *IEEE Journal of Biomedical and Health Informatics*, 27(1), 123–135.
26. Zhang, S., Li, Y., Zhang, S., Shahabi, F., Xia, S., Deng, Y., & Alshurafa, N. (2022). Deep learning in human activity recognition with wearable sensors: A review on advances. *Sensors*, 22(4), 1476.

Part III

Industry 5.0 in Practice

Case Studies, Tools, and Strategic Insights

9 Corporate Social Responsibility in Sustainable Agriculture Investment in Taiwan

Fitria Dewi Navisa and Hsing-Hao Wu

9.1 Introduction

In light of climate change and the growing demand for food worldwide, sustainable agriculture has emerged as a key problem for many nations, including Taiwan. In Taiwan, the agriculture industry plays a significant part in preserving environmental sustainability and food security. However, conventional farming methods frequently violate sustainability standards, particularly when it comes to waste management and the use of natural resources. A solution to this problem that might have beneficial effects on the economy, society, and environment is the corporate social responsibility (CSR) approach. CSR becomes a crucial tool for promoting ethical investing, particularly in the field of sustainable agriculture (Hsu, 2021). CSR in sustainable agriculture aims not only to create economic benefits for companies but also to promote environmental protection and improve the welfare of local communities. These policies, among other things, provide incentives for companies that adopt environmentally friendly technologies and sustainable farming methods. This is done to achieve long-term sustainability goals that have a broad impact on food security and environmental sustainability (Sulandjari & Suparwata, 2023).

The role of CSR in the sustainable agriculture sector is very important because companies have a responsibility to contribute to sustainable economic, social, and environmental development. In Taiwan, companies are encouraged to not only pursue economic benefits but also ensure that their agricultural practices reduce environmental impacts and improve people's well-being. One form of CSR implementation that is growing in Taiwan is the management of agricultural waste and its utilization as renewable energy.

CSR in sustainable agriculture investment plays a vital role in ensuring that economic activities in the agricultural sector are not only financially profitable but also environmentally friendly and socially responsible. CSR encourages companies to consider the long-term impact of their operations on soil, water, biodiversity, and ecosystems as a whole. In Taiwan, where agricultural land is limited and food demand continues to increase, the implementation of CSR helps maintain a balance between production and the conservation of natural resources.

DOI: 10.4324/9781003647621-12

Beyond environmentally friendly practices, CSR in agriculture also involves protecting farmers' rights, empowering local communities, and improving the welfare of surrounding populations. Through CSR programs, investors can provide training on sustainable farming, offer access to eco-friendly technology, and build agricultural infrastructure that supports local food security. This approach is highly relevant in Taiwan, particularly in addressing the challenges of climate change and limited resources. Furthermore, CSR adds value to the relationships among the private sector, government, and civil society in achieving the Sustainable Development Goals (SDGs). Companies that consistently implement CSR tend to gain higher public trust, enhance their brand reputation, and strengthen the sustainability of their business. In Taiwan, where consumers are increasingly aware of environmental and social issues, CSR in agricultural investment becomes a strategic factor in ensuring the continuity of the agricultural sector while generating a broader positive impact on society and the environment.

Taiwan has successfully managed various types of agricultural waste such as rice husk, straw, and livestock waste through material and energy recycling processes. According to National Statistics, the amount of agricultural waste reached more than 2.1 million tons in 2017, with the majority in the form of rice straw and livestock waste. The Taiwanese government has implemented a number of laws and policies, including the Waste Disposal Act and Agricultural Waste Recycling Management Measures, to enhance the use of this trash, which aim to encourage the use of agricultural waste as biomass fuel and biogas energy (Hsu, 2021).

This study addresses a research gap that remains underexplored regarding the specific implementation of CSR in the sustainable agriculture sector in Taiwan, particularly how local agricultural companies and foreign investors translate CSR principles into practical actions on the ground. While CSR has been widely discussed in the context of Taiwan's manufacturing and technology industries, there is limited research focusing on its application in agriculture—especially its impact on environmental sustainability and the empowerment of local farmers. This research aims to bridge that gap by analyzing CSR approaches adopted in agricultural investments, evaluating their effectiveness in improving farmers' welfare and preserving natural resources, and providing practical recommendations for developing more inclusive and context-specific CSR policies in Taiwan's agricultural sector.

As shown in Table 9.1, the implementation of CSR in Taiwan's agricultural sector often involves investments in environmentally friendly technologies such as bioenergy and agricultural waste management. These initiatives have contributed to reducing CO_2 emissions through the adoption of sustainability technologies; however, several barriers such as high operational costs, local resistance, and regulatory complexity continue to challenge broader implementation.

This chapter aims to analyze how Taiwan government policies and regulations support the implementation of CSR in investment in the sustainable agriculture sector. In addition, this chapter will also explore the legal, social, and environmental challenges companies face in implementing CSR in this sector, as well as the strategies that can be taken to overcome these obstacles. With this understanding, it is hoped that more effective policy recommendations can be provided for

Table 9.1 Aspect and Description of CSR.

Aspect	*Description*
Implementation of CSR	Investments in environmentally friendly technologies, such as bioenergy and agricultural waste management
Impact on the environment	Reduction of CO_2 emissions through the adoption of bioenergy and sustainability technologies
Barriers to implementation	High costs, local resistance, and regulatory complexity

governments, companies, and other stakeholders for encouraging responsible and sustainable investment in the agricultural sector.

9.2 Method

This study uses a normative legal research approach, which is research that focuses on the study of applicable legal rules and relevant legal principles. This study will analyze various regulations and policies that support the implementation of CSR in the sustainable agriculture sector in Taiwan.

9.2.1 Legal Approach (Statute Approach)

This approach is applied to use it to analyze laws, regulations, and policies issued by the Taiwan government regarding CSR and the sustainable agriculture sector. This research will review legal documents such as government policies that encourage sustainable agricultural practices. The main focus is on how the legal framework supports or hinders the implementation of CSR in the agricultural sector.

9.2.2 The Conceptual Approach

The conceptual approach is carried out to understand the legal concepts underlying the application of CSR, especially in the agricultural sector. This study will outline the concept of CSR as part of corporate legal responsibility, by highlighting sustainability principles applied in agricultural sector investments. This approach will also compare the application of CSR concepts in Taiwan with international standards, such as Global GAP and ISO 26000, to understand how practices in Taiwan relate to global standards.

9.2.3 Data Sources

This research will use secondary data obtained from various sources, including legislation, law journals, textbooks, and reports from relevant national and international organizations. Using this normative legal method, the study aims to provide a clear picture of the relationship among CSR policies, government regulations, and sustainable agricultural practices in Taiwan, and provide relevant recommendations for policymakers and other stakeholders.

9.3 Discussion

The idea of CSR emphasizes that businesses have obligations beyond generating profits for shareholders. The economic, social, and environmental obligations that a business must take on in the course of its operations are all included in CSR. Economically speaking, the firm must be able to operate economically and efficiently in order to maintain long-term viability (Hung et al., 2019). In sustainable agriculture, the experience of PT. Pertamina RU II Dumai through the DEWIGATRA program shows how the transformation of structures and processes can improve the welfare of peatland farming communities. The program faces major challenges, including vulnerability to forest fires, water access constraints, and high production costs (Chang & Lee, 2020). In this case, sustainable agriculture is not only about environmentally friendly farming methods but also includes the economic empowerment of local communities and risk management.

References related to the DEWIGATRA program show the importance of integrating various strategies in achieving sustainable agriculture, from risk management and technology use to economic empowerment and community capacity building. In Taiwan, similar challenges in agricultural waste management can be addressed with an integrated and sustainable approach, which focuses not only on production efficiency but also on environmental protection and people's well-being (Fatkhullah et al., 2021).

CSR and sustainable agriculture have a close relationship as both aim to create a balance between economic benefits and social and environmental responsibility. In sustainable agriculture, companies are expected to not only pursue profit but also ensure that their agricultural practices are environmentally friendly and provide social benefits to the local community (Chiou & Shu, 2019). For example, companies that invest in renewable energy through agricultural waste management not only help reduce carbon emissions but also create clean energy sources and strengthen national energy security. In addition, through the implementation of CSR, the company can support local farmers by providing access to modern agricultural technology, training, and infrastructure needed to increase productivity sustainably (Van Hong et al., 2021).

The implementation of CSR in sustainable agriculture in Taiwan also involves supporting local farmers. Companies can provide access to modern agricultural technology, training, and infrastructure that supports sustainable productivity. Thus, through the implementation of CSR, the company helps build food security, improve the welfare of farmers, and encourage broader economic sustainability. In addition, investment in sustainable agricultural practices has the potential to expand access to international markets that increasingly prioritize environmentally friendly products (Huang, 2020).

Starbucks coffee company offers a clear illustration of how marketing tactics and CSR can be combined to promote environmental sustainability. In order to reduce adverse effects on the environment, Starbucks, a multinational corporation, employs a green marketing strategy that consists of green product, green price, green place, and green promotion. Some of the initiatives launched by Starbucks include the following:

1. "Take a Mug Pledge" program: Educates consumers to use reusable eco-friendly packaging, such as Starbucks on-the-go tumblers, and provides discounts for consumers who use those tumblers.
2. Use of 10% recycled paper for product packaging: Starbucks is committed to using more environmentally friendly packaging materials, reducing dependence on plastic packaging.
3. Grounds for Hope Program: Processing coffee grounds into compost through collaboration with Bali Compost Crafters, which reduces waste and contributes to organic waste management.
4. Water for Change Program: Building AquaTower with Planet Water Foundation to provide access to clean water for people who have difficulty getting clean water.

Starbucks' CSR programs show how companies can use their social responsibility to support environmental sustainability and improve the company's image. The integration of CSR with marketing strategies creates added value for products and builds engagement with consumers, which in turn supports sustainable environmental development (Li et al., 2021). Thus, the CSR carried out by Starbucks and other companies implementing similar principles demonstrates the importance of the company's role in supporting sustainable agriculture and environmental protection (Shidiq et al., 2019).

Discussions on CSR and sustainable agriculture in Taiwan provide valuable insights when compared to findings from Sukabumi Regency. In Taiwan, CSR is widely applied in the agricultural sector to support sustainable agriculture and reduce the environmental impact of agricultural practices. Companies in Taiwan are expected to focus not only on economic profitability but also on their social and environmental responsibilities. This includes waste management, the use of renewable energy such as biogas and biomass, and support for local farming communities in adopting new technologies and sustainable farming practices.

9.3.1 Analysis of Findings from Sukabumi Regency

The findings from Kabupaten Sukabumi show that farmers in the region face a variety of challenges, including limited infrastructure, knowledge, and access to technology. The demographic diversity of farmers in Sukabumi adds to the complexity of the problems faced in the application of modern agricultural technology and sustainable agricultural practices. Therefore, a well-designed CSR can play an important role in supporting smallholders in Sukabumi. For example, CSR can offer the training and technology transfer needed to adopt modern agricultural practices such as vertical farming and bioinformatics.

By comparing the implementation of CSR in Taiwan and Sukabumi, we can see how CSR can be used to address the challenges of sustainable agriculture and technology in both circumstances. Successful CSR will include support for clean technologies, renewable energy, training for smallholders, and strategies to address the impact of trade on the environment. The Taiwan model that emphasizes CSR and environmental responsibility can provide guidance for more effective efforts in supporting sustainable agriculture in Sukabumi (Wang et al., 2020). CSR regulations in Taiwan operate within a legal framework that encourages companies to meet high standards in social and environmental responsibility. The Taiwanese government has implemented policies that support the implementation of CSR, especially in the agricultural sector (Tsai, 2020).

9.3.2 Corporate Social Responsibility Regulations in Taiwan and Their Impact on Agriculture and the Environment

CSR regulations in Taiwan operate within a legal framework that encourages companies to meet high standards in social and environmental responsibility. The Taiwanese government has implemented policies that support the implementation of CSR, especially in the agricultural sector. As presented in Table 9.2, Taiwan's CSR

Table 9.2 CSR Regulations in Taiwan and Their Impact on Agriculture and the Environment.

Aspect	*Description*
CSR regulatory framework in Taiwan CSR	Regulation in Taiwan encourages companies to meet high standards in social and environmental responsibility. The government supports this policy, especially in the agricultural sector
National policy	Obligation for companies to report on CSR practices, ensuring business activities are economically, socially, and environmentally beneficial
Rules related to foreign investment	Requirements such as environmental impact evaluation before starting a project and waste management in a sustainable manner to ensure economic benefits without harming the environment
Taiwan's total factor productivity	TFP reached 1,033, with an average increase of 3.3% in agricultural productivity during 2002–2010, higher than Japan and Korea.
	Taiwan has a high technical and environmental efficiency score (1.00) for most years, indicating technically efficient and environmentally sound agricultural practices
CO_2 emission reduction potential	Taiwan lacks the CO_2 emission reduction potential of the agricultural sector, thanks to the integration of environmentally friendly agricultural practices and advanced technologies
Contribution of CSR regulations	Strict CSR regulations and advanced agricultural technologies contribute to high efficiency and good environmental management, promoting the achievement of sustainable agricultural productivity
Model for other countries	Taiwan can be an example for other countries in the implementation of effective CSR and sustainable agricultural management, showing the relationship between CSR and agricultural technology

framework includes mandatory reporting requirements, environmental impact assessments for foreign investments, and waste management obligations that ensure economic growth aligns with ecological protection.

The results showed that Taiwan had a positive change in total factor productivity (TFP) during the period 2002–2010, with the highest TFP compared to other countries such as Japan and Korea. Taiwan's TFP reached 1,033, reflecting an average increase of 3.3% in agricultural productivity. These data highlight the important role of technology and efficiency in improving Taiwan's agricultural productivity. The analysis of technical and environmental efficiency also showed that Taiwan, along with Japan and Korea, had high technical and environmental efficiency throughout the study period. Taiwan achieved a perfect technical and environmental efficiency score (1.00) for most of the year, showing that agricultural practices in Taiwan are not only efficient from a technical point of view but also in terms of environmental impact. This is inversely proportional to other countries such as Thailand, which have significantly lower technical and environmental efficiency scores.

Furthermore, Table 9.2 illustrates that Taiwan achieved a perfect technical and environmental efficiency score (1.00) for most years, indicating that its agricultural practices are not only technically efficient but also environmentally sound. This finding contrasts with other countries, such as Thailand, which show significantly lower technical and environmental efficiency scores.

Energy security and sustainable agriculture are two important aspects that are interrelated with the management of energy and natural resources. Energy security, which includes aspects of energy security, affordability, and environmental sustainability, is crucial to ensuring a stable and affordable energy supply while minimizing environmental impacts. The concept of energy trilemma promoted by the World Energy Council suggests that the priority of each country in responding to this trilemma depends on the level of economic growth and welfare of its people. Developed countries tend to prioritize environmental sustainability, while developing countries focus more on energy security and affordability (Tseng et al., 2019).

Energy security and sustainable agriculture have a close relationship that cannot be separated with the management of energy and natural resources. Energy security includes three main aspects: energy security, energy affordability, and environmental sustainability. These three aspects must be considered to ensure that the energy supply remains stable and affordable while reducing adverse environmental impacts. The concept of the energy trilemma proposed by the World Energy Council indicates that the priorities in dealing with this problem vary depending on the level of economic growth and well-being of the people of each country. Developed countries tend to emphasize environmental sustainability, while developing countries focus more on energy security and affordability (Chung, 2021).

The sociocultural aspect is no less important. In the development of tourism destinations, relationships with local communities and support for local culture are key to success. In sustainable agriculture, the recognition of the rights of local farmers and the adoption of practices that support the well-being of communities are also essential to achieve sustainability (Khan et al., 2021). Therefore, in both tourism and agriculture, a holistic and integrative approach to sustainability—encompassing

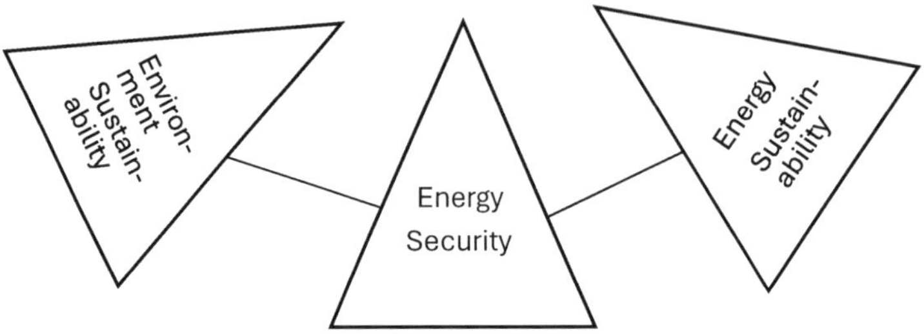

Figure 9.1 Energy security and sustainable agriculture.

environmental, economic, and sociocultural aspects—is key to achieving successful sustainability goals. The implementation of sustainability indicators covering a wide range of stakeholder perspectives can assist in formulating effective strategies to achieve energy security and sustainable agriculture (Lee et al., 2021).

The implementation of CSR by the company adopts the principle of double commitment that prioritizes the balance between economic success and social progress, in line with the thinking of its founder, Tirto Utomo. Since 2010, the company has been implementing CSR, which can be grouped into four main areas:

1. Social field: One of the main activities in the social field is the social mapping carried out from March to July 2017. Social mapping aims to obtain information about the composition of stakeholders, environmental issues, community needs, and recommendations for the development of social programs that are in accordance with the company's mission.
2. Economic sector: Community-based conservation, such as Avocado Field School, is an initiative implemented since July 2015. The program supports learning about sustainable agriculture through the process of crop experimentation and analysis, as well as mitigating problems facing communities.
3. Environment: The Kehati Park Program, which began in 2014, aims to develop green open spaces and the conservation of flora and fauna. Kehati Park serves as a collection of local endemic plants and supports life systems through soil quality guards and groundwater storage.
4. Community empowerment: PKBI West Sumatra, as a CSR implementing partner, carries out empowerment programs in Nagari Batang Barus with a focus on reproductive health and family planning. This program includes clean water, improving clean and healthy living behavior (PHBS), and community-based total sanitation (STBM), with support from Aqua's CSR.

The success of a CSR program lies in its impact on the community, where community participation and stakeholder engagement are essential. CSR programs implemented by PT. Tirta Investama is expected to provide significant benefits to the local community, create mutually beneficial relationships, and support the principles of good governance (Risa, 2020).

The Taiwanese government has implemented various policies to support the international expansion of microchip manufacturers. The regulations implemented include quality standards and international certifications, which help manufacturers understand and comply with the requirements of the global market. The policy aims to strengthen the position of Taiwan's microchip industry by increasing its competitiveness in the international market. The growth of this sector promotes an increase in GDP and job creation, as well as investment in research and development. However, the high dependence on this industry makes Taiwan's economy vulnerable to global market fluctuations. Therefore, risk mitigation strategies need to be implemented to address the potential for economic volatility (Vrabcová & Urbancová, 2021).

On the environmental side, the semiconductor industry in Taiwan faces major challenges related to energy and water consumption. The process of making microchips consumes very high energy and can result in water shortages, with potential impacts on the agricultural and infrastructure sectors. TSMC, Taiwan's top producer, has committed to using renewable energy by 2040, but rapid industrial growth raises concerns about Taiwan's ability to balance economic growth and environmental preservation. Effective CSR implementation is crucial to managing environmental impacts while maximizing the economic benefits of this international expansion policy (Putri et al., 2024).

9.3.3 CSR Implementation in Sustainable Agricultural Investment

This section presents examples of CSR implementation by companies in Taiwan. A number of businesses in Taiwan have shown a strong dedication to sustainable agriculture through CSR (Azhar, 2025). The Taiwan Sugar Corporation is a well-known illustration. They have incorporated CSR concepts by making investments in ecologically friendly agricultural technologies and putting in place training programs for regional farmers. To create crop varieties that are more resilient to disease and climate change, the business uses biotechnology. In order to boost crop yields while lessening their negative effects on the environment, they also teach farmers sustainable agricultural methods, such as the application of biofertilizers and soil conservation strategies. By taking these actions, local farmers' welfare is supported as agricultural output increases.

Environmental sustainability in the agricultural industry is becoming increasingly important in Taiwan. According to a study conducted on Dongshan tea in Dongshan Township, Yilan County, Taiwan, the primary factors affecting the product's carbon footprint were energy use and nitrogen fertilizer use. The consumer use phase accounted for the most carbon emissions, primarily from the usage of energy to boil water. According to the investigation, the energy consumption and nitrogen fertilizer use during the tea-making process have a major influence on both human health and climate change. Taiwan shows that a comprehensive approach to agriculture may lower carbon footprints, improve sustainability, and enhance the well-being of local populations by fusing green technologies with CSR activities that support sustainability (Hu et al., 2019).

Adoption of advanced technologies in agriculture is an integral part of CSR implementation. Companies in Taiwan are increasingly adopting innovations such as vertical farming, smart water management, and renewable energy technologies as components of their CSR strategies (Hsu & Chen, 2021). For example, Taiwanese agri-tech companies have invested in vertical farming systems that enable crop production in a controlled environment, reduce land and water use, and minimize carbon footprint. In addition, they apply advanced water management technologies to optimize the use of water resources, reduce waste, and ensure efficient distribution to farmers. Renewable energy technologies, such as solar panels installed on agricultural facilities, also play an important role in reducing dependence on fossil energy sources and promoting sustainability (Ilyas et al., 2022). Through the application of these technologies, companies not only fulfill their CSR commitments but also contribute to the wider sustainability of agriculture.

Based on available data, the adoption of advanced technologies in agriculture in Taiwan shows a significant relationship among knowledge, the importance of technology, and adoption behavior. For example, automated environmental control systems and agricultural management applications are the technologies that are most recognized by and important for farmers, with high average scores for knowledge and importance (Jnr et al., 2021). This reflects that the understanding and assessment of the technologies greatly affect the adoption rate. Data shows that knowledge of advanced agricultural technologies and perceptions of the importance of such technologies are positively related to the rate of technology adoption. For example, the advanced technology adoption score shows that a 1% increase in knowledge and perception of the importance of technology can increase adoption rates by 0.93% and 0.81%, respectively (Ilyas et al., 2022).

However, the results of the analysis show that although knowledge and the importance of technology have a positive relationship with adoption, factors such as farm size, annual turnover, and demographic characteristics also play an important role. Farmers with larger farm sizes or higher business volumes tend to adopt technology faster. Therefore, it is important for policies and training programs to focus on improving market access and systematic training on applications of technologies such as Internet of Things (IoT) and big data analytics to help farmers engage more effectively in advanced farming practices. Overall, the application of these technologies in CSR strategies not only reflects CSR but also plays a role in promoting broader agricultural sustainability, with significant contributions to resource management and environmental impact reduction (Chuang et al., 2020).

9.4 Positive Social and Environmental Impact

The implementation of CSR in the agricultural sector brings significant social and environmental benefits. First, reduction in carbon emissions can be achieved through the use of environmentally friendly technologies and sustainable agricultural practices. The use of biofertilizers and soil conservation techniques helps in the absorption of carbon dioxide and reduces the carbon footprint of agricultural activities (Dauda, 2024).

Second, the adoption of efficient water management systems and vertical farming techniques helps conserve water resources, minimize excessive usage, and enhance resilience to drought. Third, farmer training and capacity-building initiatives contribute to improved livelihoods by equipping farmers with advanced agricultural knowledge and practices, ultimately increasing their income and quality of life. Moreover, the impact on food security becomes more evident as modern technologies and sustainable farming methods boost crop productivity and reinforce the stability of food systems. Therefore, these innovations not only strengthen the implementation of CSR but also play a crucial role in promoting sustainable agriculture while delivering extensive benefits to both society and the environment (Bilyay-Erdogan et al., 2024).

Reduction of carbon emissions, water management technologies, training of farmers, and contribution to food security are some of the key benefits of CSR in agriculture. Data related to the impact of solar panel capacity on agricultural land prices in Tainan city may clarify some additional effects of CSR.

Effect of solar panel capacity on agricultural land prices: Based on the spatial lag model (SLM) estimation results, the following are the impacts of solar panel capacity on agricultural land prices in various divisions:

1. Division I (Tainan city and surrounding areas):
   - Each kilowatt (kW) of solar panel capacity within a 2-km radius increases the price of agricultural land by \$0.68 per m^2.
   - This reflects that in Division I, where average solar panel capacity is still low, there is an expectation that land prices will increase if solar panel capacity increases around agricultural land.
2. Division II (central):
   - Higher solar panel capacity is likely to lower the price of agricultural land. This may be because Division II is a core development zone in Tainan city, with most of the land to be converted into industrial parks. The existence of solar panels can inhibit the change of land function into an industrial park.
3. Division III (north):
   - The price of agricultural land increases by \$1.71 per m^2 with each kilowatt of solar panel capacity. This suggests that in Division III, which is more oriented toward agricultural production, solar panels can increase the value of land.
4. Division IV (coastal):
   - The price of agricultural land increased by \$0.78 per m^2 with the capacity of solar panels. Despite the improvement, this value is lower compared to Division III.

5. Division V (mountains):
    - Solar panel capacity had little effect on farmland prices, with an increase of only \$0.14 per m^2. This suggests a more limited impact within mountainous areas.

9.5 Effect of Distance from Farmland to the Nearest Solar Panel

- Division I shows that proximity to the land with solar panels impacts the price of agricultural land, which decreases with increasing distance from the land with solar panels. Hence, farmers in these urban areas prefer a reasonable distance from solar panels, as land prices increase the farther away they are from the land with solar panels.
- Division II shows that land prices tend to decrease as the distance from solar panel installations increases, since land in this area is more likely to be converted into industrial parks. Furthermore, the combined impact of solar panel capacity and spacing influences both land-use patterns and energy efficiency, where optimal configurations can maximize power generation while minimizing negative effects on surrounding land values.
- Division I showed a total positive impact of \$5.32 per m^2, or 3.4% of the average price of agricultural land. This means that solar panels with a certain capacity within a radius of 2 km can increase land prices.
- Division II showed a negative impact with a decrease in land prices by \$11.19 per m^2, or 11.77% of the average price. This reflects the decline in land value in an industrial development zone with many solar panels in the vicinity.
- Divisions III, IV, and V show an increase in land prices by various percentages, depending on the capacity of the solar panels and the distance from the nearest land to the solar panels.

The development of sustainable agricultural systems requires collaboration among governments, farmers, scientists, research institutions, and communities. By implementing this approach, agricultural systems can become more sustainable, produce enough food, and preserve the environment for generations to come. Food sustainability, which includes food security, sustainable agriculture, food diversity, food loss reduction, and sustainable consumption, is key in achieving this goal. By adopting a holistic and collaborative approach, Taiwan can overcome challenges and obstacles in the implementation of CSR in the sustainable agriculture sector, thereby creating a positive impact on the environment, society, and economy as a whole (Siregar, 2023).

CSR initiatives in sustainable agriculture are often intertwined agrarian regulations and licensing processes. Data related to bioenergy production and emission reductions provide a clear picture of how regulations and policies can affect the implementation of such initiatives. First, the influence of regulations on bioenergy production is very real. A significant reduction in green bond rates could lead to a drastic increase in biofuel and biopower production, as seen from the increase in biofuel production from 160 million liters to 295 million liters at an interest rate of 6.5%. However, strict regulations on land use and licensing processes often slow or even limit the implementation of bioenergy projects if they are not aligned with government policies.

Second, complexity of the licensing process is another big challenge. Although low interest rates can increase bioenergy production, a complicated and unsupportive licensing process can hinder the achievement of desired results. Sustainable agriculture projects often face difficulties if the licensing process is inefficient or too complex. In addition, compliance with strict environmental standards can be an obstacle for companies lacking the resources or expertise to comply with such regulations.

Third, green bond policies play an important role in influencing bioenergy production. A reduction in green interest rates, as the data shows, could significantly increase biofuel and biopower production and reduce emissions. This suggests that proper regulation of supportive policies, including financial incentives, can facilitate the implementation of CSR and sustainable agriculture. Finally, the impact on land use cannot be ignored either. The decrease in green interest rates is related to the increase in land use for energy crops, which can improve the social well-being of the agricultural sector. Regulations supporting the development of energy plants will provide greater social and economic benefits from bioenergy projects.

Overall, although the data shows great potential for increased bioenergy production and emission reductions with supportive policies, the successful implementation of CSR in sustainable agriculture largely depends on regulatory and policy support that facilitates licensing and compliance processes. Supportive regulations and efficient licensing procedures can accelerate the implementation of CSR initiatives, while strict regulations or complicated processes can be significant barriers (Kung et al., 2022).

Increased public education and awareness of the benefits of CSR and sustainable technologies are also key interventions to address local resistance and lack of understanding. This can be done through training programs, public awareness campaigns, and effective counseling. In addition, the provision of funding and technical support, such as financial aid programs or grants, can help reduce the high costs often associated with CSR implementation. By strategically implementing these solutions, companies in Taiwan can be more effective in implementing CSR in the sustainable agriculture sector (Wang et al., 2019).

9.6 Conclusion

The use of renewable energy technology, including solar panels, has a major impact on the cost of agricultural land, according to Taiwan's CSR in sustainable agricultural investment. The effect of solar panel capacity on land prices varies by division, according to estimations from the SLM. Prices increase in more agriculturally oriented areas like Divisions III and IV, while prices decrease in zones that would be turned into industrial parks, like Division II. These results suggest that depending on local conditions and land usage, solar panels may increase the value of property in some locations.

In Taiwan's agriculture sector, the application of environmentally friendly technologies as part of CSR implementation has the dual benefits of boosting environmental sustainability and increasing the economic value of land. However, there are also issues with CSR implementation, such as high expenses, local opposition, and farmers' ignorance of sustainable technologies. Legal and administrative barriers are also major issues, particularly when it comes to complicated licensing procedures and agricultural rules. To overcome these barriers, partnerships between the private sector and the government are needed, which can provide incentives and regulatory policies that further support sustainable agricultural practices. Increased education and awareness campaigns on the benefits of CSR and sustainable technologies are also very important. With a holistic and collaborative approach, CSR in Taiwan's agricultural sector has the potential to create long-term positive impacts, for the environment, society, and economy.

References

Azhar, Z. (2025). *The Role of Ethical Leadership in HRM-Driven Corporate Social Responsibility-CSR*. SSRN. https://papers.ssrn.com/sol3/papers.cfm?abstract_id=5184673

Bilyay-Erdogan, S., Danisman, G. O., & Demir, E. (2024). ESG performance and investment efficiency: The impact of information asymmetry. *Journal of International* www.sciencedirect.com/science/article/pii/S1042443123001877

Chang, J. I., & Lee, C. Y. (2020). The effect of service innovation on customer behavioral intention in the Taiwanese insurance sector: The role of word of mouth and corporate social responsibility. *Journal of Asia Business Studies*, *14*(3), 341–360. https://doi.org/10.1108/JABS-06-2018-0168

Chiou, C. L., & Shu, P. G. (2019). How does foreign pressure affect a firm's corporate social performance? Evidence from listed firms in Taiwan. *Journal of Multinational Financial Management*, *51*(510), 1–22. https://doi.org/10.1016/j.mulfin.2019.04.006

Chuang, J. H., Wang, J. H., & Liou, Y. C. (2020). Farmers' knowledge, attitude, and adoption of smart agriculture technology in Taiwan. *International Journal of Environmental Research and Public Health*, *17*(19), 1–8. https://doi.org/10.3390/ijerph17197236

Chung, H. S. (2021). Taiwan's offshore wind energy policy: From policy dilemma to sustainable development. *Sustainability (Switzerland)*, *13*(18). https://doi.org/10.3390/su131810465

Dauda, S. (2024). Dynamics of corporate social responsibility (CSR) during mineral exploration: Evidence from Eastern region, Ghana. *Journal of Environmental Management*. www.sciencedirect.com/science/article/pii/S0301479724020103

Fatkhullah, M., Mulyani, I., & Imawan, B. (2021). Strategi Pengembangan Masyarakat Petani Lahan Gambut melalui Program Tanggung Jawab Sosial Perusahaan: Analisis Pendekatan Penghidupan Berkelanjutan. *Journal of Social Development Studies*, *2*(2), 15–29. https://doi.org/10.22146/jsds.2186

Hsu, E. (2021). Cost-benefit analysis for recycling of agricultural wastes in Taiwan. *Waste Management*, *120*(xxxx), 424–432. https://doi.org/10.1016/j.wasman.2020.09.051

Hsu, F. J., & Chen, S. H. (2021). US quantitative easing and firm's default risk: The role of corporate social responsibility (CSR). *The Quarterly Review of Economics and Finance*. www.sciencedirect.com/science/article/pii/S1062976921000727

Hu, A. H., Chen, C. H., Huang, L. H., Chung, M. H., Lan, Y. C., & Chen, Z. (2019). Environmental impact and carbon footprint assessment of Taiwanese agricultural products: A case study on Taiwanese Dongshan tea. *Energies*, *12*(1), 1–13. https://doi.org/10.3390/en12010138

Huang, C. C. (2020). Cocreating social innovations between an agro-food company and rice farmers in Taiwan: Exploring the process mechanisms. *British Food Journal*, *122*(12), 3837–3851. https://doi.org/10.1108/BFJ-10-2019-0759

Hung, S. W., Li, C. M., & Lee, J. M. (2019). Firm growth, business risk, and corporate social responsibility in Taiwan's food industry. *Agricultural Economics (Czech Republic)*, *65*(8), 366–374. https://doi.org/10.17221/339/2018-AGRICECON

Ilyas, M., Mian, R. U., & Suleman, M. T. (2022). Economic policy uncertainty and firm propensity to invest in corporate social responsibility. *Management Decision*. https://doi.org/10.1108/MD-06-2021-0746

Jnr, A., Kukah, A. S. K., & Akotia, J. (2021). Assessing perception and practice of corporate social responsibility (CSR) by energy sector firms in Ghana. *International Journal of Real Estate* https://intrest.utm.my/index.php/intrest/article/view/87

Khan, A., Chen, C. C., Suanpong, K., Ruangkanjanases, A., Kittikowit, S., & Chen, S. C. (2021). The impact of CSR on sustainable innovation ambidexterity: The mediating role of sustainable supply chain management and second-order social capital. *Sustainability (Switzerland)*, *13*(21). https://doi.org/10.3390/su132112160

Kung, C. C., Lan, X., Yang, Y., Kung, S. S., & Chang, M. S. (2022). Effects of green bonds on Taiwan's bioenergy development. *Energy*, *238*, 121567. https://doi.org/10.1016/j.energy.2021.121567

Lee, T. H., Jan, F. H., & Liu, J. T. (2021). Developing an indicator framework for assessing sustainable tourism: Evidence from a Taiwan ecological resort. *Ecological Indicators*, *125*, 107596. https://doi.org/10.1016/j.ecolind.2021.107596

Li, Z., Jin, M., & Cheng, J. (2021). Economic growth of green agriculture and its influencing factors in China: Based on emergy theory and spatial econometric model. *Environment, Development and Sustainability*, *23*(10), 15494–15512. https://doi.org/10.1007/s10668-021-01307-1

Putri, A. A. M. N., Lestari, A. S., Wijaya, N. P., & Alfaizi, R. (2024). Implikasi Kebijakan Penjualan Microchip Taiwan Dalam Pasar Ekonomi Global: Analisis Poliheuristik. *POLITEIA: Jurnal Ilmu Politik*, *16*(2), 74–81.

Risa, Y. (2020). Pemberdayaan Masyarakat Sebagai Wujud Tanggung Jawab Sosial Perusahaan (Studi Kasus Pt. Tirta Investama Plant Solok) Di Nagari Batang Barus Kabupaten Solok. *Jurnal Usm Law Review*, *3*(1), 199. https://doi.org/10.26623/julr.v3i1.2378

Shidiq, A. F., Purnama, F. H., Raharjo, S. T., Humaedi, S. (2020). Strategi Tanggung Jawab Sosial Starbuck Dalam Mewujudkan Lingkungan Yang Berkelanjutan. *Pros. Penelit. dan Pengabdi. Kpd. Masy*, *6*, 192–201, https://doi.org/10.24198/jppm.v6i3.26207

Siregar, F. (2023). Pengembangan Sistem Petanian Berkelanjutan Untuk Mencapai Keberlanjutan Pangan. *Jurnal Universitas Medan Area*, 1–11.

Sulandjari, K., & Suparwata, D. O. (2023). Inovasi Dalam Pertanian Berkelanjutan di Kabupaten Sukabumi: Dari Pertanian Vertikal hingga Bioinformatika Pertanian Mewujudkan Sumber Pangan yang Lebih Berkelanjutan. *Jurnal Multidisiplin West Science*, *2*(09), 780–789. https://doi.org/10.58812/jmws.v2i09.667

Tsai, W. T. (2020). Turning food waste into value-added resources: Current status and regulatory promotion in Taiwan. *Resources*, *9*(5). https://doi.org/10.3390/RESOURCES9050053

Tseng, M. L., Tan, P. A., Jeng, S. Y., Lin, C. W. R., Negash, Y. T., & Darsono, S. N. A. C. (2019). Sustainable investment: Interrelated among corporate governance, economic performance and market risks using investor preference approach. *Sustainability (Switzerland)*, *11*(7), 1–15. https://doi.org/10.3390/su11072108

Van Hong, P., Nguyen, N. T., Huy, D. T. N., Thuy, N. T., & Huong, L. T. T. (2021). Evaluating several models of quality management and impacts on lychee price applying for

Vietnam agriculture products value chain sustainable development. *Alinteri Journal of Agricultural Sciences*, *36*(1), 122–130. https://doi.org/10.47059/alinteri/v36i1/ajas21018

Vrabcová, P., & Urbancová, H. (2021). Approaches of selected organisations in the Czech Republic to promoting the concept of sustainable development and corporate social responsibility. *Agricultural Economics (Czech Republic)*, *67*(7), 255–265. https://doi.org/10.17221/8/2021-AGRICECON

Wang, L., Su, C. W., Ali, S., & Chang, H. L. (2020). How China is fostering sustainable growth: The interplay of green investment and production-based emission. *Environmental Science and Pollution Research*, *27*(31), 39607–39618. https://doi.org/10.1007/s11356-020-09933-4

Wang, Y., Xie, L., Zhang, Y., Wang, C., & Yu, K. (2019). Does FDI promote or inhibit the high-quality development of agriculture in China? An agricultural GTFP perspective. *Sustainability (Switzerland)*, *11*(17), 1–19. https://doi.org/10.3390/su11174620

10 Analysis of Drivers of Industry 5.0 Using MCDM Tool

A Case Study

Ganesan Muruganantham, S. Vinodh, Saumendra Das, and Udaya Sankar Patro

10.1 Introduction

The manufacturing industry is going through a big change. It is moving from an efficiency-driven view of Industry 4.0 to a more human-centered and value-driven one of Industry 5.0. Unlike its predecessor, which focused largely on automation and smart technologies, Industry 5.0 places people back in the spotlight. It emphasizes collaboration between humans and intelligent technology, hence facilitating a more equitable and significant relationship with robots (Agerskans et al., 2025; Plooy et al., 2024). The technique is about leveraging technology to help individuals be more creative, comprehend things better, and make better choices.

At the same time, there is an increasing need to address problems with the environment. Manufacturers all around world are now considering sustainability as the top priority because of strict restrictions, limited resources, and climate change. Companies are working hard to reduce waste, minimize pollution, and enable effective resource utilization (Mulongo, 2024). Eco-innovation is very essential because it helps firms create greener technologies and processes that balance their financial aims with their environmental obligations (Majid et al., 2023; Santhi & Muthuswamy, 2023). These forward-thinking projects are good for the planet and make businesses more competitive and adaptable in a world that is changing quickly.

Industry 5.0 presents a special chance to unite all these objectives. Companies can now center sustainability in their operations using smart systems and modern digital tools (Bhat & Parvez, 2024; Ghobakhloo et al., 2022). Manufacturers are finding the ideal mix of performance and environmental care by means of technologies offering real-time data, predictive insights, and improved decision-making (Adel, 2022). Thanks to digital innovation, ideas like circular economy, where resources are reused and recycled instead of becoming disposable waste, are increasingly realistic. Industry 5.0 is in many dimensions aiming to reconsider what manufacturing should represent.

Adopting Industry 5.0, however, involves understanding the true drivers of effective change, beyond simply implementing new technology. Leadership support, workforce readiness, environmental goals, and consumer expectations all have roles (Atif, 2023; Bhat & Parvez, 2024; Dacre et al., 2024; Santhi & Muthuswamy, 2023). Researchers have investigated some of these characteristics, but there is

DOI: 10.4324/9781003647621-13

still no clear picture of which ones have the biggest effect on a complicated business like manufacturing, where people, processes, and technology need to work together.

As people start to agree on the basic rationale behind Industry 5.0, a closer look shows that there are disagreements on how to put the technologies into practice and which pillars should be given the most importance. For instance, some researchers point out how digital technologies like artificial intelligence AI and Internet of Things (IoT) could change things by making them more efficient and encouraging new ideas (Mulongo, 2024). Others, on the other hand, caution against focusing only on technology, saying that the real success of Industry 5.0 depends on carefully combining human elements and placing well-being ahead of productivity (Santhi & Muthuswamy, 2023). Also, when people discuss on implementing Industry 5.0, they often have different ideas about the most important enablers. Some studies focus on technological infrastructure, while others underline the relevance of organizational culture and retraining the workforce. It is still hard to find a balance among technical progress, better human roles, and ambitious environmental goals. This shows how important it is to have frameworks that can fully assess these interconnected areas.

The present study addresses critical factors influencing the transition to Industry 5.0, with a particular emphasis on eco-innovation and green manufacturing. Employing multi-criteria decision-making (MCDM) technique, these drivers are assessed and ranked to ascertain their practical impact. The objective is to equip industry leaders, researchers, and policymakers with the required insights to make informed decisions, supporting a transition toward a future that is technologically advanced, sustainable, and centered on human values. This study is guided by the conceptual framework illustrated in Figure 10.1, which maps the relationships between key Industry 5.0 drivers and its desired outcomes.

There is a lot to explore in Industry 5.0 in both business and academia, but there is still a big gap in the literature when it comes to a systematic, data-driven framework for fully assessing and prioritizing its many adoption drivers, especially when it comes to human-centered and sustainability outcomes. The main goals of this study are to (1) find and group the main factors that affect the adoption of Industry 5.0 in the industry sector and (2) apply the TOPSIS method to create and use a MCDM framework to objectively evaluate and rank these factors.

A conceptual framework illustrating Industry 5.0. On the left, four drivers are listed: technology-driven, human-centric, resilience, and sustainability, all leading toward the central concept of Industry 5.0. On the right, four outcomes are shown: personalization, circular economy, social well-being, and human–robot collaboration.

This chapter outlines factors influencing Industry 5.0 adoption. Section 10.2 reviews the literature on theoretical foundations, technologies, sustainability, and human-centric aspects. Section 10.3 details the research methodology using TOPSIS-based MCDM. Section 10.4 presents a case study evaluating the drivers through expert inputs and modeling. Section 10.5 discusses the findings and implications. Section 10.6 concludes with the contributions, limitations, and future research directions.

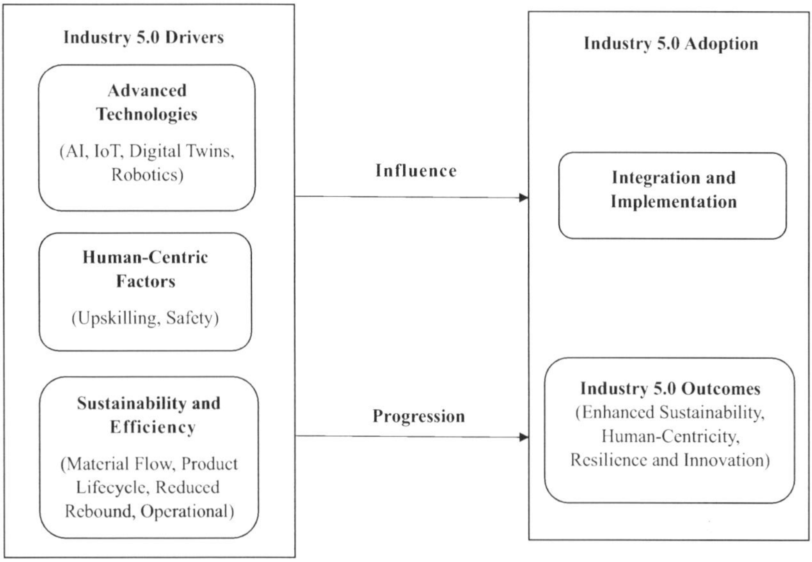

Figure 10.1 Conceptual framework of Industry 5.0 drivers and outcomes.

10.2 Review of Literature

The shift from Industry 4.0 to Industry 5.0 marks a change that goes beyond typical technological growth (Mulongo, 2024; Zaidi et al., 2024). Industry 4.0 had smart machines, real-time data, and systems that could coordinate with each other, but it didn't always pay attention to people. Industry 5.0 seeks to correct the present challenges faced by the industries. Moktadir et al. (2024) clarified that the subsequent phase goes beyond mere operational acceleration or data aggregation. It stresses the need to link new ideas to people's needs, social aspirations, and health of the planet.

The idea behind Industry 5.0 is that technology should be used to serve people, not the other way around. Collaborative robots (cobots), smart interfaces, and systems that place people first are all meant to aid people with new ideas, solving problems, and so on (Demir et al., 2019). Zafar et al. (2024) explained that when workplaces are built with empathy and respect for people's capacities, people are happier and more innovative. The concept looks forward to a time when machines and people work together to get things done (Ghobakhloo et al., 2021).

Stakeholders are starting to see sustainability as more than just a procedural formality; it is becoming a key idea that guides how people make decisions today. Industry 5.0 needs to help solve the unprecedented environmental concerns that are hurting our planet. Aggarwal et al. (2022) stressed that industrial processes are relying more on ideas like circular economy, with the focus on better resource utilization: making things that can be used again, and regenerative systems can assist save money and make sure that things last for a long time.

Akundi et al. (2022) said that modularity helps companies make goods that are good for the environment, flexible, and can be upgraded. It boosts consumer value while promoting sustainability and reducing waste. For example, consider a mobile phone designed for modular upgrades, extending its usability well beyond the typical 2-year cycle. It depicts wonderful examples of the possibilities in design motivated by both long-term strategic planning and creative inspiration. These developments are made possible in great part by emerging technologies. In addition, AI, extended reality (XR), and IoT now essentially revolutionize daily industrial processes. Chu and Pan (2024) addressed the use of XR tools in remote problem resolution, design simulation, and people training. These technologies improve personal information and involvement levels by creating safer and more intelligent surroundings (Ghobakhloo et al., 2021). While human supervision is still vital, AI solutions help ensure better decision-making processes by means of supply chain risk prediction and process optimization.

Moktadir et al. (2025) suggested printed electronics as an eco-friendly option to traditional systems for saving energy and reducing trash. Santhi and Muthuswamy (2023) showed that AI can estimate the amount of energy that will be needed in real time, hence making industrial operations more sustainable and efficient. These adjustments are not simply small steps; they're huge steps toward making industry less destructive to the environment and more cost-effective. It is vital to admit that Industry 4.0 had certain problems. As Awouda et al. (2024) pointed out, even if it improved efficiency, it usually neglected the human component of industrial operations. Not regularly addressed were areas such as mental health, inclusiveness, and ethical behavior. Rising to these shortcomings, Industry 5.0 advocates a reorientation of business toward human dignity and society benefit. The paradigm shift highlights the change in viewing workers as mere resources to acknowledging them as essential collaborators in innovation.

Three basic dimensions, namely, advanced technology, sustainability imperatives, and a clear human-centric focus, help us move toward Industry 5.0. Technology innovations such as cobots, digital twins, AI, machine learning, IoT, and cyber–physical systems (CPSs) help shape more intelligent and flexible manufacturing ecosystems (Adel, 2022; Ivanov, 2023). In terms of sustainability, aspects such as lowered rebound effects, material flow efficiency, and extended product lifecycles are crucial in lessening environmental impact and therefore supporting circular economy practices (Ghobakhloo et al., 2022, 2023). Central to Industry 5.0 is a human-centric strategy that stresses employee empowerment, upskill, occupational safety, job happiness, and operational efficiency, thereby ensuring that technical developments are inclusive and significant (Ghobakhloo et al., 2021). Table 10.1 presents the drivers of Industry 5.0 systematically and the several yet linked forces.

Given the numerous dynamic components such as technology, human factors, environmental considerations, and ethical implications, it is evident that a systematic approach to decision-making is essential. MCDM approaches, like the Technique for Order of Preference by Similarity to Ideal Solution (TOPSIS), efficiently fulfil the objective. These methods help businesses and researchers deal

Table 10.1 Drivers of Industry 5.0.

Sl. No.	*Drivers of Industry 5.0*	*References*
1	Cobots	Demir et al. (2019), Mulongo (2024)
2	Digital twins	Adel (2022), Bhat & Parvez (2024)
3	AI	Ivanov (2023)
4	Machine learning	Ivanov (2023)
5	IoT	Ivanov (2023)
6	CPSs	Ivanov (2023)
7	Big data analytics	Adel (2022)
8	Industrial robots and automation	Adel (2022), Mulongo (2024), Zaidi et al. (2024)
9	Smart materials	Ghobakhloo (2020), Turner et al. (2022)
10	Smart manufacturing	Ghobakhloo (2020), Turner et al. (2022)
11	Reduced rebound effect	Ghobakhloo et al. (2021, 2023)
12	Material flow efficiency	Ghobakhloo et al. (2022)
13	Prolonged product lifecycle	Ghobakhloo et al. (2022)
14	Employee empowerment	Ghobakhloo et al., (2021)
15	Upskilling	Ghobakhloo et al. (2021)
16	Workplace safety	Ghobakhloo et al. (2021)
17	Job satisfaction	Ghobakhloo et al. (2021)
18	Operational efficiency	Ghobakhloo et al. (2021)

Source: Compiled by Authors.

with complicated situations by analyzing multiple factors, comparing different scenarios, and figuring out the best way forward.

Industry 5.0 research mostly focuses on human-centered methods, sustainability, and the use of cutting-edge technology, but it does not focus enough at the aspects that affect how these ideas can be put into practice. This study fills up this research gap by ranking 18 drivers across seven areas using the TOPSIS MCDM approach. TOPSIS is a data-driven framework for organizations to identify the key drivers of Industry 5.0. It does this using expert opinion and a wide range of data to look at complex human, technological, and environmental factors that make up Industry 5.0. Section 10.3 outlines the methodology employed in this study, drawing insights from the literature. This involves the application of the TOPSIS-based MCDM approach to evaluate and rank the key drivers of Industry 5.0.

10.3 Methodology

The TOPSIS technique is used in this study to systematically evaluate and rank the most important factors that affect the adoption of Industry 5.0. The evaluation is based on seven important criteria: resilience, resource consumption, environmental impact, technology utilization, creative and innovative skills, new business opportunities, and product and process innovation. These were chosen because they are important to the main goals of Industry 5.0, which are to promote sustainability, technological progress, and innovation that place people at the center. Eighteen drivers are considered, including AI, IoT, digital twins, and enablers like human–robot collaboration and employee upskilling. Three industry experts rated every driver across all seven criteria using a specified number scale. They were chosen because they had a lot of professional experience and had directly worked on implementing industrial transformation technologies that are in line with Industry 4.0 and 5.0. Each specialist had been working in manufacturing, industrial automation, or technological consulting for more than 10 years. Their background gave them a comprehensive awareness of both the drivers and evaluation criteria, which led to evaluations that were based on real-world situations and practical knowledge. Each expert assigned the drivers a score on all parameters using a defined scoring scale. These individual assessments were averaged to form a consolidated choice matrix. Then, vector normalization was used to fix any differences in scale and ensure that the comparison analysis was consistent.

Reflecting their respective significance in Industry 5.0 environment, weights were allocated to each criterion following normalization depending on either expert consensus or relevant literature. The seven evaluation criteria were identified through a literature review of Industry 5.0, sustainability, and technological innovation, and then validated by domain experts. The final weights were determined through expert consensus for TOPSIS analysis: [0.15, 0.15, 0.15, 0.15, 0.15, 0.10, 0.15], corresponding to resilience, resource consumption, environmental impacts, technology utilization, creative and innovative skills, new business opportunities, and product and process innovation.

The weighted normalized decision matrix was obtained by multiplying normalized values by these weights. The ideal solution that performs best on every criterion and the negative-ideal solution that performs worst were found by means of this matrix. Next, Euclidean distances were calculated for every alternative from both ideal and negative-ideal solutions. The relative closeness of every driver to the optimal solution was determined using these distances, which finally helped rank the drivers. The alternatives were sorted in declining order of their closeness coefficients, therefore demonstrating their priority and impact in guiding Industry 5.0 transition. A methodologically rigorous, expert-informed strategy guarantees analytically sound rankings that are helpful in fostering industrial innovation and informing successful governmental decisions.

10.3.1 TOPSIS Methodology

TOPSIS approach is used to rank the main forces of Industry 5.0 based on expert inputs. The steps are described as follows:

Step 1: Assignment of Ratings to Drivers and Criteria

Considering the inherent subjectivity associated with expert-based evaluation, this study employs an aggregation strategy by calculating the average of the scores assigned by the three domain experts for each driver across all criteria. This approach facilitates the capture of a consensus perspective and mitigates potential bias of individual evaluators, thereby providing a more reliable basis for analysis.

Let $A = \{A_1, A_2, \ldots, A_m\}$ be the set of *alternatives* (*drivers*) and

$C = \{C_1, C_2, \ldots, C_n\}$ be the set of *evaluation criteria*.

Each expert provides a score $x_{ij}^{(k)}$ for alternative A_i with respect to criterion C_j, where

$k = 1, 2, 3, \ldots K$ (number of experts). The average rating is computed as

$$x_{ij} = \frac{1}{K}\sum_{k=1}^{K} x_{ij}^{(k)}$$

This results in the aggregated decision matrix:

$$X = \begin{bmatrix} x_{11} & x_{12} & x_{13} & x_{14} \\ x_{21} & x_{22} & x_{23} & x_{24} \\ \vdots & \vdots & \ddots & \vdots \\ x_{m1} & x_{m2} & \cdots & x_{mn} \end{bmatrix}$$

Step 2: Construction of the Decision Matrix

The decision matrix X captures the aggregated expert scores for all alternatives against each criterion.

Step 3: Identify Maximum and Minimum Values Based on the Nature of the Criterion

The criteria are classified as

- *Benefit criteria* (to be maximized)
- *Cost criteria* (to be minimized)

For each criterion C_j:

$$x_j^{max} = \max_i \left(x_{ij}\right)$$

$$x_j^{min} = \min_i \left(x_{ij}\right)$$

Step 4: Normalization of the Decision Matrix

Normalization is a mathematical procedure that converts the scores of various criteria into a standardized scale, typically ranging from 0 to 1. This process ensures that no single criterion exerts an undue influence on the outcome due to its larger numerical values. In this study, vector normalization was employed to maintain comparability across all drivers and criteria, thereby establishing an equitable basis for evaluation.

Criteria are normalized by

$$r_{ij} = \frac{x_{ij}}{\sum_{j=1}^{m} x_{ij}}$$

The resulting *normalized matrix* $R = \left[r_{ij} \right]$ is

$$R = \begin{bmatrix} r_{11} & r_{12} & \cdots & r_{1n} \\ r_{21} & r_{22} & \cdots & r_{2n} \\ \vdots & \vdots & \ddots & \vdots \\ r_{m1} & r_{m2} & \cdots & r_{mn} \end{bmatrix}$$

Step 5: Construction of the Weighted Normalized Decision Matrix

Let $W = \left[w_1, w_2, \ldots, w_n \right]$ be the weight vector for criteria, derived by averaging expert-assigned weights. Each element of the weighted normalized matrix $V = \left[v_{ij} \right]$ is computed as

$$v_{ij} = w_j \cdot r_{ij},$$

resulting in

$$V = \begin{bmatrix} v_{11} & v_{12} & \cdots & v_{1n} \\ r_{21} & r_{22} & \cdots & v_{2n} \\ \vdots & \vdots & \ddots & \vdots \\ r_{m1} & r_{m2} & \cdots & r_{mn} \end{bmatrix}.$$

Step 6: Determination of Ideal and Negative-Ideal Solutions

Positive ideal solution (PIS):

$$V^{+} = \left\{ max\left(v_{ij} \right) \mid j \in J_{benefit}; min\left(v_{ij} \right) \mid j \in J_{cost} \right\}$$

Negative ideal solution (NIS):

$$V^{-} = \left\{ \min\left(v_{ij} \right) \mid j \in J_{benefit}; \max\left(v_{ij} \right) \mid j \in J_{cost} \right\}$$

Step 7: Separation Measures from Ideal and Negative-Ideal Solutions

Euclidean distance of each alternative A_i from the PIS and NIS is calculated as

$$S_i^{+} = \sqrt{\sum_{j=1}^{n} \left(v_{ij} - v_j^{+} \right)^2}, S_i^{-} = \sqrt{\sum_{j=1}^{n} \left(v_{ij} - v_j^{-} \right)^2}$$

Step 8: Calculation of Relative Closeness Coefficient

The relative closeness to the ideal solution for each alternative is given by

$$C_i = \frac{S_i^{-}}{S_i^{+} + S_i^{-}}, 0 \le C_i \le 1$$

Step 9: Ranking of Alternatives

The alternatives (drivers) are ranked in descending order of C_i. A higher value of C_i indicates a greater preference and priority in the context of Industry 5.0 adoption. Building upon the established methodological framework, Section 10.4 presents a case study, illustrating the practical implementation of the TOPSIS method to evaluate and rank identified drivers of Industry 5.0.

10.4 Case Study

Table 10.1 summarizes the key drivers that will assist in the shift to Industry 5.0. This is the basis for the empirical evaluation in this case study. A total of 18 drivers were identified by focusing on both academic research and the real-world business scenario. The alternatives include technological progress (AI, machine learning, IoT, and digital twins), environmental concerns (material flow efficiency and reduced rebound effect), and human-centered factors (employee empowerment, upskilling, and workplace safety). This set of drivers shows the main idea of Industry 5.0: innovation that is socially inclusive, ecologically friendly, and technologically advanced.

Step 1: Assignment of Ratings to Drivers and Criteria

Cobots, digital twins, AI, machine learning, the IoT, CPSs, big data analytics, industrial robots and automation, smart materials, smart manufacturing, reduced rebound effect, material flow efficiency, prolonged product lifecycle, employee empowerment, upskilling, workplace safety, job satisfaction, and operational efficiency are *alternatives* (*drivers*), and resilience, resource consumption, environmental impacts, technology utilization, creative and innovative skills, new business opportunities, and product and process innovation are the *evaluation criteria*.

Each expert provides a score $x_{ij}^{(k)}$ for alternative A_i with respect to criterion C_j, where

$k = 1, 2, 3, \ldots, K$ (number of experts). The average rating is computed as

$$x_{ij} = \frac{1}{K}\sum_{k=1}^{K} x_{ij}^{(k)}.$$

This results in the aggregated decision matrix

$$X = \begin{bmatrix} x_{11} & x_{12} & x_{13} & x_{14} \\ x_{21} & x_{22} & x_{23} & x_{24} \\ \vdots & \vdots & \ddots & \vdots \\ x_{m1} & x_{m2} & \cdots & x_{mn} \end{bmatrix}.$$

Step 2: Construction of the Decision Matrix

The decision matrix X captures the aggregated expert scores for all alternatives against each criterion. The construction of the decision matrix is presented in Table 10.2.

Table 10.2 Aggregated Decision Matrix.

Criterion / *Driver*	*Resilience*	*Resource Consumption*	*Environmental Impacts*	*Technology Utilization*	*Creative and Innovative Skills*	*New Business Opportunities*	*Product and Process Innovation*
D1	7.33	6.33	6.67	8.67	8.67	8.67	8.67
D2	7.67	7.33	7.33	8.67	8.00	8.67	8.00
D3	7.67	7.67	8.33	7.67	8.67	7.67	8.00
D4	7.67	8.33	7.67	8.67	8.33	7.67	8.67
D5	7.33	8.33	8.33	7.67	8.33	8.33	8.00
D6	6.67	6.00	6.33	8.67	7.67	8.33	7.67
D7	7.33	7.33	7.00	7.33	8.33	8.67	8.33
D8	6.67	8.00	8.00	7.67	8.33	7.67	8.00
D9	7.33	8.33	8.00	8.00	7.00	7.33	8.33
D10	7.67	8.00	7.67	9.00	7.67	6.67	8.00
D11	7.67	8.00	7.67	6.67	8.33	7.33	7.67
D12	7.33	7.00	8.33	7.33	8.67	7.67	8.67
D13	6.67	7.33	7.67	7.67	7.67	8.33	8.33
D14	6.67	7.33	7.67	7.67	8.00	7.67	8.00
D15	6.67	7.33	7.67	8.67	7.33	8.00	8.00
D16	6.00	6.67	6.67	6.67	7.33	7.33	7.67
D17	7.00	7.33	7.00	8.33	8.33	7.33	8.00
D18	7.67	7.67	7.67	8.00	8.33	7.67	7.33

Source: Compiled by Authors.

Table 10.3 Benefit or Cost Criteria.

Criterion	*Type*	*Ideal = Max or Min*	*Ideal Value*	*Anti-Ideal Value*
Resilience	Benefit	Max	7.67	6
Resource consumption	Cost	Min	6	8.34
Environmental impacts	Benefit	Max	8.34	6.33
Technology utilization	Cost	Min	6.67	9
Creative and innovative skills	Benefit	Max	8.66	7
New business opportunities	Benefit	Max	8.66	6.67
Product and process innovation	Benefit	Max	8.66	7.33

Source: Compiled by authors.

Step 3: Benefit or Cost Criteria
The criteria are classified as

- *Benefit criteria* (to be maximized)
- *Cost criteria* (to be minimized)

For each criterion C_j:

$$x_j^{max} = \max_i \left(x_{ij}\right)$$

$$x_j^{min} = \min_i \left(x_{ij}\right)$$

Table 10.3 shows the criteria for costs and benefits that were considered in the analysis.

Step 4: Vector Normalization
Normalization is carried out by

$$r_{ij} = \frac{x_{ij}}{\sum_{j=1}^{m} x_{ij}}.$$

The resulting *normalized matrix* $R = \left[r_{ij}\right]$ is

$$R = \begin{bmatrix} r_{11} & r_{12} & \cdots & r_{1n} \\ r_{21} & r_{22} & \cdots & r_{2n} \\ \vdots & \vdots & \ddots & \vdots \\ r_{m1} & r_{m2} & \cdots & r_{mn} \end{bmatrix}.$$

Table 10.4 presents the vector normalization of the decision matrix.

Table 10.4 Normalized Decision Matrix.

Driver	*Resilience*	*Resource Consumption*	*Environmental Impacts*	*Technology Utilization*	*Creative and Innovative Skills*	*New Business Opportunities*	*Product and Process Innovation*
Cobots	0.057	0.047	0.049	0.061	0.060	0.061	0.060
Digital twins	0.059	0.055	0.054	0.061	0.055	0.061	0.055
AI	0.059	0.057	0.061	0.054	0.060	0.054	0.055
Machine learning	0.059	0.062	0.057	0.061	0.057	0.054	0.060
IoT	0.057	0.062	0.061	0.054	0.057	0.059	0.055
CPSs	0.052	0.045	0.047	0.061	0.053	0.059	0.053
Big data analytics	0.057	0.055	0.052	0.051	0.057	0.061	0.057
Industrial robots and automation	0.052	0.060	0.059	0.054	0.057	0.054	0.055
Smart materials	0.057	0.062	0.059	0.056	0.048	0.052	0.057
Smart manufacturing	0.059	0.060	0.057	0.063	0.053	0.047	0.055
Reduced rebound effect	0.059	0.060	0.057	0.047	0.057	0.052	0.053
Material flow efficiency	0.057	0.052	0.061	0.051	0.060	0.054	0.060
Prolonged product lifecycle	0.052	0.055	0.057	0.054	0.053	0.059	0.057
Employee empowerment	0.052	0.055	0.057	0.054	0.055	0.054	0.055
Upskilling	0.052	0.055	0.057	0.061	0.051	0.057	0.055
Workplace safety	0.047	0.050	0.049	0.047	0.051	0.052	0.053
Job satisfaction	0.054	0.055	0.052	0.058	0.057	0.052	0.055
Operational efficiency	0.059	0.057	0.057	0.056	0.057	0.054	0.050

Source: Compiled by authors.

Step 5: Weighted Normalized Matrix
Let W = [0.15,0.15,0.15,0.15,0.15,0.1,0.15] be the weight vector for criteria, derived by averaging expert-assigned weights. Each element of the weighted normalized matrix $V = \left[v_{ij}\right]$ is computed as

$$v_{ij} = w_j \cdot r_{ij}.$$

This results in

$$V = \begin{bmatrix} v_{11} & v_{12} & \cdots & v_{1n} \\ r_{21} & r_{22} & \cdots & v_{2n} \\ \vdots & \vdots & \ddots & \vdots \\ r_{m1} & r_{m2} & \cdots & r_{mn} \end{bmatrix}.$$

Table 10.5 presents a weighted normalized matrix derived from the normalized decision matrix and corresponding criteria weights.

Step 6: Determination of Ideal and Non-Ideal Solutions
Positive ideal solution (PIS):

$$V^+ = \left\{max\left(v_{ij}\right) \mid j \in J_{benefit}; min\left(v_{ij}\right) \mid j \in J_{cost}\right\}$$

Negative ideal solution (NIS):

$$V^- = \left\{\min\left(v_{ij}\right) \mid j \in J_{benefit}; \max\left(v_{ij}\right) \mid j \in J_{cost}\right\}$$

As shown in Table 10.6, the ideal and non-ideal solutions are derived by selecting the best and worst values for each criterion, forming the foundation for distance and closeness calculations in the subsequent steps.

Step 7: Distances from Ideal Solutions and NIS and Step 8: Calculation of Relative Closeness Coefficient
The Euclidean distance of each alternative A_i from PIS and NIS is calculated as

$$S_i^+ = \sqrt{\sum_{j=1}^{n}\left(v_{ij} - v_j^+\right)^2}, S_i^- = \sqrt{\sum_{j=1}^{n}\left(v_{ij} - v_j^-\right)^2}$$

The relative closeness to the ideal solution for each alternative is given by

$$C_i = \frac{S_i^-}{S_i^+ + S_i^-}, 0 \le C_i \le 1$$

As shown in Table 10.7, Step 7 involves calculating the Euclidean distances from the ideal solution and NIS, while Step 8 determines the relative closeness coefficient, which aids in ranking the alternatives.

Table 10.5 Weighted Normalized Matrix.

Driver	*Resilience*	*Resource Consumption*	*Environmental Impacts*	*Technology Utilization*	*Creative and Innovative Skills*	*New Business Opportunities*	*Product and Process Innovation*
Cobots	0.0085	0.0071	0.0074	0.0091	0.0090	0.0061	0.0089
Digital twins	0.0089	0.0082	0.0081	0.0091	0.0083	0.0061	0.0083
AI	0.0089	0.0086	0.0092	0.0080	0.0090	0.0054	0.0083
Machine learning	0.0089	0.0093	0.0085	0.0091	0.0086	0.0054	0.0089
IoT	0.0085	0.0093	0.0092	0.0080	0.0086	0.0059	0.0083
CPSs	0.0078	0.0067	0.0070	0.0091	0.0079	0.0059	0.0079
Big data analytics	0.0085	0.0082	0.0077	0.0077	0.0086	0.0061	0.0086
Industrial robots and automation	0.0078	0.0089	0.0088	0.0080	0.0086	0.0054	0.0083
Smart materials	0.0085	0.0093	0.0088	0.0084	0.0072	0.0052	0.0086
Smart manufacturing	0.0089	0.0089	0.0085	0.0094	0.0079	0.0047	0.0083
Reduced rebound effect	0.0089	0.0089	0.0085	0.0070	0.0086	0.0052	0.0079
Material flow efficiency	0.0085	0.0078	0.0092	0.0077	0.0090	0.0054	0.0089
Prolonged product lifecycle	0.0078	0.0082	0.0085	0.0080	0.0079	0.0059	0.0086
Employee empowerment	0.0078	0.0082	0.0085	0.0080	0.0083	0.0054	0.0083
Upskilling	0.0078	0.0082	0.0085	0.0091	0.0076	0.0057	0.0083
Workplace safety	0.0070	0.0074	0.0074	0.0070	0.0076	0.0052	0.0079
Job satisfaction	0.0081	0.0082	0.0077	0.0087	0.0086	0.0052	0.0083
Operational efficiency	0.0089	0.0086	0.0085	0.0084	0.0086	0.0054	0.0076

Source: Compiled by authors.

Table 10.6 Determination of PIS and NIS.

	Resilience	*Resource Consumption*	*Environmental Impacts*	*Technology Utilization*	*Creative and Innovative Skills*	*New Business Opportunities*	*Product and Process Innovation*
PIS	0.0089	0.0093	0.0092	0.0094	0.0090	0.0061	0.0089
NIS	0.0070	0.0067	0.0070	0.0070	0.0072	0.0047	0.0076

Source: Compiled by authors.

Step 9: Final Rankings based on Closeness coefficient (C_i)

The alternatives (drivers) are ranked in the descending order of C_i. A higher value of C_i indicates a greater preference and priority in the context of Industry 5.0 adoption. The final prioritization of alternatives, determined through the relative closeness coefficients, is presented in Table 10.8.

Based on the results obtained from the case study, a detailed discussion is presented in Section 10.5, interpreting the findings and highlighting their theoretical and practical implications.

10.5 Discussions

The analysis of the drivers for Industry 5.0 adoption provides significant insight into the current goals and obstacles within sustainable manufacturing. Digital technologies such as machine learning and IoT are becoming increasingly important, which shows that the industry is highly interested in smart automation and making decisions based on data (Ivanov, 2023). These technologies are considered essential for improving the efficiency, flexibility, and competitiveness of modern production systems. However, intermediate rating of employee empowerment and upskilling shows that while human-centered principles are seen as important to Industry 5.0, they may not yet be fully integrated into implementation plans (Ghobakhloo et al., 2021).

This indicates a divergence between theoretical focus on human–machine collaboration and its practical implementation on the manufacturing floor. Additionally, aspects like workplace safety and CPSs, which are placed lower, can be seen as normal attributes of Industry 4.0, These findings illustrate a prevailing trend: organizations often choose quick technology improvements over long-term investments in human capital and sustainability. The low rankings of employee empowerment and workplace safety are surprising, given their importance in Industry 5.0. This could reflect how many industries still prioritize technology over people in practice. It may also suggest a gap between theory and practice. These results highlight the need to rethink how human-centric factors are evaluated and call for future research to explore this disconnect more closely. The findings also raise significant enquiries about the potential shifts in priority across different industries, locations, or degrees of digital maturity. The present study offers a significant foundation;

Table 10.7 Distances from PIS and NIS and Calculation of Relative Closeness Coefficient.

Driver	*Resilience*	*Resource Consumption*	*Environmental Impacts*	*Technology Utilization*	*Creative and Innovative Skills*	*New Business Opportunities*	*Product and Process Innovation*	*Distances to Ideal (S+)*	*Distance to NIS (S-)*	*Closeness Coefficient (Ci)*
Cobots	0.0085	0.0071	0.0074	0.0091	0.0090	0.0061	0.0089	0.0029	0.0037	0.559
Digital twins	0.0089	0.0082	0.0081	0.0091	0.0083	0.0061	0.0083	0.0019	0.0039	0.674
AI	0.0089	0.0086	0.0092	0.0080	0.0090	0.0054	0.0083	0.0019	0.0041	0.689
Machine learning	0.0089	0.0093	0.0085	0.0091	0.0086	0.0054	0.0089	0.0011	0.0046	0.803
IoT	0.0085	0.0093	0.0092	0.0080	0.0086	0.0059	0.0083	0.0017	0.0044	0.724
CPSs	0.0078	0.0067	0.0070	0.0091	0.0079	0.0059	0.0079	0.0039	0.0026	0.403
Big data analytics	0.0085	0.0082	0.0077	0.0077	0.0086	0.0061	0.0086	0.0026	0.0033	0.554
Industrial robots and automation	0.0078	0.0089	0.0088	0.0080	0.0086	0.0054	0.0083	0.0022	0.0036	0.625
Smart materials	0.0085	0.0093	0.0088	0.0084	0.0072	0.0052	0.0086	0.0023	0.0040	0.632
Smart manufacturing	0.0089	0.0089	0.0085	0.0094	0.0079	0.0047	0.0083	0.0021	0.0042	0.672
Reduced rebound effect	0.0089	0.0089	0.0085	0.0070	0.0086	0.0052	0.0079	0.0030	0.0036	0.551
Material flow efficiency	0.0085	0.0078	0.0092	0.0077	0.0090	0.0054	0.0089	0.0024	0.0038	0.609
Prolonged product lifecycle	0.0078	0.0082	0.0085	0.0080	0.0079	0.0059	0.0086	0.0025	0.0030	0.544
Employee empowerment	0.0078	0.0082	0.0085	0.0080	0.0083	0.0054	0.0083	0.0026	0.0029	0.527
Upskilling	0.0078	0.0082	0.0085	0.0091	0.0076	0.0057	0.0083	0.0024	0.0033	0.577
Workplace safety	0.0070	0.0074	0.0074	0.0070	0.0076	0.0052	0.0079	0.0045	0.0011	0.192
Job satisfaction	0.0081	0.0082	0.0077	0.0087	0.0086	0.0052	0.0083	0.0024	0.0031	0.561
Operational efficiency	0.0089	0.0086	0.0085	0.0084	0.0086	0.0054	0.0076	0.0022	0.0037	0.631

Source: Compiled by authors.

Table 10.8 Rankings Based on Closeness Coefficient.

Driver	*Closeness Coefficient (Ci)*	*Rank*
Machine learning	0.803	1
IoT	0.724	2
AI	0.689	3
Digital wins	0.674	4
Smart manufacturing	0.672	5
Smart materials	0.632	6
Operational efficiency	0.631	7
Industrial robots and automation	0.625	8
Material flow efficiency	0.609	9
Upskilling	0.577	10
Job satisfaction	0.561	11
Cobots	0.559	12
Big data analytics	0.554	13
Reduced rebound effect	0.551	14
Prolonged product lifecycle	0.544	15
Employee empowerment	0.527	16
CPSs	0.403	17
Workplace safety	0.192	18

Source: Compiled by authors.

nevertheless, further research is essential to examine how various stakeholders perceive and react to the changing requirements of Industry 5.0.

10.5.1 Theoretical Implications

This research enhances Industry 5.0 theory by offering a systematic, prioritized perspective on essential adoption variables in sustainable manufacturing. The results highlight the increasing significance of sophisticated data-driven technologies, especially machine learning, IoT, and AI, in the development of the forthcoming generation of intelligent industrial systems. These drivers are in line with and support theoretical models that imply autonomy, adaptability, and digital integration as key ideas in Industry 5.0. The comparatively low emphasis on cobots and employee empowerment shows that even while human-centricity is widely accepted as a key component of Industry 5.0, its theoretical importance is still changing compared to other ideas that are more focused on technology. The low ranks of CPSs and workplace safety indicate that these elements, although fundamental, are now perceived as essential criteria rather than as strategic differentiators. Furthermore, the incorporation of sustainability-focused metrics, such as material flow efficiency and reduced rebound effect, exemplifies an increasing theoretical convergence between digital innovation and environmental concerns. The study enhances Industry 5.0 research through the application of the TOPSIS technique, facilitating comprehensive comparisons among technological, organizational, and social aspects.

While Ghobakhloo et al. (2021) established themes of digital innovation, sustainability, and human-centricity in Industry 5.0, the present study provides empirical evidence through the TOPSIS method's prioritization. Machine learning, IoT, and AI emerged as dominant factors, while employee empowerment and workplace safety received lower practical priority. This gap between theoretical aspirations and implementation provides a novel perspective and guides decision-makers in Industry 5.0 transition, helping align technology, human resources, and organizational objectives. To further enrich the discourse on human-centricity within Industry 5.0, it is imperative to examine the findings through the frameworks of human–technology co-evolution and socio-technical systems theory. These perspectives underscore that technological advancement and human adaptation are interdependent processes that evolve concurrently, influencing each other over time.

The present study findings indicate a relatively lower prioritization of traditionally human-centric factors such as employee empowerment and workplace safety, suggesting a transitional phase wherein organizations are still aligning sociocultural readiness with the rapid adoption of technology. By conceptualizing Industry 5.0 as a socio-technical system, it is recognized that meaningful innovation must encompass not only advanced tools and platforms but also values, needs, and capabilities of the individuals who engage with them. This theoretical foundation aids in interpreting the observed gaps and underscores the necessity of devising implementation strategies that harmonize digital transformation with human development.

10.5.2 Practical Implications

The results offer practical insights for industry leaders, policymakers, and decision-makers driving Industry 5.0 initiatives. The focus on machine learning, IoT, and AI shows how important it is to invest right away and keep investing in smart, adaptable technologies that help real-time optimization of production processes and make predictions. Increasing the focus on smart manufacturing and digital twins makes it even more important to use simulation-based, cyber–physical designs that improve efficiency, transparency, and responsiveness. These findings indicate that for organizations seeking to migrate to Industry 5.0 paradigms, digital infrastructure should serve as the primary focus of their implementation strategies. At the same time, mid-level drivers like employee empowerment, upskilling, and job satisfaction show the importance of ensuring the synchronization between technological progress and human capital growth.

A well-prepared and engaged workforce is essential for ensuring the long-term success of Industry 5.0 initiatives. Although workplace safety and prolonged product lifecycle are not as important, they should not be ignored in policies or operations. These findings provide policymakers a strong reason to create targeted support systems, such as incentives, subsidies for training, and changes to the rules that promote balanced and inclusive technological change. To ensure human-centric Industry 5.0 implementation, policymakers must prioritize worker well-being alongside digital transformation through national upskilling programs, workplace safety incentives, and human-focused performance metrics. Government should establish regulatory frameworks for ethical AI and robotics use, emphasizing transparency and accountability to ensure technology enhances rather than displaces human experience. The study provides a prioritized list of factors that encourage adoption, which helps stakeholders coordinate their technical work with human and environmental goals in a step-by-step, evidence-based manner. Based on key insights discussed, Section 10.6 concludes the chapter, summarizing the main findings and outlining future research directions.

10.6 Conclusions

The study uses the TOPSIS technique to systematically assess and prioritize key factors driving Industry 5.0 adoption in sustainable manufacturing. The analysis emphasizes intelligent systems, particularly machine learning, IoT, and AI, as the most significant facilitators of Industry 5.0, thereby underscoring the critical role of digital technologies in the forthcoming industrial paradigm.

The inclusion of human-centered variables such as employee empowerment, upskilling, and job satisfaction in the middle tier of the ranking underscores the significance of both technological innovation and human integration. This split is at the heart of Industry 5.0, which aims to connect digital development with the well-being of people and the environment. The research contributes to theoretical discourse by validating established concepts and introducing a novel analytical perspective, while concurrently providing practical guidance to industry leaders and policymakers. As industrial settings keep changing, the successful adoption of Industry 5.0 will need to align technology, organizations, and people in a strategic way.

10.6.1 Limitations and Future Research Directions

A limitation of this study is the small expert panel of three evaluators, although they were selected for their Industry 4.0 and 5.0 expertise. A larger, diverse panel could improve generalizability. Future research could include experts from various industries and regions to validate the prioritization of drivers. The priorities of Industry 5.0 drivers vary by location: a European tech factory might prioritize AI and automation, while a developing country firm may focus on sustainability. Future studies could examine how rankings differ across sectors and digital maturity levels. While expert judgment was valuable, case studies and secondary data from actual Industry 5.0 implementation could further validate results and provide practical insights. The low prioritization of workplace safety and employee empowerment, despite their theoretical importance, suggests a gap between strategy and implementation that warrants further qualitative research. This study excluded ethical considerations in evaluating drivers; future research should incorporate ethical frameworks with MCDM methods to ensure responsible adoption. Additionally, environmental impact assessments could be refined using lifecycle analysis to better understand drivers' ecological performance and help evaluate trade-offs between innovation and environmental responsibility.

References

Adel, A. (2022). Future of industry 5.0 in society: Human-centric solutions, challenges and prospective research areas. *Journal of Cloud Computing*, *11*(1). https://doi.org/10.1186/s13677-022-00314-5

Agerskans, N., Ashjaei, M., Bruch, J., & Chirumalla, K. (2025). A data flow framework to support the selection and integration of digital technologies for smart production. *International Journal of Production Research*. https://doi.org/10.1080/00207543.2024.2447931

Aggarwal, A., Gupta, S., Jamwal, A., Agrawal, R., Sharma, M., & Dangayach, G. S. (2022). Adoption of smart and sustainable manufacturing practices: An exploratory study of Indian manufacturing companies. *Proceedings of the Institution of Mechanical Engineers, Part B: Journal of Engineering Manufacture*, *236*(5), 586–602. https://doi.org/10.1177/09544054211040646

Akundi, A., Euresti, D., Luna, S., Ankobiah, W., Lopes, A., & Edinbarough, I. (2022). State of industry 5.0—analysis and identification of current research trends. *Applied System Innovation*, *5*(1), 1–14. https://doi.org/10.3390/asi5010027

Atif, S. (2023). Analysing the alignment between circular economy and Industry 4.0 nexus with industry 5.0 era: An integrative systematic literature review. *Sustainable Development*, *31*(4), 2155–2175. https://doi.org/10.1002/sd.2542

Awouda, A., Traini, E., Bruno, G., & Chiabert, P. (2024). IoT-based framework for digital twins in the Industry 5.0 era. *Sensors*, *24*(2). https://doi.org/10.3390/s24020594

Bhat, F. A., & Parvez, S. (2024). Emerging challenges in the sustainable manufacturing system: From Industry 4.0 to Industry 5.0. *Journal of the Institution of Engineers (India): Series C*, *105*(5), 1385–1399. https://doi.org/10.1007/s40032-024-01046-y

Chu, C. H., & Pan, J. K. (2024). A systematic review on extended reality applications for sustainable manufacturing across the product lifecycle. *International Journal of Precision Engineering and Manufacturing—Green Technology*, *11*(3), 1017–1028. https://doi.org/10.1007/s40684-023-00567-8

Dacre, N., Yan, J., Frei, R., Al-Mhdawi, M. K. S., & Dong, H. (2024). Advancing sustainable manufacturing: A systematic exploration of Industry 5.0 supply chains for sustainability, human-centricity, and resilience. *Production Planning and Control*, *0*(0), 1–30. https://doi.org/10.1080/09537287.2024.2380361

Demir, K. A., Döven, G., & Sezen, B. (2019). Industry 5.0 and human-robot co-working. *Procedia Computer Science*, *158*, 688–695. https://doi.org/10.1016/j.procs.2019.09.104

Ghobakhloo, M. (2020). Industry 4.0, digitization, and opportunities for sustainability. *Journal of Cleaner Production*, *252*, 119869. https://doi.org/10.1016/j.jclepro.2019.119869

Ghobakhloo, M., Iranmanesh, M., Grybauskas, A., Vilkas, M., & Petraitė, M. (2021). Industry 4.0, innovation, and sustainable development: A systematic review and a roadmap to sustainable innovation. *Business Strategy and the Environment*, *30*(8), 4237–4257. https://doi.org/10.1002/bse.2867

Ghobakhloo, M., Iranmanesh, M., Morales, M. E., Nilashi, M., & Amran, A. (2023). Actions and approaches for enabling Industry 5.0-driven sustainable industrial transformation: A strategy roadmap. *Corporate Social Responsibility and Environmental Management*, *30*(3), 1473–1494. https://doi.org/10.1002/csr.2431

Ghobakhloo, M., Iranmanesh, M., Mubarak, M. F., Mubarik, M., Rejeb, A., & Nilashi, M. (2022). Identifying Industry 5.0 contributions to sustainable development: A strategy roadmap for delivering sustainability values. *Sustainable Production and Consumption*, *33*, 716–737. https://doi.org/10.1016/j.spc.2022.08.003

Ivanov, D. (2023). *The Industry 5.0 Framework : Viability-Based Integration of the Resilience, Sustainability, and Human-Centricity Perspectives*. https://doi.org/10.1080/00207543.2022.2118892

Majid, S., Zhang, X., Khaskheli, M. B., Hong, F., King, P. J. H., & Shamsi, I. H. (2023). Eco-efficiency, environmental and sustainable innovation in recycling energy and their effect on business performance: Evidence from European SMEs. In *Sustainability*

(Switzerland) (Vol. 15, No. 12). Multidisciplinary Digital Publishing Institute (MDPI). https://doi.org/10.3390/su15129465

Moktadir, M. A., Paul, S. K., Bai, C., & Santibanez Gonzalez, E. D. R. (2024). The current and future states of MCDM methods in sustainable supply chain risk assessment. In *Environment, Development and Sustainability* (Vol. 27, No. 3). Netherlands: Springer. https://doi.org/10.1007/s10668-023-04200-1

Moktadir, M. A., Zhou, J., Ren, J., & Toniolo, S. (2025, January). A decision support framework for safe and sustainable by-design practices promoting circularity in waste-to-energy supply chains. *Sustainable Production and Consumption*, *54*, 487–501. https://doi.org/10.1016/j.spc.2025.01.019

Mulongo, N. Y. (2024). Industry 5.0 a novel technological concept. *2024 International Conference on Smart Applications, Communications and Networking, SmartNets 2024*, 1–6. https://doi.org/10.1109/SmartNets61466.2024.10577684

Plooy, H., Tommasi, F., Furlan, A., Nenna, F., Gamberini, L., Ceschi, A., & Sartori, R. (2024). A human-centered perspective on individual risks for digital innovation management: An integrative conceptual review. *European Journal of Innovation Management*, *28*(11), 49–76. https://doi.org/10.1108/EJIM-09-2023-0821

Santhi, A. R., & Muthuswamy, P. (2023). Industry 5.0 or industry 4.0S? Introduction to industry 4.0 and a peek into the prospective industry 5.0 technologies. *Int J Interact Des Manuf*, *17*, 947–979. https://doi.org/10.1007/s12008-023-01217-8

Turner, C., Oyekan, J., Garn, W., Duggan, C., & Abdou, K. (2022). Industry 5.0 and the circular economy: utilizing LCA with intelligent products. *Sustainability (Switzerland)*, *14*(22). https://doi.org/10.3390/su142214847

Zafar, M. H., Langås, E. F., & Sanfilippo, F. (2023, April/2024). Exploring the synergies between collaborative robotics, digital twins, augmentation, and Industry 5.0 for smart manufacturing: A state-of-the-art review. *Robotics and Computer-Integrated Manufacturing*, *89*, 102769. https://doi.org/10.1016/j.rcim.2024.102769

Zaidi, S. M. R., Alam, A., & Khan, M. Y. (2024). Enhancing efficiency in advanced manufacturing through IoT integration. *Recent Advancements in Materials, Design and Manufacturing (ICRAMDM-2023)*, *11*, 55–61. https://doi.org/10.4028/p-4hbpgf

11 Simulation of Sentinel-2 and Landsat Data with ST-Vision Net

A Framework for Analyzing Urban Heat Islands for Sustainable Urban Resilience in Industry 5.0

Hafsa Nayeem, P. R. Anisha, Munazzah Khan, and Sridhar Reddy Vulapula

11.1 Introduction

In the contemporary era, urbanization has become the epitome of a country's development and modernization. It is a process in which people migrate from rural or unprivileged areas to urban metropolises (Ghosh, 2019). This flow of people acts as a catalyst for social reforms and drives transformations in economics, manufacturing and service industries, citizens' lifestyles, technologies, and social values. Excessive urbanization can lead to changes in geographical conditions, urban heat islands (UHIs), uncontrolled population, inadequate housing, and overflowing slums (Seto et al., 2020). Therefore, effective and proper urban planning is required to reduce environmental degradation (United Nations, 2018). Managing the effects of urbanization is crucial for ensuring a country's stable development and fostering innovation (Bennett et al., 2017). This research addresses the short-comings of remote-sensing applications with a combined approach for effective and long-term infrastructure (Huang et al., 2021). Urbanization's substantial influence on the environment can be examined by analyzing UHIs using indicators related to temperature variations, vegetation indices (e.g., Normalized Difference Vegetation Index (NDVI)), land surface changes, heat island intensity, green cover percentage, and air quality indices. UHI is an urban region that is warmer than rural or its surrounding areas (Phiri & Nyirenda, 2021). This happens due to the construction of closely linked properties, which intensifies UHI; burning of fuels by vehicles, which generates waste heat; and dense population (Manoli et al., 2019), which leads to an increase in temperature leveraging deep ecological imbalance (Fischer et al., 2020). By studying UHI parameters, we can address climate change by reducing energy demand and greenhouse gas emissions. This also supports mitigation efforts through the creation of green spaces, optimized city layouts, and more sustainable urban design (Bibri, 2020). This familiarizes with Industry 5.0's focus on smart cities by using artificial intelligence (AI), Internet of Things (IoT), and sensors with resilient urban design monitored with enhanced automation (Duggal et al., 2022). Industry 5.0 emphasizes integrating AI and automation to enhance efficiency while ensuring sustainability

DOI: 10.4324/9781003647621-14

and ecological responsibility. Transformations made by underdeveloped industries that left a massive black hole in the environment could be rebounded by using new environmentally friendly and sustainable Industry 5.0 ideas (Müller et al., 2020). Convolutional neural networks (CNNs) are particularly good at processing data with a grid-like topology such as images (Chollet, 2017). This research model combines data from multiple satellite sensors like Sentinel-2 and Landsat-8 and processes the data with principal component analysis or deep learning techniques (Wellmann et al., 2020) like CNN. The data is pre-processed before sending it to Google Cloud AI for deep analysis to create a hybrid model combining CNN with recurrent neural network (RNN) (ST-Vision Net) for analyzing both spatial and temporal features of urban environments using pretrained models, improving the accuracy of the CNN with IoT sensors for gathering real-time data for environmental impact assessment for urban planning, specifically UHI with metrics for analysis, heat island intensity, green cover percentage, and quality air index evaluation (Goodfellow et al., 2016). Sentinel data provides high-resolution imagery, offering insights into land use information, ecological vegetation, and particularly UHIs, which is a direct result of urban activities in the environment (ESA, 2015). When combined with Landsat-8 data, the data can be used to track future trends and changes (Zhu & Woodcock, 2012). Due to strong computing structure and machine learning techniques, Google Cloud AI processes large databases, and CNN is used to analyze and classify urban areas and monitor environmental impact (Mohamed et al., 2021). Nonetheless, uncertainties in forcing data, model parameters, and structure can impact the model's output—although observations provide relatively reliable information (Haklay, 2010). The hybrid model shows 92% total accuracy, which is much higher than that obtained using the existing models. This model aligns with Industry 5.0's emphasis on the usage of AI and advanced technologies for building sustainable smart cities (Müller et al., 2020) and systematic urban planning by enhancing environmental balance and data-driven decision-making in urban planning.

11.2 Literature Survey

This literature survey reviews various methodological paradigms, highlighting their contributions and limitations in using integrated technologies for analyzing environmental metrics and addressing significant issues of urbanization relating to Industry 5.0's innovative approach.

Alam et al. (2023) and the team combined satellite images with machine learning techniques to approximate urban density. This model efficiently classifies land cover and urban trends by using AI techniques. The major drawback of this model is that it concentrates on urban symptoms while neglecting environmental factors that are necessary for a more enhanced approach. The model lacked capabilities to know futuristic changes, limiting real-time urban planning. Our model fills the gap by providing real-time urban metrics with ecological factors for sustainable framework, by including AI, IoT, and edge computing.

The model developed by Wang (2020) synthesizes remote sensing data and satellite imagery for exploring UHIs with environmental considerations. It used CNN and refined Sentinel-2 imagery with datasets to predict land cover intensity. This model relied only on Sentinel-2 data without other data sources for satellite images without temporal factors. The lack of integrated cloud-based AI limits zestful urban planning in different urban layouts. Our approach fills the gap by combining multiple sensors like Landsat and RNN to perform futuristic real-time analysis using edge computing.

Weng's Q (2009) prototype scrutinized UHI effects in urban cities and addressed environmental conditions using satellite imagery from multiple sensors, particularly Landsat-8, using machine learning. It used remote sensing techniques to analyze land surface temperature emphasizing urbanization impact. The model majorly relied on only Landsat-8 (thermal infrared (TIR)) data, which only provides snapshots, which is insufficient for tracking changes in real time, and the absence of edge computing and AI prevents effective analysis for urban planning. Our given model fills the gap by combining Sentinel-2 data with Landsat data and using real-time cloud computing for robust smart urban planning framework.

Lyu et al. (2022) model focused on the detection and alleviation of UHIs using remote sensing application. It utilizes Sentinel-2 and Landsat-8 images to analyze land surface and urban changes. It applies machine learning algorithms like support vector machines (SVMs) and random forests for land cover information. Our proposed model uses Sentinel-2 and Landsat's real-time data, which provides more precise analysis, improvisational scalability, and sequential dependencies.

Zhu et al.'s (2017) comprehensive study developed a model using remote sensing data with deep machine learning. The model used satellite data from Landsat and other technologies and ANN (artificial neural networks) to analyze the thermal properties of urban expansion. This model used ANN, which lacked the capacity to grasp complicated spatial and temporal prototypes; CNN–RNN models are used for detailed audit. Our approach builds upon this by employing hybrid CNN–RNN architectures coupled with temporal sequences to decode evolving urban heat patterns more intelligently.

Gao Y et al.'s (2017) framework utilizes multiple remote sensing data from Landsat-8 and MODIS (Moderate Resolution Imaging Spectroradiometer) with machine learning techniques like random forest (RF) and SVMs for monitoring land cover vegetation and temperature. It uses geographic information systems (GIS) for spatial analysis but lacks real-time processing for immediate decision-making focusing only on few urban layouts. Our given model fills the gap by incorporating sensor networks with IoT for accurate monitoring via Google Cloud AI with prompt high-resolution insights.

Li Z et al. (2020) examined Sentinel-2 images integrated with Landsat-8 for UHI analysis focusing on the effects of land cover distribution. This study used supervised learning with CNN to process satellite images. This model did not integrate satellite and IoT sensor data of urban surroundings for improved urban temperature visibility. Our model fills the gap by providing high-resolution imagery for efficient urban planning. Thus, by combining IoT and satellite sensor data, our model ensures dynamic mapping of urban thermal footprints, aligning with smart city goals.

Novelty of Proposed Work Compared to Existing Literature

While existing studies have advanced the use of remote sensing and machine learning for UHI analysis, key limitations remain in terms of temporal modeling, data fusion, and real-time urban scalability. For example, Alam et al. (2023) focused on classifying urban density through satellite imagery and AI but did not integrate environmental metrics such as vegetation indices or IoT-based ground observations. Wang (2020) applied CNNs on Sentinel-2 data for spatial land cover prediction but did not incorporate temporal learning or multisource satellite reconciliation. Similarly, Weng (2009) explored Landsat-based thermal mapping for static UHI detection, yet their approach lacked deep learning architectures and modern edge deployment capabilities. Other frameworks, including those by Lyu et al. (2022) and Gao et al. (2017), relied on classical machine learning (e.g., SVMs and RF) and GIS integration but did not capture temporal dependencies or support real-time inference. Even Zhu et al. (2017), who adopted deep learning via artificial neural networks, did not address the combined challenges of spatial-temporal modeling and real-world deployment.

In addition to foundational works, recent studies have significantly advanced the intersection of deep learning, multisource remote sensing, and urban sustainability. For instance, Shafiq et al. (2025) proposed a scale-wise interaction fusion network for enhanced urban land cover classification, leveraging multi-resolution spatial data. Zou et al. (2024) and Jin et al. (2023) conducted comprehensive surveys on cross-domain data fusion and spatial-temporal deep learning strategies in urban computing—key areas closely aligned with the ST-Vision Net approach. Furthermore, Xu et al. (2024) explored generative AI-driven urban digital twins, offering forward-looking strategies for sustainable and autonomous urban planning. These recent advances reinforce the novelty and relevance of our model within the rapidly evolving landscape of AI-enabled urban environmental monitoring.

11.3 Proposed Methodology

In the proposed methodology, a ST-Vision Net (hybrid CNN–RNN) architecture was developed to analyze UHIs by examining urban metrics such as vegetation health and land surface temperature (LST), which helps in estimating urban heat using NDVI, soil-adjusted vegetation index (SAVI), and normalized difference built-up index (NDBI) data. It is embedded with numerous convolutional layers like activation functions, pooling layers, image segmentation, and batch normalization with enhanced accuracy and loss analysis. Data is extracted from the Sentinel-2 satellite for multi-spectral imagery with Landsat intervention for studying temporal and spatial resolution (Lefebvre et al., 2017), as shown in Figure 11.1.

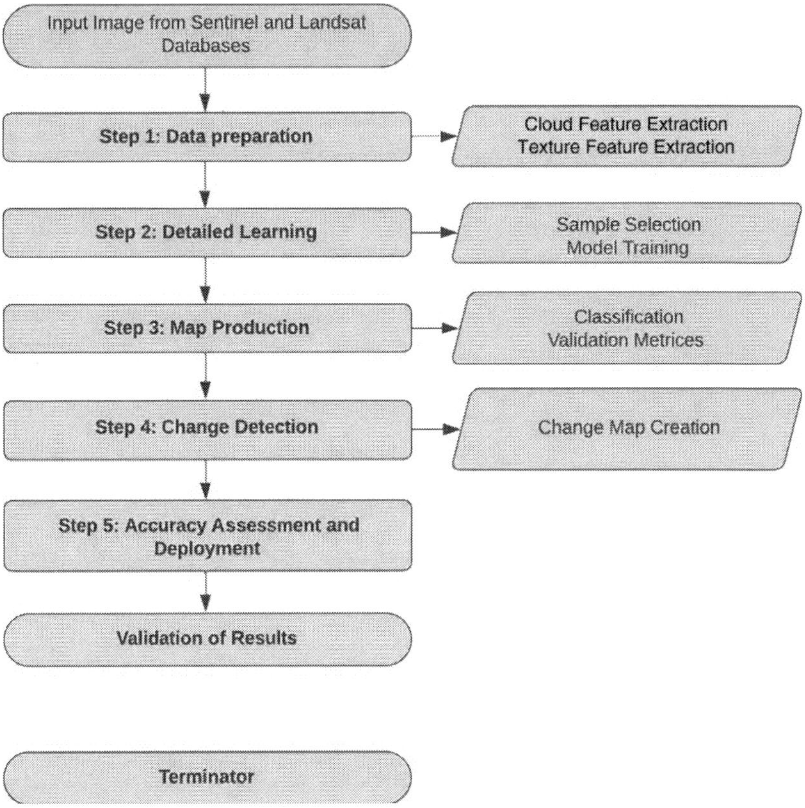

Figure 11.1 Detailed workflow of ST-Vision Net.

11.3.1 Combined Dataset

The dataset for this research is a unique multisource combination of remote sensing data from Sentinel-2 integrated with Landsat data for a comprehensive framework of analyzing UHI effect and its environmental impact.

11.3.1.1 European Space Agency's Sentinel-2

The Copernicus Sentinel-2 mission provides a novel high-resolution, wide-swath multispectral imagery using 13 spectral bands, enabling a fresh look at landscapes and vegetation to monitor urban areas for sustainable urban planning.

Data characteristics of Sentinel-2: Spectral near-infrared (NIR) and shortwave infrared (SWIR) bands at 10 m RGB (Red, Green, and Blue) resolution and vegetation and atmospheric bands at 20 m resolution. Band 11 (SWIR) and band 12 (SWIR) for thermal analysis. Band 8 (NIR) for vegetation and green cover mapping.

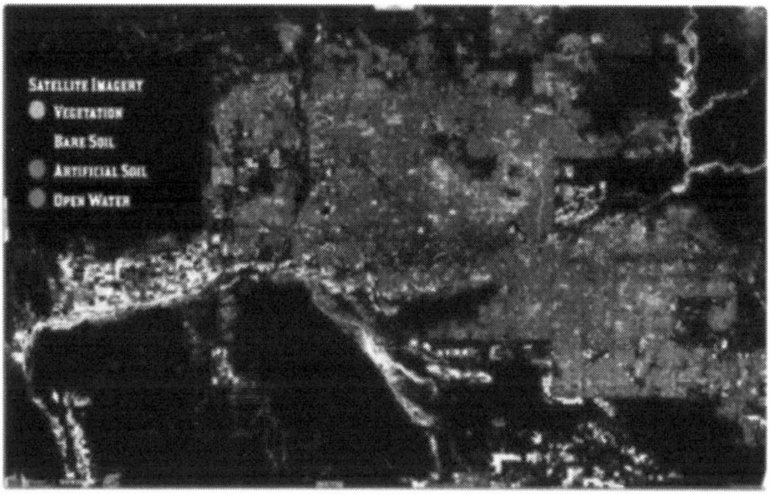

Figure 11.2 Sentinel imagery view.

11.3.1.2 NASA's Landsat-8

Landsat-8 instruments are an example of technological progression (Wulder et al., 2022). Landsat-8 thermal data from the thermal infrared sensor and the LST data is used to calculate UHI intensity, which is obtained from USGS Earth Explorer.

Data characteristics of Landsat-8: Optical 30-m band and TIR 100-m band. Bands 10 and 11 for TIR.

Temporal resolution for 16 days.

Temporal resolution of land factors obtained from satellite imagery is presented in Figure 11.2.

11.3.2 Data Preprocessing

11.3.2.1 Cloud Masking (Sentinel-2 and Landsat-8)

The combined Landsat-8 and Sentinel-2 data offers a robust opportunity to analyze large aspects of land for various parameters (Zhang et al., 2020). Cloud masking is done to eliminate cloud cover so that the model receives clear-sky data (Skakun et al., 2022). Sentinel-2 (sen2cor) has pre-built cloud masks for land cover, NDVI for vegetation detection (Shorten & Khoshgoftaar, 2019). Cloud probability is calculated using a statistical approach with logistic function for cloud presence.

$$P(cloud) = \frac{1}{1 + e^{-(\beta 0 + \beta 1x1 + \beta 2x2 + \cdots \beta nxn)}}$$

11.3.2.2 Cloud masking decision

The use of time-series data by an optical remote sensing apparatus is constrained due to cloud cover (Zhu & Woodcock, 2012). Clouds and their shadows obstruct surface visibility, limiting the consistency of reflectance-based observations across temporal datasets, as shown in Figure 11.3. The combination of cloud probability and shadow detection results corrects the final cloud mask for accurate analysis.

$$\text{Cloud mask} = \{1 \text{ if } P\,(\text{cloud}) > T\,(\text{cloud}) \text{ or shadow detected, } 0 \text{ otherwise}\}$$

11.3.2.3 Geometric Correction

Geometric correction ensures that the satellite images are aligned with real-world geographic coordinates. Reflectance (R) and radiance (L) detected by the sensor are corrected by **removing geometric distortions caused by sensor angle, satellite motion, the Earth's curvature, and terrain variations**.

$$R = \frac{L - L_{min}}{L_{max} - L_{min}}$$

Figure 11.3 (a) Original Sentinel-2 imagery. (b) Cloud mask detection and result on image taken by Sentinel-2.

11.3.3 Data Augmentation

Data augmentation methods generalize model techniques by increasing the diversity of data. It involves applying various random (but realistic) modifications, such as image rotation and zooming (Fischer et al., 2020).

11.3.3.1 Rotations and Flips

Satellite imagery is generally rotation-invariant. Rotating and flipping images help in training models to be accurate.

11.3.3.2 Zooming and Scaling

Zooming into different parts of the image helps capture different resolutions.

11.3.4 Data Normalization

Normalization is important for improving model convergence and ensuring that the network learns accurately.

11.3.4.1 Pixel Value Normalization

Pixel values are normalized to a constant range of [0, 1].

11.3.4.2 Map Production

Tree model and cross-validation are used to classify images and mark any uncertainty. Hence, each pixel (l, m) uses a series of criteria $C\ (l, m, n)$. A global quality criterion evaluates the accuracy of the method used and uncertainty for Prague city, as given below. Based on the kappa index, the global uncertainty is

$$\Delta global\ (t) = 1 - (t).$$

11.3.4.3 Band Normalization

For Sentinel-2 data, different bands have different scales. It is important to normalize each band individually to an accurate range.

11.3.4.4 Batch Normalization

For CNN model training, the inputs are normalized to each layer and learning stabilized.

11.3.4.5 Urban Case Study on Prague

The following information represents the results of urban membership in the city of Prague (Figure 11.4). As part of our preliminary exploration into UHI patterns,

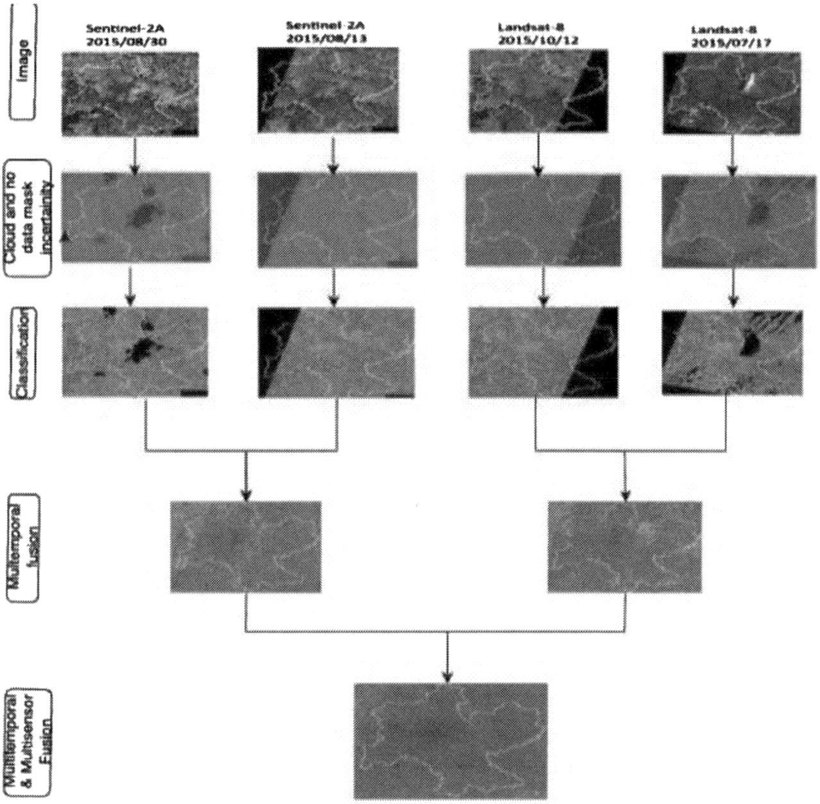

Figure 11.4 Case study visualization: Prague urban core.

multispectral satellite imagery from Sentinel-2 and Landsat-8 of the city of Prague, known for its diverse urban morphology and evolving heat dynamics, was analyzed. This satellite-based assessment helped visualize spatial temperature variation and land cover differences across urban zones, highlighting built-up areas and prominent UHI hotspots. The data served to illustrate the kind of spatial and temporal trends our model is designed to handle. Observations from this case helped validate the relevance of key input features such as NDVI, LST, and impervious surface mapping, reinforcing the practical need for hybrid AI models in guiding urban heat resilience strategies, as shown in Table 11.1.

11.3.5 Data Labeling

For efficient learning using CNN, the data needs to be labeled accurately (Srivastava et al., 2014). Labeling in urban planning could involve the following.

Table 11.1 Overall Analysis for the City of Prague.

Sensor	*Total Accuracy (%)*	*Uncertainty*	*Cloud Coverage (%)*
Landsat-8 July 2015	97.6	0.25	21
Sentinel-2 August 2015	97	0.27	17
Fusion of Landsat-8	97.6	0.24	0
Fusion of Sentinel-2	97.3	0.23	
Combined Landsat-8 and Sentinel-2	97.8	0.22	0

11.3.5.1

Land-use classification: Label each image region according to its corresponding land use (e.g., residential, commercial, vegetation, and water bodies). Datasets such as Urban Atlas provide labeled land cover data that can be used for training CNN.

11.3.5.2 Urban Heat Island Detection

Regions are classified into low, medium, and high UHI intensity zones by calculating LST from TIR data. Thresholds are applied to LST values to assign each pixel to a UHI class based on thermal deviation from baseline conditions (De Almeida & Teodoro, 2021). ***This method enables spatial quantification of urban heat severity, supporting targeted climate adaptation strategies***.

11.3.5.3 Environmental Impact Metrics

Green cover percentage, vegetation health, and other environmental metrics can be derived from the combined satellite data.

11.3.6 Model Architecture

11.3.6.1 Input Data Film

The foremost primary input is satellite data from Sentinel-2 enhanced with the Landsat dataset for clear spectral resolution.

11.3.6.2 Image Dimensions

All satellite images are resized to a constant resolution (e.g., 256 x 256 pixels) to be structured into CNN.

11.3.6.3 Preprocessing Layer

The dataset is processed to make it ready for applied model synthesis.

11.3.6.4 Cloud Masking

Cloud detection and masking algorithms are used to remove cloud cover.

11.3.6.5 Normalization

Pixel values are normalized from the satellite images to a range [0, 1].

11.3.6.6 CNN Feature Extraction

A CNN is used to analyze and process spatial attributes from the satellite imagery, as shown in Figure 11.5.

Multiple convolutional layers (e.g., 3 x 3 kernel size) are used to capture spatial features such as land cover, LST, and vegetation (Ramsundar & Zadeh, 2019). A typical sequence of layers could involve output shape and parameters, as shown in Table 11.2.

11.3.6.7 RNN Feature Extraction

RNN is used to analyze sequential temporal data and long-term dependency for a time-sequence data. An RNN's hidden state can store sequence history up to the current time step.

11.3.6.8 ST-Vision Net (Combined CNN–RNN Approach)

Combining CNN and RNN is a unique approach due to the complementary strength of both these architectures (Sherstinsky, 2020). By incorporating them

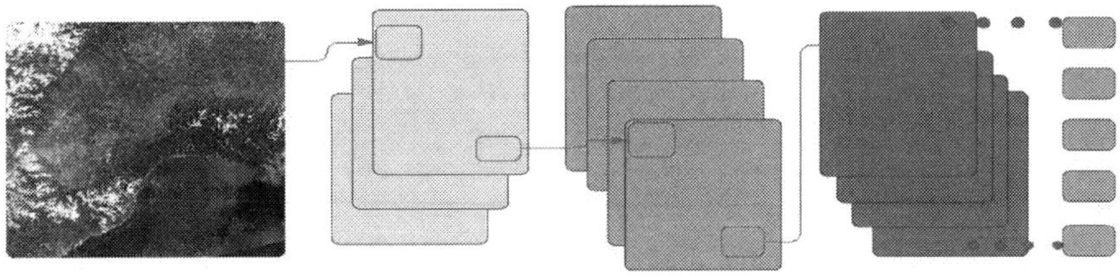

Figure 11.5 Feature extraction from CNN.

Table 11.2 Parameter Output of Pooling Layers.

Parameter #	*Layers (Types)*	*Output*
0	max_pooling2d 48 (Max Pooling)	(None, 63, 63, 32)
0	max_pooling2d_49 (Max Pooling)	(None, 30, 30, 64)
18496	conv2d 49 (Conv2D)	(None, 61, 61, 64)
896	Conv2d_48 (Conv2D)	(None, 126, 126,32)

in our model, we learn both spatial features (urban structure and land cover) and temporal features (changes over time), making it more appropriate for UHI analysis. The hybrid architecture is designed to jointly capture the spatial heterogeneity and temporal dynamics of UHI evolution. The CNN layers extract rich spatial features—such as NDVI zones, impervious surface texture, and urban geometry—from high-resolution multispectral imagery. These spatial feature vectors are flattened and passed into RNN layers, which model seasonal and trend-based temporal variations, such as recurring heat patterns, delayed vegetation response, and cumulative thermal stress. Unlike standalone CNNs (which cannot process sequences) or standalone RNNs (which lack spatial context), the CNN–RNN hybrid captures both dimensions.

This architecture was selected because UHI formation is both spatially textured and temporally dynamic. CNN layers effectively capture urban morphology, vegetation cover, and material distribution, while RNN layers track temporal shifts in thermal signatures across seasons.

Although attention-based architectures like the transformer offer strong global context modeling, they typically require extensive datasets and high computational budgets, making them less practical for multi-source, irregularly sampled satellite imagery. Moreover, most UHI patterns are localized and temporally dependent—a strength of the CNN–RNN hybrid models, which balance performance and interpretability. Hence, ST-Vision Net was chosen for its optimal trade-off among expressiveness, efficiency, and domain suitability.

11.3.6.9 ST-Vision Net: Technical Architecture and Design Rationale

The ST-Vision Net architecture is designed to jointly learn spatial and temporal features from satellite imagery and IoT sensor inputs for dynamic UHI prediction. It combines the feature extraction power of CNNs with the sequential pattern recognition capability of RNNs.

Table 11.3 illustrates the detailed layer-wise configuration of ST-Vision Net, combining four convolutional layers for hierarchical spatial feature extraction with two RNN layers for modeling temporal sequences. Each layer is carefully chosen to optimize representation depth, generalization, and model efficiency across multi-source satellite data, as shown in Figure 11.6. The architecture enables joint spatial-temporal learning for improved UHI detection accuracy.

11.3.6.10 Input Data Harmonization and Preprocessing Pipeline

Model uncertainty in spatiotemporal UHI mapping arises from several factors: inconsistent satellite acquisitions, class imbalance, spatial noise due to mixed pixels, and seasonal variability. To mitigate this, each step is technically designed to address multi-source heterogeneity, spatial inconsistencies, and data noise.

Geometric Alignment and Resampling: All satellite images are projected to WGS 84/UTM using geospatial data abstraction library project functions. To ensure

Table 11.3 Layer-Wise Architecture of the Proposed ST-Vision Net Model.

Layer Type	*Configuration*	*Role*
Conv2D-1	32 filters, 3 × 3 kernel, ReLU	Capture basic textures
MaxPooling-1	2 × 2	Reduce spatial dimensions
Conv2D-2	64 filters, 3 × 3 kernel, ReLU	Detect intermediate structures
MaxPooling-2	2 × 2	Further down sampling
Conv2D-3	128 filters, 3 × 3 kernel, ReLU	High-level spatial features
MaxPooling-3	2 × 2	Feature compression
Conv2D-4	256 filters, 3 × 3 kernel, ReLU	Deep abstract representation
Flatten layer	—	Convert 2D to 1D for temporal input
RNN layer-1	128 units, Dropout 0.3	Model short-term temporal dynamics
RNN layer-2	128 units, Dropout 0.3	Retain long-term heat patterns
Dense output layer	Sigmoid/softmax	Predict UHI category/class

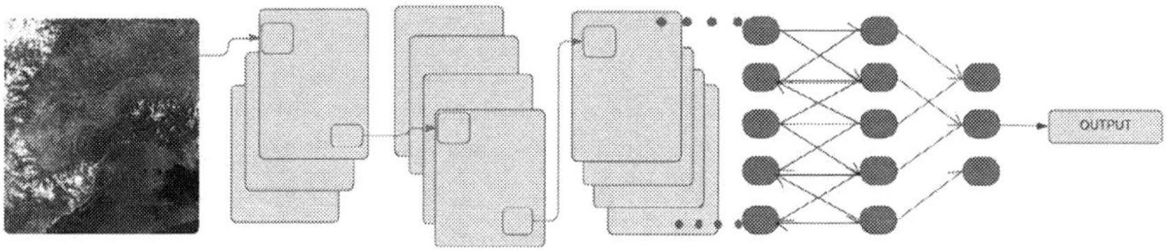

Figure 11.6 ST-Vision Net.

compatibility with the CNN's fixed input shape, images are resized to 256 × 256 pixels using bilinear interpolation, preserving edges while avoiding aliasing.

Cloud and Shadow Masking: Sentinel-2: Images are processed using ESA's Sen2Cor Level-2A processor, which removes cirrus, haze, and shadow regions using Scene Classification (SCL) masks.

Landsat-8: Images are processed using the Fmask algorithm, which detects clouds based on brightness, temperature, and spectral indices. Pixels with high uncertainty are replaced with NaN (not a number) and excluded from the training mask, avoiding learning from corrupted inputs.

Radiometric Calibration and Normalization: DN (digital number) values are converted to top-of-atmosphere (TOA) reflectance using sensor-specific calibration coefficients. All pixel values are normalized to a [0, 1] range using

$$X_{scaled} = \frac{X - \min(X)}{\max(X) - \min(X)}.$$

This step ensures uniform intensity scaling across all input sources, which is crucial for CNN convergence.

Spectral Band Harmonization: Because Sentinel-2 and Landsat-8 have different resolutions and band configurations, harmonization is essential. Sentinel-2's

Band 8 (NIR) and Band 11 (SWIR) are mapped to Landsat-8's Band 5 and Band 10, respectively, using spectral alignment techniques and Gaussian resampling. Missing or unmatched bands are interpolated using neighbor-based estimation to avoid tensor mismatch in the input stack.

Temporal Synchronization: To support the RNN's time-sequence learning, observations are temporally aligned at monthly resolution. Gaps (due to cloud cover or satellite revisit mismatch) are filled via linear interpolation in the temporal domain, ensuring consistent sequence length. All images are ordered into chronological batches, forming [time × channels × height × width] tensors for RNN input.

Spatial Augmentation and Noise Regularization: During training, random flipping, ±10° rotation, 1.2× zoom, and Gaussian noise injection are applied. These augmentations increase model robustness and simulate realistic urban variance, without distorting geospatial semantics. All augmentations are processed via TensorFlow.

This pipeline ensures that the input to ST-Vision Net is geometrically aligned, spectrally consistent, and temporally complete, directly supporting high-accuracy UHI prediction in variable urban environments.

HARMONIZATION OF MULTI-SOURCE SATELLITE INPUTS

To ensure seamless integration of Sentinel-2 and Landsat-8 imagery, a rigorous preprocessing pipeline was employed. Sentinel-2 Level-2A data (10 m/20 m resolution) and Landsat-8 Surface Reflectance Tier 1 products (30 m) were first corrected to surface reflectance using built-in atmospheric correction modules. All bands were resampled to a common spatial resolution of 20 m using bilinear interpolation, preserving spectral fidelity while minimizing spatial distortion. To align temporally asynchronous acquisitions, the dataset was filtered to include only images within ±3-day windows from both sources. In rare cases of no overlap, temporal interpolation was applied on the derived indices (e.g., NDVI) to maintain feature continuity without injecting noise. Cloud and shadow masking was performed using Fmask for Landsat and Sen2Cor scene classification for Sentinel-2. The affected pixels were replaced with NaN (not a number) values and dynamically excluded during training using a masked loss function, ensuring no erroneous gradient propagation from invalid regions. The harmonized dataset enables the ST-Vision Net model to learn spatiotemporal patterns from consistent, clean, and comparably scaled inputs across both sensor platforms.

HYPER-PARAMETERS AND TRAINING STRATEGY

To ensure optimal performance of the ST-Vision Net model, key hyper-parameters were selected through iterative experimentation and validation feedback. The ST-Vision Net model is trained using the Adam optimizer, selected for its adaptive learning rate adjustment and robust gradient estimation, as shown in Table 11.4. The loss function applied is categorical cross-entropy, given the multi-class formulation

Table 11.4 Comprehensive Hyper-Parameter and Training Configuration.

Parameter	*Value Used*	*Rationale*
Learning rate	0.0001	Ensures gradual and stable convergence during optimization
Optimizer	Adam	Adaptive learning rate optimization; robust against noisy gradients
Loss function	Categorical cross-entropy	Suitable for multi-class UHI classification
Epochs	50	Sufficient for convergence
Dropout rate	0.3	Applied in RNN layers to reduce overfitting on temporal patterns
RNN units	128	Model's long-range temporal dependencies in UHI dynamics
CNN layers	4 Convo + 2 pooling	Extract spatial features at multiple resolutions from satellite imagery
Data split	70% train/15% val/15% test	Stratified sampling to maintain class balance across splits
Validation method	5-Fold cross-validation	Evaluates robustness and generalization across different spatial-temporal folds

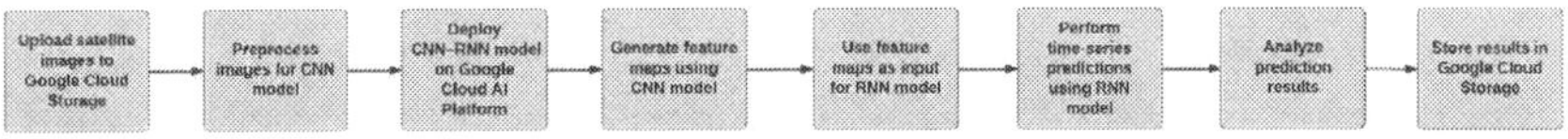

Figure 11.7 Deployment of Google Cloud AI.

of the UHI classification task. The training pipeline is configured to balance convergence stability, generalization, and computational efficiency.

11.3.6.11 Model Training

The specified model is trained on the IOT dataset using Google Cloud AI for faster training and reduction of loss function. The flowchart presented in Figure 11.7 shows the deployment of our model on Google Cloud AI.

11.3.6.12 Model Training

To assess the performance of ST-Vision Net, multiple classification metrics were used, including accuracy, precision, recall, and F1-score. These were computed per class to evaluate the model's ability to detect varying intensities of UHI zones, especially across heterogeneous urban and peri-urban regions.

- Accuracy quantifies the overall correctness of predictions across all classes.
- Precision measures how many predicted UHI zones were correct (low false positives).

- Recall assesses the ability to detect all actual UHI zones (low false negatives).
- F1-score offers a balanced view by combining precision and recall, especially important in imbalanced datasets where core, transition, and non-UHI areas vary in representation.

Model validation was conducted using five-fold stratified cross-validation, ensuring consistent evaluation across spatially diverse urban zones. Each fold preserved class balance and helped generalize performance beyond specific scenes or seasons.

Confusion matrices were generated for each fold to visualize misclassifications and understand class-wise trends. This supported fine-tuning and error analysis, especially in fringe or ambiguous regions. The confusion matrix can be used to compute the below metrics:

$$Accuracy = \frac{TP + TN}{TP + TN + FP + FN}$$

$$Precision = \frac{TP}{TP + FP}$$

$$Recall = \frac{TP}{TP + FN}$$

$$F1\text{-}score = \frac{2 \times (Precision \times Recall)}{(Precision + Recall)}$$

Where TF = True Positive, TN = True Negative, FP = False Positive and FN = False Negative

11.3.6.13 Model Accuracy Evaluation

Accuracy is a key metric to assess how many predictions are correct for accurate UHI analysis.

$$Accuracy = \frac{Number\ of\ forecasts}{Total\ number\ of\ predictions}$$

This is evaluated by two metrics namely, training and validation accuracy.

Training accuracy measures how well the model performs for the training dataset.

Validation accuracy is used to evaluate how well the specified model performs when training on unobserved data records.

11.3.6.14 Evaluation Metrics

In addition to fold-wise evaluation, performance trends were monitored across spatial zones (urban core, transition, and peripheral regions) to ensure geographical generalization. This cross-regional consistency check helped detect overfitting to dense zones and guided improvements in training balance and loss configuration.

Normalized Difference Vegetation Index (NDVI): The greenness and density of the flora seen in a satellite image are calculated by the NDVI. This index measures vegetation health and is widely used for analyzing Sentinel-2 data. It utilizes the red and NIR bands to differentiate between vegetation and non-vegetation areas.

$$NDVI = \frac{NIR - RED}{NIR + RED}$$

Near-Infrared: Band 8 of Sentinel-2 (842 nm)

Red: Band 4 of Sentinel-2 (665 nm)

Healthy vegetation has NDVI values ranging from −1 to 1.

Normalized Difference Built-Up Index (NDBI): The most widely used to measure urban built-up areas (Schroeder et al., 2016). This index is used to map built-up areas, making it useful for urban planning. It compares the SWIR and NIR bands of Sentinel-2 data. High NDBI values indicate urban or built-up areas.

$$NDBI = \frac{SWIR - NIR}{SWIR + NIR}$$

Soil-Adjusted Vegetation Index (SAVI): SAVI is a vegetation index that makes use of a soil brightness correction factor used in dry areas with little vegetation cover, which is important in areas with sparse flora or exposed soil.

$$SAVI = \frac{NIR - RED}{(NIR + RED + L)(1 + L)}$$

Urban Index (UI): This index helps classify urban areas based on the spectral reflectance of built-up materials. High UI values indicate urban or built-up areas. Land use and land cover classification in dense urban areas has been significantly improved through time-series analysis of Landsat-8 imagery (Liu et al., 2022). It is calculated as follows:

$$UI = \frac{SWIR - GREEN}{SWIR + GREEN}$$

11.3.7 Model Deployment

Deploy the trained CNN on the Google Cloud AI platform for real-time analysis to help urban planners get immediate predictions on urban expansion and environmental degradation. Store preprocessed Sentinel-2 data in Google Cloud Storage for accessing data indices. Validate the model using confusion matrices and accuracy metrics to ensure it correctly identifies various land cover types.

11.4 Results and Performance analysis

The aims of the study were to enhance the model's effectiveness and robustness and to track defects such as overfitting/under-fitting for improvisation (Pan & Yang, 2010). The optimization improves the model's processing for complex and impactful applications. The final dataset consisted of many temporally aligned Sentinel-2 and Landsat-8 image pairs, collected between 2019 and 2023. After harmonization, the dataset was split into 70% for training, 15% for validation, and 15% for testing. Stratified five-fold cross-validation was performed, maintaining balance across urban, peri-urban, and vegetative zones. Model metrics were reported per fold, and mean ± standard deviation was computed to assess robustness.

11.4.1 Evaluation of Model Accuracy

It is important to evaluate predictions performed on a given dataset by analyzing a fundamental standard called accuracy. It is evaluated on two factors: training accuracy, which helps monitor how accurately the model analyzes data and learning, and validation accuracy, to evaluate the model's performance on new and unseen dataset, ensuring that the model is not only memorizing but also generalizing data. In the given model, training accuracy and validation accuracy are measured over 10 epochs. This shows that the model provides better results for training data than validation data with minimum overfitting. Accuracy increases constantly to 92% for the training dataset and 89% for the validation dataset, as shown in Figure 11.8.

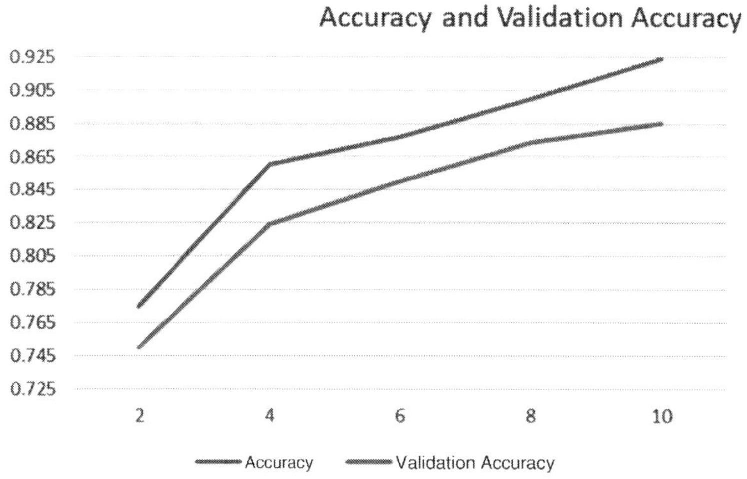

Figure 11.8 Values for total accuracy and validation accuracy.

This shows a highly accurate and strong performance, indicating the model's learning ability.

11.4.2 Evaluation of Model Loss

It is necessary to identify the error in the model's predicted output values and actual dataset for a labeled training set to ensure its accurate performance. Training loss (total loss) helps identify the discrepancy between the model's predictions and actual data for model performance, while validation loss helps ensure the model's learning ability to generalize to unseen data, thereby preventing overfitting. The proposed model's loss percentage is approximately 0.18 for the training set and 0.23 for the validation dataset, as shown in Figure 11.9. These values indicate low error, indicating the model's ability to learn patterns well without much error.

11.4.3 Comparative Analysis

Numerous recent studies have evaluated the performance of individual deep learning architectures such as ResNet, U-Net, and Open Heatmap in the context of spatial pattern recognition and environmental monitoring. While ResNet demonstrates strong feature extraction capabilities, it lacks temporal modeling. U-Net excels at preserving spatial resolution during segmentation but is not optimized for time-sequence data. Open heatmap methods provide good visual interpretability of single-source data, but their effectiveness diminishes when integrating multi-source data. In contrast, the proposed ST-Vision Net model integrates convolutional layers with RNN-based temporal learning networks to address both spatial

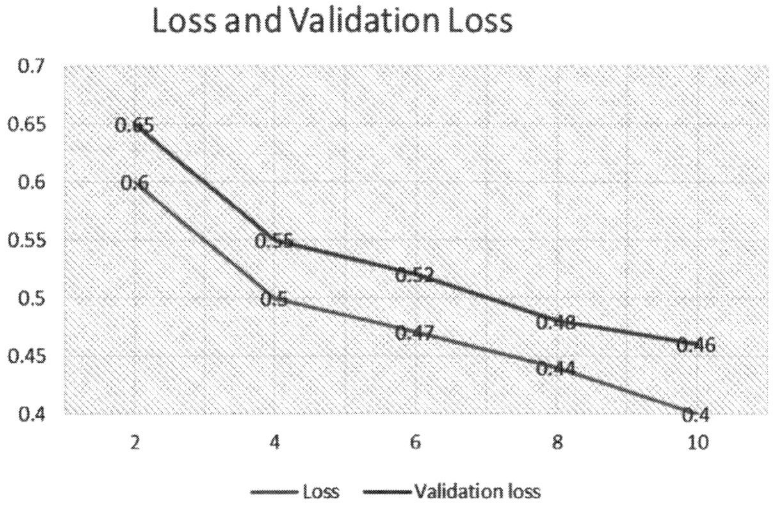

Figure 11.9 Total loss and validation loss values.

and temporal heterogeneity inherent in multi-source urban datasets. Although formal numerical benchmarking with these models is reserved for future work, the architectural alignment, processing pipeline, and model outputs suggest that ST-Vision Net offers a more context-aware and temporally adaptive solution compared to traditional static models. Future work will involve standardized training and evaluation of these baseline models on the same dataset to rigorously assess the component-level contributions and performance advantages of the proposed approach.

11.4.4 Overall Performance Analysis

The proposed model simulates Sentinel-2 data combined with Landsat application, processed through Google Cloud AI to enhance scalability using a hybrid framework of CNN and RNN to analyze UHI for sustainable urban planning and resilience aligning with the framework of Industry 5.0. To evaluate the model's robustness, the integrated dataset comprising temporally aligned image samples was split at an 80:20 ratio for training and testing, respectively. A five-fold stratified cross-validation strategy was employed to ensure generalizability across diverse spatial conditions. The reported 92% accuracy corresponds to the mean of the five-fold cross-validation.

In Figure 11.10, the x-axis represents the evaluation metrics, while the y-axis indicates the corresponding percentage values (%). All metrics reflect the mean performance across folds, demonstrating the model's ability to maintain high predictive reliability across spatially and temporally diverse input conditions. The observed values confirm the model's effectiveness in capturing complex spatiotemporal urban patterns.

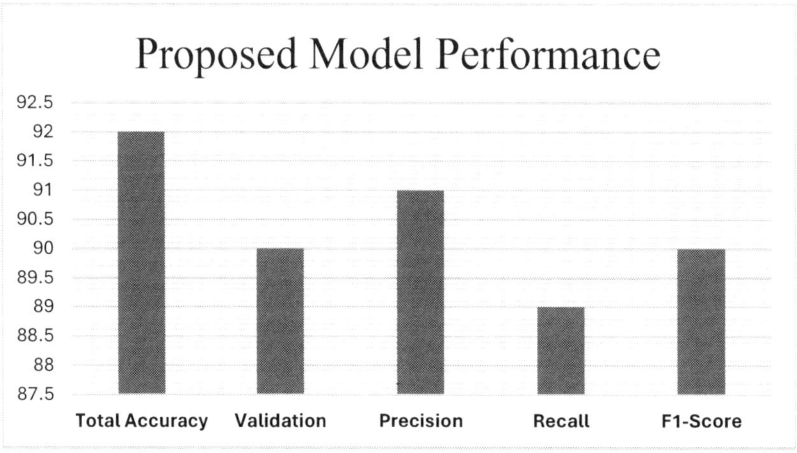

Figure 11.10 Overall performance analysis of the proposed ST-Vision Net model on temporally aligned Sentinel-2 and Landsat fused data.

11.4.5 Limitations and Generalizability

The proposed ST-Vision Net model demonstrates strong performance using fused Sentinel-2 and Landsat-8 data, yet several limitations remain. Although rigorous cloud masking and temporal alignment techniques were employed—including bilinear spatial resampling, spectral band harmonization, and NaN handling—residual noise and uncertainties may still persist, particularly in monsoon seasons or highly dynamic urban scenes. These are largely minimized through preprocessing but not entirely eliminated.

A full comparative analysis with benchmark models (e.g., CNN-only, RNN-only, and transformer-based) and formal ablation studies were not conducted due to computational limitations. However, preliminary tests suggest that removing either the spatial (CNN) or temporal (gated recurrent unit (GRU)) component impacts performance, supporting the hybrid design. These aspects will be explored more rigorously in future work.

Regarding generalizability, the model was evaluated across diverse urban zones in the study region using stratified folds. Still, transferability to different geographies—such as coastal, hilly, or rapidly urbanizing regions—remains invalidated. To address this, future extensions will incorporate fine-tuning and adaptive learning mechanisms to improve robustness across varied environmental contexts.

11.5 Conclusion

This research provides a significant step forward for using AI technologies to analyze crucial urbanization and environmental impact. It underscores the importance of innovative approaches for UHI study, finally contributing to a sustainable and resilient urban atmosphere in the era of Industry 5.0. By leveraging multi-source satellite data—particularly Sentinel-2 and Landsat-8 imagery—in addition to hybrid CNN–RNN architectures, the model effectively captures both spatial and temporal features to improve UHI classification accuracy. The model combines satellite data from Sentinel-2 and Landsat-8 and processes the data using hybrid CNN–RNN to analyze spatial and temporal features for increased accuracy metrics, contributing to UHI dynamics and making it an easy for urban planners. Clubbing the model with Google Cloud AI enriches real-time data analysis to help plan smart cities and sustainable urban cover through the intersection of AI and Industry 5.0. This study acknowledges the demands and limitations of data integration, making the model reliable and timely. It not only fills the existing gaps in current research modules but also opens the way for future incorporation by taking full advantage of satellite data and cloud-based AI integration.

References

Alam, M. G., Waleed, M., Sajjad, M., & Acheampong, A. O. (2023). Towards sustainable and livable cities: Leveraging remote sensing, machine learning, and geo--information modelling to explore and predict thermal field variance in response to urban growth-. *Sustainability, 15*(2), 1416. https://doi.org/10.3390/su15021416

Bennett, E. M., Cumming, G. S., & Peterson, G. D. (2017). Guidelines for the management of socio-ecological systems in urban areas: A review of challenges and opportunities. *Urban Ecosystems*, *20*(6), 1221–1238. https://doi.org/10.1007/s11252-017-0682-0

Bibri, S. E. (2020). Data-driven smart sustainable cities of the future: An evidence synthesis approach to a comprehensive state-of-the-art literature review. *Sustainable Cities and Society*, *60*, 102301. https://doi.org/10.1016/j.scs.2020.102301

Chollet, F. (2017). Deep Learning with Python. Manning Publications.

De Almeida, C. R., & Teodoro, A. C. (2021). Study of urban heat island (UHI) using remote sensing data/techniques: A systematic review. *Remote Sensing Applications: Society and Environment*, *22*, 100504. https://doi.org/10.1016/j.rsase.2021.100504

Duggal, A. S., Malik, P. K., Gehlot, A., Singh, R., Gaba, G. S., Masud, M., & Al-Amri, J. F. (2022). A sequential roadmap to Industry 6.0: Exploring future manufacturing trends. *IEEE Access*, *10*, 83657–83678. https://doi.org/10.1109/ACCESS.2022.3184915

ESA (European Space Agency). (2015). Sentinel-2 User Handbook. https://sentinel.esa.int/documents/247904/685211/Sentinel-2_User_Handbook

Fischer, G., Hizsnyik, E., Prieler, S., Shah, M., & van Velthuizen, H. (2020). Linking land use and socio-economic dynamics: New methods and data sources. *Ecological Economics*, *169*, 106488. https://doi.org/10.1016/j.ecolecon.2019.106488

Gao, Y., Chen, X., Zhang, L., & Liu, X. (2017). *Data fusion for vegetation and temperature monitoring. Remote Sensing, 9*(10), 1034. https://doi.org/10.3390/rs9101034

Ghosh, I. (2019). 70 Years of Urban Growth in 1 Infographic. Visual Capitalist. www.visualcapitalist.com/urbanization-in-70-years/

Goodfellow, I., Bengio, Y., & Courville, A. (2016). Deep Learning. MIT Press.

Haklay, M. (2010). How good is volunteered geographical information? A comparative study of OpenStreetMap and Ordnance Survey datasets. *Environment and Planning B: Planning and Design*, *37*(4), 682–703. https://doi.org/10.1068/b35097

Huang, C., Wang, Y., Yang, J., & Liu, Y. (2021). Integrating multi-source remote sensing data for land cover mapping: A review of current methods and future directions. *Remote Sensing of Environment*, *260*, 112466. https://doi.org/10.1016/j.rse.2021.112466

Jin, G., Liang, Y., Fang, Y., Shao, Z., Huang, J., Zhang, J., & Zheng, Y. (2023). *Spatio-Temporal Graph Neural Networks for Predictive Learning in Urban Computing: A Survey*. https://doi.org/10.48550/arXiv.2303.14483

Lefebvre, A., Sannier, C., & Thomas, I. (2017). Monitoring urban areas with Sentinel-2A data: Application to the update of the Copernicus high resolution layer imperviousness degree. *Remote Sensing*, *9*(4), 342. https://doi.org/10.3390/rs9040342

Li, Z., Shen, H., Li, H., Xia, G., Gamba, P., & Zhang, L. (2020). Multi-feature combined cloud and cloud shadow detection in GaoFen-1 wide field of view imagery. *ISPRS Journal of Photogrammetry and Remote Sensing*, *166*, 254–268. https://doi.org/10.1016/j.isprsjprs.2020.06.006

Liu, Z., Deng, Z., He, G. et al. (2022). Challenges and opportunities for carbon neutrality in China. *Nature Reviews Earth & Environment*, *3*, 141–155. https://doi.org/10.1038/s43017-021-00244-x

Lyu, F., Wang, S., Han, S. Y. et al. (2022). An integrated cyberGIS and machine learning framework for fine--scale prediction of urban heat island using satellite remote sensing and urban sensor network data-. *Urban Informatics, 1*(1), 6. https://doi.org/10.1007/s44212-022-00002-4

Manoli, G., Fatichi, S., Schläpfer, M., Yu, K., Crowther, T. W., Meili, N., Burlando, P., Katul, G. G., & Bou-Zeid, E. (2019). Magnitude of urban heat islands largely explained by climate and population. *Nature*, *573*(7772), 55–60. https://doi.org/10.1038/s41586-019-1512-9

Mohamed, N., Rao, L. S., Sharma, M., Rajasekaran, S. B., & Alfurhood, B. S. (2021). In-depth review of integration of AI in cloud computing. *Journal of Cloud Computing*, *10*, 33. https://doi.org/10.1186/s13677-021-00241-1

Müller, J. M., Kiel, D., & Voigt, K.-I. (2020). Industry 4.0 and Industry 5.0: Towards an integrated framework for the future of work. *Journal of Industrial Information Integration, 20*, 100168. https://doi.org/10.1016/j.jii.2020.100168

Pan, S. J., & Yang, Q. (2010). A survey on transfer learning. *IEEE Transactions on Knowledge and Data Engineering, 22*(10), 1345–1359. https://doi.org/10.1109/TKDE.2009.191

Phiri, D., & Nyirenda, V. R. (2021). Sentinel-2 data for land cover/use mapping: A review. *Remote Sensing, 13*(13), 2305. https://doi.org/10.3390/rs13132305

Ramsundar, B., & Zadeh, R. (2019). TensorFlow for Deep Learning: From Linear Regression to Reinforcement Learning. O'Reilly Media.

Schroeder, W., Oliva, P., Giglio, L., & Csiszar, I. A. (2016). Remote sensing for the sustainable development goals: A review of applications and challenges. *Remote Sensing of Environment, 179*, 134–152.

Seto, K. C., Frolking, S., Mahtta, R., Milliman, T., & Esch, T. (2020). Global urban structural growth shows a profound shift from spreading out to building up. *Nature Sustainability, 3*, 305–314. https://doi.org/10.1038/s41893-019-0468-6

Shafiq, M., Obidallah, W. J., Fan, Q., Bilal, A., & Alduraywish, Y. A. (2025). Scale-wise interaction fusion network for land cover classification of urban scene imagery. *IET Image Processing, 19*(1). https://doi.org/10.1049/ipr2.70139

Sherstinsky, A. (2020). Fundamentals of recurrent neural network (RNN) and long short-term memory (LSTM) network. *Physica D: Nonlinear Phenomena, 404*, 132306. https://doi.org/10.1016/j.physd.2019.132306

Shorten, C., & Khoshgoftaar, T. M. (2019). A survey on image data augmentation for deep learning. *Journal of Big Data, 6*, 60. https://doi.org/10.1186/s40537-019-0197-0

Skakun, S., Wevers, J., Brockmann, C., Doxani, G., Aleksandrov, M., Batič, M., Frantz, D., & Louis, J. (2022). Cloud Mask Intercomparison eXercise (CMIX): An evaluation of cloud masking algorithms for Landsat 8 and Sentinel-2. *Remote Sensing of Environment, 274, 112990. https://doi.org/10.1016/j.rse.2022.112990*

Srivastava, N., Hinton, G., Krizhevsky, A., Sutskever, I., & Salakhutdinov, R. (2014). Dropout: A simple way to prevent neural networks from overfitting. *Journal of Machine Learning Research, 15*, 1929–1958.

United Nations. (2018). *World Urbanization Prospects: The 2018 Revision.*

Wang, X. (2020). Remote sensing applications to climate change. *Remote Sensing, 12*(2), 340.

Wellmann, T., Lausch, A., Andersson, E., Knapp, S., Cortinovis, C., Jache, J., Scheuer, S., Kremer, P., Mascarenhas, A., Kraemer, R., Haase, A., Schug, F., & Haase, D. (2020). Remote sensing in urban planning: Contributions towards ecologically sound policies. *Landscape and Urban Planning, 204*, 103921.

Weng, Q. (2009). Thermal infrared remote sensing for urban climate and environmental studies: Methods, applications, and trends. *ISPRS Journal of Photogrammetry and Remote Sensing, 64*(4), 335–344. https://doi.org/10.1016/j.isprsjprs.2009.03.007

Wulder, M. A., Roy, D. P., Radeloff, V. C., Loveland, T. R., Anderson, M. C., & Johnson, D. M. (2022). Fifty years of Landsat science and impacts. *Remote Sensing of Environment, 280*, 113195.

Xu, H., Omitaomu, F., Sabri, S., Zlatanova, S., Li, X., & Song, Y. (2024). *Leveraging Generative AI for Urban Digital Twins: A Scoping Review on the Autonomous Generation of Urban Data, Scenarios, Designs, and 3D City Models for Smart City Advancement.* https://doi.org/10.48550/arXiv.2405.19464

Zhang, J., Feng, Y., & Li, C. (2020). Data fusion for land cover mapping: A review of methods and applications. *International Journal of Applied Earth Observation and Geoinformation, 93*, 102193.

Zhu, X. X., Tuia, D., Mou, L., Xia, G.-S., Zhang, L., Xu, F., & Fraundorfer, F. (2017). Deep learning in remote sensing: A comprehensive review and list of resources. IEEE

Geoscience and Remote Sensing Magazine, 5(4), 8–36. https://doi.org/10.1109/MGRS.2017.2762307

Zhu, Z., & Woodcock, C. E. (2012). Fmask: A flexible algorithm for cloud and cloud shadow detection in Landsat imagery. *Remote Sensing of Environment*, *118*, 83–94.

Zou, X., Yan, Y., Hao, X., Hu, Y., Wen, H., Liu, E., Zhang, J., Li, Y., & Liang, Y. (2024). *Deep Learning for Cross-Domain Data Fusion in Urban Computing: Taxonomy, Advances, and Outlook*. https://doi.org/10.48550/arXiv.2402.19348

12 Eco-Innovation and Circularity in Industry 5.0

Pathways to Resilient Industrial Futures

Nasir Assad

12.1 Introduction

The history of industrial systems over the last two centuries has been marked by accelerated technological progress, first driven by mechanization and, more recently, by digitization (Szeszák et al., 2025). The latest development, known as, Industry 4.0, has opened a new era of smart technologies, such as artificial intelligence (AI), the Internet of Things (IoT), and cyber–physical systems, changing manufacturing into a highly competitive, efficient, and data-driven enterprise (Kumar et al., 2024a). Industry 4.0 has succeeded in making the operations more productive, but automated and digitalized industry plus efficiency in operation has come at the expense of environmental friendliness and inclusiveness of humans. In its turn, the new Industry 5.0 paradigm strives to overcome these failures by focusing on the notions of sustainability, humanistic values, and population welfare in the process of industrial innovation. *Industry 5.0* refers to the next evolution of industrialization that emphasizes human–machine collaboration, sustainability, and resilience, going beyond the automation and efficiency focus of Industry 4.0 (Golovianko et al., 2023). *Sustainability* denotes the capacity to meet present needs without compromising the ability of future generations to meet their own, encompassing environmental protection, social equity, and economic viability (Mensah, 2019).

Industry 5.0 is more about a change of attitude: replacing its technology as a means of achieving maximum efficiency with technology as a means of attaining ecological care, ethical aim, and generational sustainability (Rame et al., 2024). The chapter presents the main conceptual premises of Industry 5.0, especially its guiding idea of sustainability. In contrast to Industry 4.0, where sustainability was commonly considered a secondary effect of optimizing operations, Industry 5.0 integrates it proactively across all levels of the industrial value chain, including design and production, consumption, and recovery after use. It requires a redesign of industrial ambitions in line with planetary boundaries and social justice. The growing global environmental problems underscore the necessity of such a transition. The effects of climate change, pollution, water scarcity, deforestation, and biodiversity loss are no longer sometime in the future, but they are here and getting worse. The pressure on resources is becoming increasingly sharp, particularly in areas with high industrialization activities but low regulatory systems. Additionally,

DOI: 10.4324/9781003647621-15

social disparities, including automation-related job losses and unequal access to the benefits of technology, have increased with the advent of Industry 4.0. These facts emphasize the necessity of new industrial systems that will be not only high-tech but also environmentally sound and socially fair (Ajaj et al., 2024).

Industry 5.0 here provides a multidimensional model of sustainability with three main dimensions involved, namely, environmental regeneration, social inclusions, and technological human symbiosis. One of the fundamental foundations of this framework is the implementation of circular economy (CE) activities. *Circular economy* is a regenerative economic model aimed at minimizing waste and making the most of resources by prioritizing reuse, recycling, and sustainable design (Morseletto, 2020). The linear industrial production system, which is essentially based on extraction, manufacturing, consumption, and disposal, is unsustainable. CE systems, on the other hand, focus on resource efficiency, waste minimization, and regenerative design to maintain material use for as long as possible, as well as extract value at each stage of the product lifecycle. This involves working with disassembly, reuse, recycling, remanufacturing, and developing closed-loop production setups (Zeb et al., 2024).

These principles require technological developments. Industry 5.0 refers to the transformation of traditional factories into innovative and auto-correcting ecosystems, taking into account both the operational efficiency of those factories and the environmental sustainability of those operations. AI can monitor energy consumption in real time and predict the best setup in the cloud to minimize emissions. IoT-enabled sensors can monitor waste generated and even track wasteful components of the production process. Blockchain enables the tracing of complex supply chains, allowing materials to be sourced ethically and sustainability claims to be verified. These products transform factories into intelligent, self-correcting ecosystems that support environmental initiatives without compromising productivity. In addition to material and process novelties, Industry 5.0 focuses on the human role in industrial systems. Its goal is to bring the human touch back in manufacturing, not as a replacement for automation but as a combination with human creativity, intelligence, and ethics. Not simply "smarter," these algorithms aim to create factories that are more responsive to the values that matter to humans and the needs of society. It is a human-centered approach in that we design technology to empower, rather than replace, the workforce; promote human co-creation with machines; and make innovation as inclusive as possible. Industry 5.0 also advocates for a healthy, meaningful, and rewarding job that is safe, recognizing that sustainability is not only about the environment but also about developing a fair and equitable social fabric (Sah & Shaikh, 2024).

The consumer role is the third critical field of change in Industry 5.0, which is consumer behavior. With increasing sensitivity to ecological issues, there is a growing demand for transparency, responsibility, and an ethical profile from brands and products. Industry 5.0 addresses these calls for action through digital traceability systems, product labeling, and lifecycle transparency tools, which provide consumers with the information they need to make informed decisions. In the process, it generates positive feedback: sustainable consumer choices strengthen sustainable producer choices and help drive green innovation into the mainstream of business. At the policy level, the shift to Industry 5.0 depends on a conducive rule of law. The role of governments is central in facilitating sustainability, with interventions such as green tax breaks, carbon pricing, public procurement standards, and research support for eco-innovation. Embedded in this framework, industries must align with national and international environmental goals, such as the United Nations' Sustainable Development Goals (SDGs) and climate targets outlined in the Paris Agreement. Multilevel governance, i.e., cooperation among local, national, and global actors, will also be necessary for steering the equitable application of Industry 5.0 (Bartuś et al., 2025).

However, Industry 5.0 also entails significant challenges, some of which will be discussed in this chapter. These are seen in regulatory fragmentation, where divergent standards among countries block sustainable innovation; financial obstacles, particularly for small and medium-sized enterprises (SMEs) that cannot afford to implement green technologies; and difficulties in technological integration, especially in the case of technological upgrading of obsolete systems. Moreover, organizational inertia against transformation, incompetence in technology, and a lack of sound metrics that measure sustainability performance are factors that frequently suppress or distort the deployment of Industry 5.0 strategies. These are obstacles that highlight the need to establish enabling ecosystems in which knowledge is shared, risks are shared, and the capacities of civil society organizations are developed. While previous literature has explored aspects of Industry 5.0 such as automation, digitalization, or isolated sustainability practices, there is limited comprehensive analysis integrating environmental sustainability, CE principles, ethical governance, and human-centric innovation within a single framework. This chapter addresses that gap by offering a multidimensional review that connects these elements and examines the systemic challenges and opportunities they present. The novelty of this work lies in its holistic exploration of how Industry 5.0 can operationalize sustainability across technological, environmental, and social dimensions, helping guide both academic understanding and practical implementation.

12.1.1 Conceptual Framework

This chapter is anchored in an integrated framework that combines three well-established theories:

1. Triple Bottom Line (TBL): Emphasizes environmental integrity, social equity, and economic viability (Elkington, 1997).
2. CE Loops: Design, reuse, remanufacture, and recycle loops that retain material value (Kirchherr et al., 2017).
3. Socio-Technical Multilevel Perspective (MLP): Interactions across niches, regimes, and landscape levels that govern technology transitions (Geels, 2012).

In this model (Figure 12.1), *Industry 5.0 technological enablers* (AI, IoT, and human–machine collaboration) form the core "niche" layer that drives innovation.

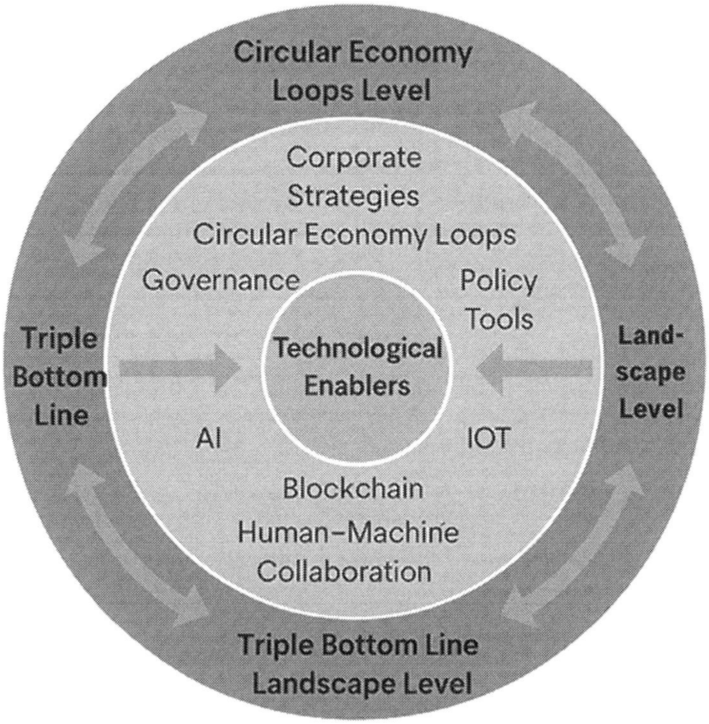

Figure 12.1 Integrated framework linking TBL objectives, CE loops, and the socio-technical MLP within an industry 5.0 context. The innermost layer represents technological enablers (AI, IoT, and cobots); the middle layer depicts organizational and policy regimes embedding circular practices; and the outer landscape layer shows macro-drivers (climate goals and SDGs). The arrows illustrate feedback among technology, circularity, and socioeconomic outcomes.

These enablers interact with regime elements—corporate strategies, supply chain structures, and policy instruments—to embed CE loops and deliver TBL outcomes (resource efficiency, decent jobs, and long-term profitability). The landscape layer comprises global sustainability drivers (e.g., SDGs and Paris Agreement) that exert pressure for large-scale transition. By situating Industry 5.0 within these overlapping lenses, the chapter shows *how* technological progress, circular resource flows, and social–ethical considerations reinforce one another, offering a robust scientific basis for the analyses that follow.

12.2 Sustainability as a Core Principle

Sustainability has evolved into a primary requirement, replacing secondary consideration within the context of modern industrial development, and specifically within the context of Industry 5.0. This new era of industrialization is as much about technological progress as it is about sustainability, ethical practices, and social equality. Climate change, natural resource depletion, and socioeconomic inequalities place increasing pressure on global industries, and sustainability becomes the key organizing principle in Industry 5.0, whereby technologies and production practices are oriented toward long-term viability and ecological harmony (Thanasi-Boçe & Hoxha, 2025).

With its approach, Industry 5.0 redefines the dynamic between humans and machines, naturally considering other organisms by fostering a harmonious, synergistic relationship between economic relationship advocates and organic relationship advocates. While Industry 4.0 had a strong focus on cyber–physical systems, automation, and digitalization toward technical efficiency, Industry 5.0 features a human-centered and environmentally friendly approach. Sustainability in Industry 4.0 was often viewed as a disadvantage, as it was perceived to be a trade-off between increased efficiency and reduced waste through automation. However, in Industry 5.0, it is a focus from the outset. Technologies are assessed based not only on their performance but also on their environmental footprint, material and energy consumption, and potential for reuse and regeneration, perhaps most notably in the trend toward CE practices, which reimagine the linear "take–make–dispose" model to prioritize reuse, recycling, and regenerative systems. Industry 5.0 takes into account the entire lifecycle of products and materials—from their creation through to the end of their useful life and potential for reuse, repurposing, or safe decomposition. It alleviates dependence on finite natural resources and minimizes waste and pollution (Sheikh et al., 2024).

Industry 5.0 also fosters the transition toward green technologies, including renewable energy systems, biodegradable materials, low-emission manufacturing processes, and carbon-neutral logistics. It is often supported at scale by purpose-built digital tools—such as AI, IoT, blockchain, and big data analytics—designed to further sustainability objectives. AI optimizes energy consumption to reduce the environmental impact in smart factories. IoT devices monitor and track emissions and resource use in real time, and blockchain provides complete and transparent tracking from raw materials to final products, ensuring the ethical sourcing and sustainability of supply chain practices. The most fundamental difference between Industry 4.0 and Industry 5.0 lies in their value orientation. Industry 4.0 prioritized efficiency and productivity, focusing on aspects such as machine autonomy and cost savings. By contrast, Industry 5.0 innovators give greater weight to the impacts of industrial activity on ecosystems, communities, and future generations, leading to more purpose-driven innovation. Innovating technologies today means not only increasing speed or reducing costs but also achieving our goals through approaches that are regenerative, inclusive, and resilient. Now, an additional point: Industry 5.0 sustainability necessarily encompasses social and ethical dimensions. Human–machine collaboration is intended to enhance human creativity and decision-making (or perhaps the best aspects of human labor) rather than replace it, ensuring that technological advances liberate rather than further strain the workforce. With changing regulatory frameworks and incentives, policymakers, industries, and consumers are expected to contribute to the transition toward sustainable practices (Ghobakhloo et al., 2022).

Within the framework of Industry 5.0, overall sustainability is more than a target: it is the basis for all industrial decisions. It moves the industrial value system away from maximal value and toward one in balance with the purpose. Meaningful human-centric tech Industry 5.0 proposes a powerful future for humanity. By framing industrial development as something that must pay equal attention to ecological and social well-being, Industry 5.0 gives us a glimpse into a future where technological innovation exists in harmony with the rest of humanity and the earth. The basic dimensions of sustainability in Industry 5.0 are CE approaches, green technologies, and human-centric and ethical principles (Figure 12.2). In Industry 5.0, routine, repetitive, and data-intensive tasks—such as predictive maintenance, process optimization, and quality monitoring—are increasingly automated using AI and IoT. Meanwhile, humans remain central in areas requiring creativity, critical thinking, emotional judgment, and ethical oversight, such as in product innovation, strategic decision-making, and managing stakeholder relationships. Building on sustainability as the foundational value, the next section explores how CE practices serve as a practical and systemic pathway to embed sustainability throughout industrial processes in Industry 5.0.

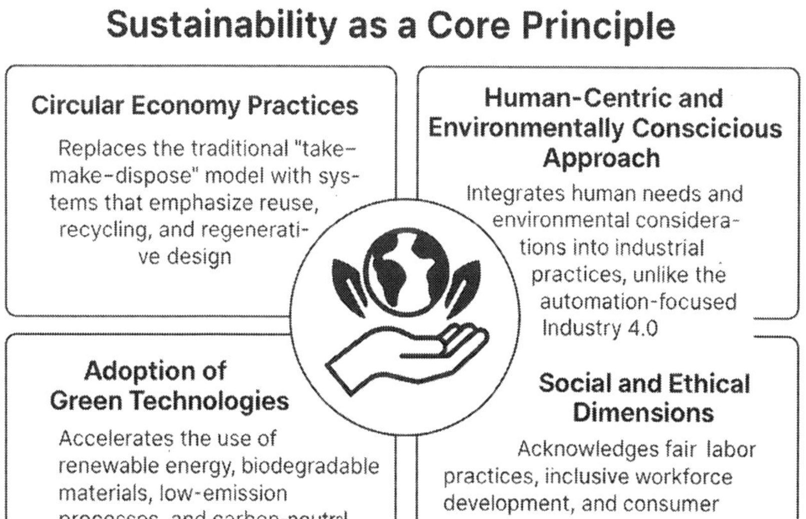

Figure 12.2 Key pillars of sustainability as a core principle in Industry 5.0, integrating environmental, technological, social, and ethical components essential to a resilient and regenerative industrial model.

12.3 Circular Economy Practices

CE practices are more than innovative solutions: CE is a pivotal vehicle for sustainable industrial transformation in the Industry 5.0 context. The old linear economic model, characterized by "take–make–use–dispose," leads to excess resource depletion, environmental destruction, and unsustainable production activities. Conversely, CE seeks to maximize the value retention of products, materials, and resources by designing closed-loop systems in all economic activities to the greatest extent possible, thereby reducing waste and pollution (Turner et al., 2022).

Waste minimization is a key component of CE. In this regard, Industry 5.0 encourages the design of products and processes for waste generation prevention at the source (Ghobakhloo et al., 2023). This spans from modular product designs that enable repair and reuse to material recovery systems that recover value from goods at the end of their life, as well as smart manufacturing practices that decrease scrap and surplus. Digital technologies—especially AI and IoT—enable the real-time tracking and control of product flow, allowing manufacturers to make adjustments that prevent overproduction and inefficiencies. This is also on the heels of resource optimization, which is the smart utilization and re-utilization of materials and energy over the lifecycle of production. Data analytics, machine learning, and predictive modeling underlie Industry 5.0 systems that further enhance supply chains, allowing resources and energy consumption to be minimized and raw material efficiency to be optimized. Doing so not only reduces production costs but also mitigates the environmental footprint caused by the mining, transportation, and disposal of raw materials. Within smart factories, materials and energy are continually monitored to enable informed, sustainable decisions (Bamakan et al., 2025).

Regenerative design represents a key principle that transcends conventional harm-reduction approaches by aiming to restore, renew, and enhance natural systems through integrative and adaptive strategies. For instance, regenerative design seeks to create products and industrial processes that are net positive (e.g., repairing ecosystems, enhancing soil and water quality, and sequestering more carbon than they emit). This approach embodies a future-driven design culture that integrates sustainability into the core of the innovation process. Materials are selected for their performance and cost, and also for their renewability, biodegradability, and safe reintegration into the environment. The significance of CE strategies in Industry 5.0 is being backed by a collaborative ecosystem that involves stakeholders ranging from suppliers and manufacturers to consumers and policymakers. Blockchain and smart contracts enable transparent and interconnected supply chains that trace materials, ensure alignment with EU legal requirements, and build trust in CE processes. In addition, consumer involvement is critical, as the shift toward sustainable products only further widens the market, making more circular products available (Payer et al., 2024).

While Industry 5.0 enhances material efficiency through its strategies and technologies, it also rethinks value creation by incorporating the principles of CE. Circular models contribute to both ecological regeneration and economic resilience by aligning industrial activities with environmental limits and social equity. The shift from linear to circular is not just a different way to operate within industries; it is an essential reimagining of industries that harmonize with the planet and with one another to create the foundation for a regenerative, inclusive, and sustainable future. Beyond circular models, technological innovation plays a crucial role in achieving environmental and economic transformation. The following section examines key eco-innovations and green technologies enabling sustainable industrial advancement.

12.4 Eco-Innovation and Green Technologies

Eco-innovation involves the development and application of novel products, processes, or systems that contribute to environmental sustainability while also delivering economic and social value (Díaz-García et al., 2015). Eco-innovation and green technologies—addressing environmental challenges and ensuring industrial competitiveness—are key enablers of sustainable transformation and Industry 5.0 development. Whereas conventional innovations tend to focus on generating profit and achieving higher performance and productivity, eco-innovation focuses on developing and implementing solutions that provide environmental benefits while creating additional economic and social value. Such innovations are crucial in driving the decarbonization of industries, reducing the consumption of natural resources, and facilitating a transition toward a circular and low-carbon economy (Ghobakhloo et al., 2023).

Eco-innovation is fundamentally the smart use of emerging digital technologies to enhance sustainability performance. The means can also include the use of technology including AI as is fundamental to maximize energy, predict the maintenance required for the equipment, and reduce the amount of waste material. AI can also be utilized to run these systems and analyze large datasets in real time to identify abnormalities, recommend optimization for energy-efficient processing, and adapt to environmental changes. It also helps reduce emissions, extends the lifespan of equipment, and reduces production costs with environmental pretensions. Green technologies are also empowered by the IoT, with smart ecosystems connected across production and supply chains. IoT devices enable the collection and transmission of accurate and real-time data on energy consumption, emission parameters, material flows, and machine parameters, allowing them to be constantly monitored and promptly controlled. In use cases such as smart factories, IoT sensors identify leaks, track air quality, and measure the greenhouse gases (GHGs) emitted, enabling corrective action to be taken without delay. Such responsiveness in real time lowers the operational inefficiencies and environmental threats (Fraga-Lamas et al., 2021).

Blockchain technology offers a transparent and accountable solution for sustainable sourcing across complex global supply chains. By tracking products from raw materials to end users, it ensures adherence to environmental standards, ethical labor, and social responsibility. Industries like apparel, electronics, and agriculture, where supply chains are often opaque, can particularly benefit from blockchain's immutable records. While some critique it for not addressing deeper socioeconomic issues, it is widely recognized as a crucial step toward long-term reform. Industry 5.0 also emphasizes the use of biodegradable and eco-friendly materials—such as plant-based plastics and recycled composites—to reduce environmental impact. These materials naturally break down without leaving toxic residues, making them ideal for packaging and disposable goods. Equally important are energy-efficient production systems that eliminate reliance on fossil fuels and lower GHG emissions. The integration of renewable energy sources—such as solar, wind, and bioenergy—with smart utility technologies and storage solutions supports a more sustainable and stable power supply. Upgrading to energy-efficient equipment and systems further reduces operational waste and boosts performance (Dai et al., 2024).

Carbon-neutral supply chains represent a critical goal in the green technology agenda. The net carbon emission of the companies is expected to be zero because of the use of renewable energy, carbon offsetting strategies, and streamlined logistics. Technologies such as route optimization software, low-emission transportation, and digital inventory management assist in reducing the environmental footprint of distribution and delivery procedures. A comparative overview of key eco-innovations and green technologies adopted in Industry 5.0, like their environmental, financial, and social impacts, is presented in Table 12.1. Collectively, these eco-innovations and green technologies define the technological backbone of Industry 5.0's sustainability vision and enable industries to operate in ways that are not only productive and competitive but also environmentally responsible and socially mindful. By seamlessly integrating these technologies into their main strategies, forward-thinking organizations can drive meaningful transformation, contributing to universal sustainability goals while safeguarding their future operations. In addition to qualitative benefits, Industry 5.0's sustainability outcomes can be tracked using quantitative performance indicators. Common metrics include reduction in carbon emissions (e.g., kg CO_2-equivalent per unit of production), energy efficiency (e.g., kilowatt-hour per unit output), material recovery rates (e.g., percentage of components recycled or remanufactured), and water usage efficiency (liters per product) (Gang et al., 2013). Frameworks such as the Global Reporting Initiative (GRI); environmental, social, and governance (ESG) metrics; and ISO 14001 environmental management standards provide structured guidelines to assess and report these outcomes, enhancing transparency and accountability (Ilori et al., 2023). Several Industry 5.0 initiatives have already begun demonstrating sustainability in practice. For instance, BMW's "Factory of the Future" in Germany integrates human-centric cobots and AI to reduce waste and optimize energy usage across its assembly lines (Tallat et al., 2023). In Japan, the company Omron Corporation uses AI-powered visual inspection systems to minimize production defects and material waste, enhancing circular practices (Alhakimi, 2024). Similarly, Procter & Gamble (P&G) has incorporated blockchain traceability in its supply chains to ensure ethical sourcing and carbon footprint tracking, aligned with ESG goals (Dua, 2025). These pilot applications illustrate the feasibility and impact of Industry 5.0 technologies when aligned with sustainability targets.

Table 12.1 Summary of Key Eco-Innovations and Green Technologies in Industry 5.0 with a Focus on Their Environmental, Economic, and Social Impact and on the Main Issues of Adoption.

Eco-Innovation/Green Tech	*Environmental Benefit*	*Economic Impact*	*Social Impact*	*Adoption Challenges*
AI-driven energy optimization	Reduces energy consumption and emissions	Lowers operational costs over time	Improves worker conditions through smarter systems	High setup and data infrastructure costs
Industrial IoT (IIoT)	Improves monitoring and minimizes resource waste	Reduces downtime and maintenance costs	Supports safer workplaces and better maintenance planning	Integration complexity with legacy systems
Blockchain traceability	Ensures sustainable and ethical sourcing	Increases consumer trust and brand value	Promotes labor rights and supply chain transparency	Lack of regulatory standards for data sharing
Biodegradable materials	Reduces landfill waste and microplastic pollution	Decreases waste management costs	Encourages sustainable consumer behavior	Higher upfront costs and limited scalability
Renewable energy integration	Decreases GHG emissions and fossil fuel reliance	Reduces energy costs, especially long-term	Enables energy access in remote areas	Grid infrastructure and intermittency issues
Carbon-neutral logistics	Lowers carbon footprint in transportation	Optimizes supply chain efficiency	Supports green job creation in the logistics sector	Coordination between multiple actors, cost of alternative fuels

While technological advancements are central to sustainability, Industry 5.0 also requires a deep reflection on its social and ethical dimensions. The next section focuses on how these innovations impact labor, consumption, and governance in a human-centric industrial paradigm.

12.5 Societal and Ethical Implications

The move to Industry 5.0 is not just technological and environmental—it is profoundly social and ethical. The labor, consumer, and governance implications are also becoming increasingly relevant as industries converge and design solutions that are sustainable, human-centered, and nature-based. Industry 5.0 values a holistic perspective on progress, where technologically driven development goes hand in hand with human dignity, civil rights, and ethical responsibility, in addition to economic and environmental goals (Longo et al., 2020).

Industry 5.0 marks a monumental turning point in our relationship with technology and the future of work. Whereas past industrial revolutions emphasized mechanization and automation that displaced many jobs, this new era envisions collaboration where machines empower rather than replace human skills. It aims to augment our creativity, nuanced decision-making, and emotional connections in digital sustainability and green design, and ensures the ethical use of AI. Emotional intelligence in Industry 5.0 refers to the ability of systems to support human-centered environments—for instance, collaborative robots (cobots) adjust to a worker's pace or AI chat interfaces that detect stress or satisfaction in customer service. Ethical AI is operationalized through "human-in-the-loop" systems, where final decisions—such as in recruitment, credit scoring, or medical diagnostics—are made or validated by human experts to ensure fairness, accountability, and alignment with human values. Still, reskilling the workforce for these evolving roles will be a significant task, requiring the adaptation of training to emerging tech-enabled careers. Beyond jobs, consumer habits are also transforming as sustainability becomes a core aspect of brands and goods. Customers increasingly demand to know how, where, and by whom products are made so that they can buy in alignment with their principles. Thanks to blockchain and barcodes, verifying lifecycles and origins is now possible, allowing shoppers to make informed choices that reflect their values. As a result, businesses must reform to meet rising consumer standards for transparency, responsibility, and sustainable goods if they hope to survive in this time of sweeping societal change (Kumar et al., 2024b).

In addition, the ethical considerations of Industry 5.0 include its regulatory and policy context. Policymakers and institutions now need to develop frameworks that generate incentives for sustainability and equity while safeguarding workers in the context of a rapidly transforming industrial economy. This means creating green legislation, offering tax incentives for low-carbon practices, providing subsidies for green technology, and implementing regulations on the responsible use of AI. These frameworks also need to consider digital ethics, data privacy, environmental justice, and access to green technologies in developing areas. There are also important ethical concerns in the design and use of technology. If AI is used in decision-making, it should be transparent, fair, and without bias that marginalizes certain communities. Similarly, green technologies should not further entrench global disparities—for example, accessing rare earths in ways that exploit vulnerable labor markets or geographic hotspots. A far more equitable BMI would have to combat technological elitism to ensure that all societies can benefit from sustainable development and contribute to it (Gamberini & Pluchino, 2024).

At its core, Industry 5.0 shifts the concept of industrial ethics to encompass not only what is technically and economically viable but also what is socially equitable and environmentally feasible. It encourages industries to reassess their profit margins and efficiency measures in light of the wider human and ecological consequences of their work. By incorporating ethical foresight into innovation, Industry 5.0 aims to shape an industrial future that's more fair, transparent, and inclusive—one where technology serves the interests of society rather than the other way around. Fundamental societal and ethical dimensions of Industry 5.0—labor environment (labor transformation), consumption (ethical consumption), governance (responsible governance), and usage (fair usage of technology)—are depicted in Figure 12.3. Addressing these social and ethical dimensions brings

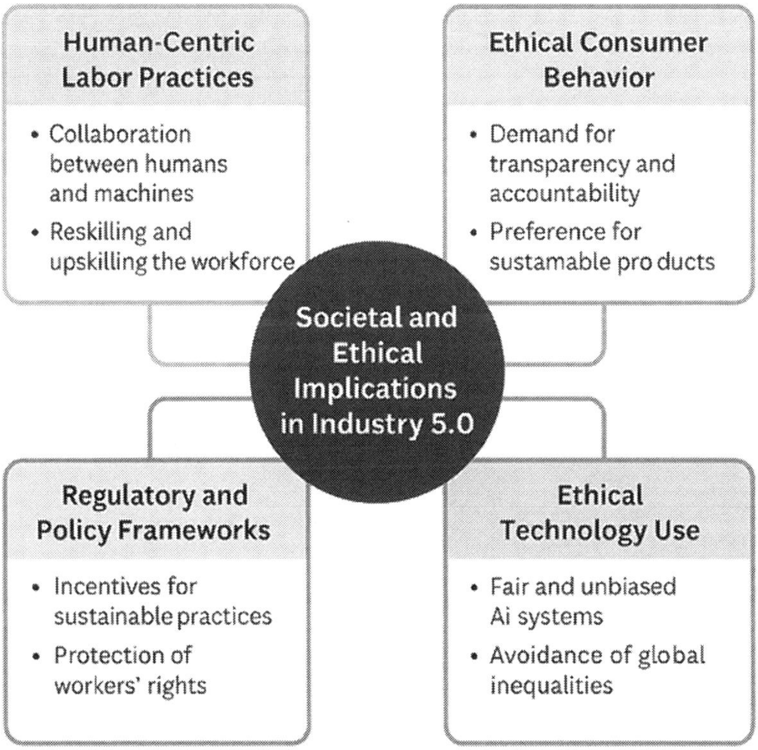

Figure 12.3 Key societal and ethical implications of Industry 5.0, highlighting the interconnections among labor practices, consumer behavior, governance frameworks, and ethical technology use within a human-centric industrial ecosystem.

Table 12.2 Comparative overview of key sustainability elements distinguishing Industry 4.0 and Industry 5.0.

Sustainability Element	*Industry 4.0*	*Industry 5.0*
Core focus	Efficiency and automation	Sustainability, resilience, and human-centric innovation
Environmental consideration	Secondary outcome	Primary design principle
Human role	Reduced (automation-centric)	Central (human–machine collaboration)
CE Integration	Minimal	Fully embedded in design and production
Ethical and social dimensions	Limited engagement	Integral to policy and practice
Technology use	AI, IoT, and cyber–physical systems	AI, IoT, blockchain, ethical AI, and cobots
Labor impact	Job displacement risk	Job augmentation and skill development
Sustainability metrics	Operational KPIs	TBL (social, environmental, and economic indicators)
Policy engagement	Low	High: alignment with SDGs and Paris Agreement

Note: The table highlights differences in environmental priorities, human roles, technology use, and alignment with global sustainability goals, emphasizing Industry 5.0's shift toward a holistic, human-centric, and regenerative industrial model.

attention to the practical barriers that industries face in adopting Industry 5.0. The next section outlines the key regulatory, financial, and organizational challenges to implementation. A comparative summary of sustainability elements distinguishing Industry 4.0 and Industry 5.0 is provided in Table 12.2 to illustrate this paradigm shift more clearly.

12.6 Challenges and Barriers

While Industry 5.0 presents a compelling vision for sustainable, human-centric, and technologically advanced industrial systems, its implementation is not without significant challenges. The transition from conventional linear models to circular, ethical, or green industrial models is hindered by technical, financial, regulatory, and behavioral barriers that need to be recognized and strategically addressed (Martini et al., 2024).

A key challenge is the diversity of regulatory systems across regions and industries. In contrast to formal regulation, which, as we have seen, is often written for linear, resource-intensive models, a regulation informed by a sustainability ethos is still emerging, and its forms or boundaries are not yet settled. Transnational companies may encounter different environmental standards and reporting requirements, as well as varying compliance systems, which can reduce barriers to new and innovative yet nonstandard, environmentally friendly technologies. Additionally, onerous bureaucracy and resistance to policy change in various countries impede the fast-tracking of changes driven by sustainability. Long-term investments and risk evaluations when industries consider investing in greener alternatives are also influenced by regulatory uncertainty (Shabur et al., 2025).

Most sustainability projects—such as renewable energy infrastructure, green materials, waste management, and digital tools like AI, IoT, and blockchain—require high upfront investment. This poses a major challenge for SMEs, which often lack access to green financing or incentive programs. Despite long-term benefits like reduced operating costs and enhanced brand value, the steep initial cost hinders adoption. This creates a need for innovative funding models, public–private partnerships, and targeted government subsidies. In addition to cost, technology integration is another barrier. Industry 5.0 promotes advanced digital tools for sustainability, but many industries still rely on outdated systems incompatible with such technologies. Upgrading to smart, energy-efficient systems demands time and expertise, and can disrupt existing workflows. Furthermore, achieving interoperability across decentralized supply chains is difficult. Building an integrated, sustainability-focused framework that aligns diverse platforms, sensors, and data systems is both technically complex and financially demanding (Katari et al., 2024).

Workforce readiness and organizational culture also play a significant role. Many organizations lack the right tools, mentality, or leadership to pursue this path of sustainable innovation (Knap-Stefaniuk, 2024). A shift in architecture toward sustainability and human–machine cooperation requires technical proficiency not only in the emerging new technologies but also in environmental science, ethical reasoning, and systemic reasoning. Change resistance, poor sustainability training, and limited employee involvement can hinder or sabotage their transformation. In addition, conventional KPIs that are inclined toward output and efficiency often fail to encompass sustainability performance, which would encourage managers to adopt greener practices. There are also ethical and operational challenges surrounding data transparency and accountability. The tracking and reporting of sustainability metrics, such as carbon emissions, material sourcing, or energy consumption, via digital technology, have implications for the accuracy of data and issues around ownership and privacy. With no established methods for measuring sustainability, it is more difficult to standardize progress across organizations and check on the authenticity of the so-called green claims, which has led to greenwashing. This mistrust further discredits the sustainability initiatives in consumer confidence. Finally, geopolitical and economic instability may create a structural deterrent to sustainable industrial growth. Politically conflicted areas, regions with inflation, resource limits, or financial difficulties can be more focused on living and financial prosperity instead of protecting the environment. Funding sustainable infrastructure, education, or a change in policy is secondary in such cases and widens the gap in the world in terms of Industry 5.0 preparedness and strengthens inequality (Romanova & Kuzmin, 2024).

In addition to regulatory, financial, and integration challenges, Industry 5.0 also poses significant socio-environmental risks (Ghobakhloo et al., 2024). Technological unemployment remains a critical concern, as increased automation and AI adoption may displace certain job categories (Kolade & Owoseni, 2022). Moreover, the digital divide, especially between the developed and developing world, can increase disparities in access to and uptake of Industry 5.0 innovations worldwide. There is also the risk of unintended environmental consequences, such as increased energy uses from digital infrastructures or the generation of e-waste from rapid technology turnover, which must be carefully managed.

Although Industry 5.0 has the potential to speed up the sustainability transition, its achievement is conditional upon the ability to overcome regulatory fragmentation, financial barriers, technological inhomogeneity, workforce shortage, and structural inequalities. These problems will have to be dealt with by concerted action of the government, industry, academia, and civil society. Only through identifying and making an effort to eliminate them will Industry 5.0 be able to deliver on its potential as a source of ecological balance and societal progress. In spite of these difficulties, there is potential future development. The final section discusses future opportunities, strategic directions, and collaborative actions needed to fully realize the potential of Industry 5.0.

12.7 Future Directions and Opportunities

Industry 5.0 opens a paradigm shift to transform the industrial system into a sustainable, inclusive, and resilient one. The integration of human values and technological innovation may serve as the guide to sustainable global economy as the world is shifting toward the abandonment of traditional models of economic growth.

One of the trends is scaling and mainstreaming green technologies. Although the implementation of AI-optimization- and IoT-based resources tracking, and the use of biodegradable materials is already present today, the future emphasis is on better cost effectiveness, accessibility, and interoperability. Private–public partnerships will play a crucial role in the further development of such technologies as green hydrogen, smart agriculture, and decentralized renewable energy.

Another priority focuses on the alignment of global policy. Climate pacts like Paris Agreement and European Green Deal push toward the pace of harmonization of regulations favorable to circularity, carbon neutrality, and ethical sourcing. Adoption can be advanced by the use of incentives such as green tax credits, emissions trading, and compulsory sustainability reporting.

It is necessary to develop workforce. The professional needs required by Industry 5.0 are associated not only with digital technologies but also with sustainable design and ethics. The educational programs should focus on green skills, interdisciplinary knowledge, and diversity to be innovative and everyone-friendly.

The cooperation among stakeholders, government, business, nonprofit organizations, academia, and communities is essential. Participatory governance is capable of establishing trust and fairness, and multi-stakeholder forums can scale down locally focused solutions, such as waste management, to community-based renewables.

New tools such as digital twins, lifecycle assessments, and predictive analytics will also enter the formula to make responsible decisions by modeling environmental impacts before taking action.

Finally, industrial success will have to be redefined, with old measures accompanied by sustainability ones, such as carbon emissions, biodiversity, and social well-being. The new model of businesses in the future should be value-focused on stakeholders, environment recovery, and sustainability.

Industry 5.0 does not only bring technological advancement but also an ethical and future-proof way to the welfare of people and our planet.

12.8 Conclusion

Industry 5.0 marks a defining moment in the evolution of industrial development—one where sustainability is not only an outcome but also a driving force of transformation. Unlike Industry 4.0, which focused on digitalization, automation, and productivity, Industry 5.0 is a paradigm shift that focuses on human-centric innovation, ecological integrity, and long-term resilience. It redefines industry as a force that is active in a global setting with respect to climate change, exhaustion of resources, deforestation, and inequality among the society. The integration among technological progress, environmental protection, and social responsibility should be the core of Industry 5.0. By implementing green technologies, CE principles, and eco-innovative policies, industries are remodeling the value creation process by moving away to more linear, extractive operating models towards more regenerative and inclusive ones. The use of AI, IoT, and blockchain technologies can provide not only increased productivity but also real-time tracking, traceability, and transparency that can be pivotal in creating sustainable and resilient supply chains. In addition, Industry 5.0 focuses on the interdependence of nature, technology, and people. It promotes an industrialization model in which machines do not substitute human creativity by enhancing it; in which products are visualized to restore rather than destroy; and in which economic development cannot be achieved at the cost of environmental and social quality. This convergence of digital smartness and environmental awareness creates new avenues to realize net-zero carbon emission, ecological consumption, and sustainable growth. Nevertheless, the path toward achieving the insignia of Industry 5.0 without limitations is not free of obstacles. High implementation costs, fragmented regulations, technological incompatibilities, and social resistance all pose serious challenges. Overcoming these barriers requires ***coordinated efforts among governments, industries, academia, and civil society***, in addition to robust policy support and inclusive innovation ecosystems. Equally critical is the investment in education and skills development to ensure that the workforce of tomorrow is prepared to lead and sustain this transformation. Looking ahead, Industry 5.0 represents more than a mere technological evolution; it offers a comprehensive vision for a sustainable industrial civilization where economic vitality, environmental preservation, and human dignity are harmoniously integrated. With sustainability embedded at its foundation, Industry 5.0 redefines the objectives of industrial advancement by aligning innovation with ecological and social imperatives. It introduces tools, frameworks, and methodologies capable of fostering resilience, equity, and long-term sustainability. In this context, Industry

5.0 emerges not only as a driver of technological progress but also as a powerful catalyst for globally inclusive and sustainable economic growth.

References

Ajaj, R., Buheji, M., & Hassoun, A. (2024). Optimizing the readiness for Industry 4.0 in fulfilling the Sustainable Development Goal 1: Focus on poverty elimination in Africa. *Frontiers in Sustainable Food Systems*, *8*, 1393935.

Alhakimi, A. M. (2024). The power of technology in Industrial Revolution 5.0. In *Utilizing Renewable Energy, Technology, and Education for Industry 5.0* (pp. 400–443). IGI Global. www.igi-global.com/chapter/the-power-of-technology-in-industrial-revolution-50/352710

Bamakan, S. M. H., Dehghan, F., & Akbarpour, S. (2025). Role of machine learning for predictive modeling for Industry 5.0 and Society 5.0. In *Human-Centric Integration of Next-Generation Data Science and Blockchain Technology* (pp. 265–285). Elsevier.

Bartuś, K., Kocot, M., & Sączewska-Piotrowska, A. (2025). Assessment and insights into the awareness and readiness of organizations to implement the assumptions of Industry 5.0: An examination of five Polish sectors. *Sustainability*, *17*(3), 903.

Dai, N. H. P., Giang, P. P., & Kiet, T. H. V. T. (2024). Readiness of enterprises to implement Industry 5.0: Challenges and security concerns. In *Modern Technologies and Tools Supporting the Development of Industry 5.0* (pp. 408–443). CRC Press.

Díaz-García, C., González-Moreno, Á., & Sáez-Martínez, F. J. (2015). Eco-innovation: Insights from a literature review. *Innovation*, *17*(1), 6–23. https://doi.org/10.1080/14479338.2015.1011060

Dua, G. K. (2025). Enhancing sustainable development through blockchain and artificial intelligence: Optimizing the supply chain and mitigating environmental footprints. In *Global Partnerships and Governance of Supply Chain Systems* (pp. 157–188). IGI Global Scientific Publishing. www.igi-global.com/chapter/enhancing-sustainable-development-through-blockchain-and-artificial-intelligence/369197

Elkington, J. (1997). The triple bottom line. *Environmental Management: Readings and Cases*, *2*(1997), 49–66.

Fraga-Lamas, P., Lopes, S. I., & Fernández-Caramés, T. M. (2021). Green IoT and edge AI as key technological enablers for a sustainable digital transition towards a smart circular economy: An Industry 5.0 use case. *Sensors*, *21*(17), 5745.

Gamberini, L., & Pluchino, P. (2024). Industry 5.0: A comprehensive insight into the future of work, social sustainability, sustainable development, and career. *Australian Journal of Career Development*, *33*(1), 5–14. https://doi.org/10.1177/10384162241231118

Gang, F., Stern, N., Edenhofer, O., Shanda, X., Eklund, K., Ackerman, F., Li, L., & Hallding, K. (2013). *The Economics of Climate Change in China: Towards a Low-Carbon Economy*. Routledge. https://books.google.com/books?hl=en&lr=&id=XdjZAAAAQBAJ&oi=fnd&pg=PP1&dq=In+addition+to+qualitative+benefits,+Industry+5.0%E2%80%99s+sustainability+outcomes+can+be+tracked+using+quantitative+performance+indicators.+Common+metrics+include+reduction+in+carbon+emissions+(e.g.,+kg+CO%E2%82%82-equivalent+per+unit+of+production),+energy+efficiency+(e.g.,+kWh+per+unit+output),+material+recovery+rates+(e.g.,+%25+of+components+recycled+or+remanufactured),+and+water+usage+efficiency+(liters+per+product).+&ots=880bSy75A1&sig=O2beZQeRAUA_3uYMr39iEv-GEgA

Geels, F. W. (2012). A socio-technical analysis of low-carbon transitions: Introducing the multi-level perspective into transport studies. *Journal of Transport Geography*, *24*, 471–482.

Ghobakhloo, M., Iranmanesh, M., Morales, M. E., Nilashi, M., & Amran, A. (2023). Actions and approaches for enabling Industry 5.0-driven sustainable industrial transformation: A strategy roadmap. *Corporate Social Responsibility and Environmental Management*, *30*(3), 1473–1494. https://doi.org/10.1002/csr.2431

Ghobakhloo, M., Iranmanesh, M., Mubarak, M. F., Mubarik, M., Rejeb, A., & Nilashi, M. (2022). Identifying Industry 5.0 contributions to sustainable development: A strategy roadmap for delivering sustainability values. *Sustainable Production and Consumption*, *33*, 716–737.

Ghobakhloo, M., Mahdiraji, H. A., Iranmanesh, M., & Jafari-Sadeghi, V. (2024). From Industry 4.0 digital manufacturing to Industry 5.0 digital society: A roadmap toward human-centric, sustainable, and resilient production. *Information Systems Frontiers*. https://doi.org/10.1007/s10796-024-10476-z

Golovianko, M., Terziyan, V., Branytskyi, V., & Malyk, D. (2023). Industry 4.0 vs. Industry 5.0: Co-existence, transition, or a hybrid. *Procedia Computer Science*, *217*, 102–113.

Ilori, O., Lawal, C. I., Friday, S. C., Isibor, N. J., & Chukwuma-Eke, E. C. (2023). A framework for environmental, social, and governance (ESG) auditing: Bridging gaps in global reporting standards. *International Journal of Social Science Exceptional Research*, *2*(1), 231–248.

Katari, V., Goyal, S., Nigam, V., Jana, M., Maitra, A., Sharma, H., & Sahu, K. K. (2024). A confluence of emerging technologies like IoT, edge & cloud computing, blockchain, Industry 4.0 & 5.0, AI & ML toward the realization of eco-friendly supercapacitors. In Ahamed, M. B., & Hussain, C. M. (Eds.), *ACS Symposium Series* (Vol. 1471, pp. 163–204). American Chemical Society. https://doi.org/10.1021/bk-2024-1471.ch007

Kirchherr, J., Reike, D., & Hekkert, M. (2017). Conceptualizing the circular economy: An analysis of 114 definitions. *Resources, Conservation and Recycling*, *127*, 221–232.

Knap-Stefaniuk, A. (2024). Exploring workforce skills in Industry 5.0 amid cultural diversity. In *Organizational Development, Innovation, and Economy 5.0* (pp. 74–85). Routledge.

Kolade, O., & Owoseni, A. (2022). Employment 5.0: The work of the future and the future of work. *Technology in Society*, *71*, 102086.

Kumar, N., Pal, S. K., Agarwal, P., Rosak-Szyrocka, J., & Jain, V. (2024b). *Human–Machine Collaboration and Emotional Intelligence in Industry 5.0*. IGI Global.

Kumar, S., Verma, A. K., & Mirza, A. (2024a). Digitalisation, artificial intelligence, IoT, and Industry 4.0 and digital society. In Kumar, S., Verma, A. K., & Mirza, A. (Eds.), *Digital Transformation, Artificial Intelligence and Society* (pp. 35–57). Singapore: Springer Nature. https://doi.org/10.1007/978-981-97-5656-8_3

Longo, F., Padovano, A., & Umbrello, S. (2020). Value-oriented and ethical technology engineering in Industry 5.0: A human-centric perspective for the design of the factory of the future. *Applied Sciences*, *10*(12), 4182.

Martini, B., Bellisario, D., & Coletti, P. (2024). Human-centered and sustainable artificial intelligence in Industry 5.0: Challenges and perspectives. *Sustainability*, *16*(13), 5448.

Mensah, J. (2019). Sustainable development: Meaning, history, principles, pillars, and implications for human action: Literature review. *Cogent Social Sciences*, *5*(1). https://doi.org/10.1080/23311886.2019.1653531

Morseletto, P. (2020). Restorative and regenerative: Exploring the concepts in the circular economy. *Journal of Industrial Ecology*, *24*(4), 763–773. https://doi.org/10.1111/jiec.12987

Payer, R. C., Quelhas, O. L. G., & Bergiante, N. C. R. (2024). Framework to supporting monitoring the circular economy in the context of Industry 5.0: A proposal considering circularity indicators, digital transformation, and sustainability. *Journal of Cleaner Production*, *466*, 142850.

Rame, R., Purwanto, P., & Sudarno, S. (2024). Industry 5.0 and sustainability: An overview of emerging trends and challenges for a green future. *Innovation and Green Development*, *3*(4), 100173.

Romanova, O., & Kuzmin, E. (2024). From Industry 4.0 to Industry 5.0: Challenges and opportunities of evolutionary transition in Russia. In Appolloni, A., Kumar, V., Kuzmin, E., & Akberdina, V. (Eds.), *The Future of Industry* (Vol. 70, pp. 55–72). Switzerland: Springer Nature. https://doi.org/10.1007/978-3-031-66801-2_5

Sah, B. P., & Shaikh, S. (2024). AI-driven IoT and blockchain integration in Industry 5.0: A systematic review of supply chain transformation. *Innovatech Engineering Journal*, *1*(1), 99–116.

Shabur, M. A., Shahriar, A., & Ara, M. (2025). From automation to collaboration: Exploring the impact of Industry 5.0 on sustainable manufacturing. *Discover Sustainability*, *6*(1), 341. https://doi.org/10.1007/s43621-025-01201-0

Sheikh, R. A., Ahmed, I., Faqihi, A. Y. A., & Shehawy, Y. M. (2024). Global perspectives on navigating Industry 5.0 knowledge: Achieving resilience, sustainability, and human-centric innovation in manufacturing. *Journal of the Knowledge Economy*, 1–36.

Szeszák, B. M., Kerékjártó, I. G., Soltész, L., & Galambos, P. (2025). Industrial revolutions and automation: Tracing economic and social transformations of manufacturing. *Societies*, *15*(4). ISSN: 2075-4698.

Tallat, R., Hawbani, A., Wang, X., Al-Dubai, A., Zhao, L., Liu, Z., Min, G., Zomaya, A. Y., & Alsamhi, S. H. (2023). Navigating industry 5.0: A survey of key enabling technologies, trends, challenges, and opportunities. *IEEE Communications Surveys & Tutorials*, *26*(2), 1080–1126.

Thanasi-Boçe, M., & Hoxha, J. (2025). Blockchain for sustainable development: A systematic review. *Sustainability*, *17*(11), 4848.

Turner, C., Oyekan, J., Garn, W., Duggan, C., & Abdou, K. (2022). Industry 5.0 and the circular economy: Utilizing LCA with intelligent products. *Sustainability*, *14*(22), 14847.

Zeb, S., Mahmood, A., Khowaja, S. A., Dev, K., Hassan, S. A., Gidlund, M., & Bellavista, P. (2024). Towards defining Industry 5.0 vision with intelligent and softwarized wireless network architectures and services: A survey. *Journal of Network and Computer Applications*, *223*, 103796.

13 Industry 5.0 and the Circular Economy

A Transformative Nexus

Md. Soleman Mollik, Rejaul Karim, Mohammad Sahabuddin, and Bilal Eneizan

13.1 Introduction

The global industrial landscape is undergoing rapid transformation as businesses and policymakers seek sustainable solutions to pressing environmental and economic challenges (Awewomom et al., 2024). Throughout history, each Industrial Revolution has experienced several transformative stages, bringing revolutionary changes to the technological landscape, production process, and societal norms, values, and structure (Mollik et al., 2024; Zong & Guan, 2024). Industry 4.0 (I4.0) emphasizes automation and digitalization using major innovative tools such as the Internet of Things (IoT), big data and advanced analytics, cloud computing, 3D printing, machine learning, cybersecurity, augmented reality (AR), and virtual reality (VR). In contrast, Industry 5.0 (I5.0) emerges as a paradigm shift by integrating and transforming these technologies with human-centric approaches, emphasizing suitability and resilience with the collaboration of human and advanced technologies like artificial intelligence (AI), machine learning (ML), deep learning (DL), robotics, and quantum computing (Li & Duan, 2025; Roshid et al., 2025c; Waaje et al., 2025b). Simultaneously, the circular economy (CE) seeks to transform production and consumption patterns by reducing waste and resource depletion (David et al., 2024). The nexus between the I5.0 and CE paradigms provides a unique opportunity for sustainable industrial growth. As a result, a comprehensive examination of how the core pillars of I5.0 might improve CE strategies and how CE practices contribute to sustainability and human welfare is presented in this chapter. To frame this discussion clearly, specific objectives and guiding questions are outlined in this chapter. The aim of the study was to investigate the transformative nexus between I5.0 and CE, examining how their integration can drive sustainable industrial practices, foster innovation, and enhance socioeconomic resilience. Specifically, the study was focused on exploring how I5.0 and CE are conceptually connected, focusing on human-centered innovation and sustainability. Additionally, a framework suggesting policy directions that use the link between I5.0 and CE to promote sustainability and resilience was developed. Moreover, the study examined how I5.0 technologies help businesses adopt CE models across industries. Finally, the study aimed to ascertain major challenges and opportunities in integrating I5.0 and CE. This chapter posits relevant questions to meet the objectives.

DOI: 10.4324/9781003647621-16

The questions include the following: what are the key conceptual links between I5.0 and CE? How do they support human-centric innovation and sustainability? How can policymakers use the synergy between I5.0 and CE to design strategies that strengthen sustainability and long-term resilience? Do I5.0 technologies help businesses transition to CE models across sectors, and in what ways? What are the leading barriers to and opportunities for integrating I5.0 and CE practices?

13.1.1 Defining Industry 5.0 and Circular Economy

Human intelligence and creativity will drive technological advancement, and workers will become closely integrated with technology. The importance of people versus technology is emphasized and demonstrated in technological approaches that enable multitalented, programmable robots. This presents a future where technology supports, rather than replaces, people driving sustainable innovation (Hashim & Mason-Jones, 2025). The CE is an economic model that prioritizes waste reduction, reuse, and resource conservation, and is characterized by re- and upcycling (compared with the take–make–dispose linear model) (Bocken et al., 2014; Noor, 2025). The complementarity of the two concepts provides strong opportunities for long-term environmental and social sustainability.

To clarify the scope of this chapter, the following frequently used concepts are defined operationally, drawing from the recent literature:

Human-centricity: It refers to design and management approaches that place human creativity, well-being, and active participation at the center of technological systems (Nahavandi, 2019; European Commission, 2021a). The I5.0 concept denotes that workers are not passive supervisors but co-creators alongside AI and robotics.

Resilience: It is the capacity of industrial systems to absorb shocks (e.g., supply chain disruptions, market volatility, and environmental events) and continue functioning while adapting and transforming to meet the new conditions (Xu et al., 2021). Here, resilience requires consideration of technological, organizational, and ecological dimensions.

Ethical innovation: It denotes the deliberate integration of ethical considerations—such as fairness, privacy, inclusiveness, and environmental responsibility—into the design and deployment of new technologies and business models (Stahl, 2012; European Commission, 2021). Within I5.0, this means ensuring AI, automation, and circular practices align with societal values.

Artificial intelligence: AI uses algorithms and ML systems to analyze data, identify patterns, and make judgments or suggestions. In I5.0, AI provides predictive maintenance, adaptive production planning, and efficient resource utilization, which are crucial for circularity.

Digital twins: Digital twins (DTs) are virtual replications of physical assets, systems, or processes, enabling real-time monitoring, modeling, and optimization. DTs help organizations reduce waste, get more out of product life, and optimize designs by anticipating how products will behave at different points of their lifecycle.

Blockchain: It is a decentralized digital ledger that provides a secure means for recording transactions transparently. Blockchain can track materials and parts throughout the supply chain, providing enhanced traceability, accurate product information, and confirmation for circular actions such as recycling and reuse.

These terms are defined to clarify their conceptual meaning and serve as a common ground for studying how integrating I5.0 with CE principles creates sustainable, human-oriented industrial systems. Furthermore, these emerging technologies become human-centric enablers if they can be designed and deployed to facilitate human decision-making, ethical innovation, and sustainability, as opposed to situational automation or cost savings.

13.1.2 Context: Moving from Industry 4.0 to Industry 5.0

The paradigm shift from I4.0 to I5.0 is that industrial systems start to recognize and consider humans as an integral part of the systems so that it no longer has only the automation, digitization, and efficiency at the heart of those systems but also the human. I4.0 revolutionized manufacturing by introducing cyber–physical systems, the IoT, and AI-powered automation. Nevertheless, there are discussions that this approach is not fully inclusive and environmentally responsible (Hashim & Mason-Jones, 2025; Wuyts, 2025). Conversely, I5.0 emphasizes the collaboration between humans and machines to drive creativity, personalization, and ethical innovation (Aiduang et al., 2025).

The transition from I4.0 to I5.0 is part of a broader technological, societal, and environmental shift. Automation and connectivity were the dominant features of I4.0, ensuring that advancements in technology serve the sustainability of society and maintain ecological balance. However, emerging global challenges, such as climate change, AI's ethical implications, and shifting working dynamics, have led to an understanding that the industry must become more human-centered. This transition also reflects the growing need for more sustainable, resilient, and personalized production, driven in part by more informed customers and cross-border regulatory constraints.

Although I4.0 and I5.0 are closely related, they have quite different focuses, purposes, and expected goals. Table 13.1 presents a structured comparison based on key criteria: technological focus, human role, sustainability orientation, innovation model, and social impact.

This critical contrast clarifies that I.5.0 aims to redefine the industrial paradigm to prioritize human-centric values and sustainability in addition to improving efficiency. I5.0 seeks to create systems that are adaptive, regenerative, and ethically aligned with societal and environmental needs by explicitly embedding CE principles.

Through this lens, the human-centric emphasis is not simply philosophical: it becomes a strategic enabler for achieving the broader goals of resilience, resource circularity, and social inclusion—essential elements missing or underdeveloped in the I4.0 model.

Table 13.1 Comparisons between I4.0 and I5.0 based on Fundamental Criteria.

Criteria	*I4.0*	*I5.0*
Technological focus	Automation, cyber–physical systems, IoT, data analytics, and AI for efficiency and productivity	Integration of advanced technologies (AI, robotics, and quantum computing) with human creativity and ethical values
Human role	Primarily viewed as system operators or overseers of automated processes	Central, active co-creators and decision-makers in collaboration with intelligent systems
Sustainability orientation	Efficiency-driven resource optimization; indirect contribution to sustainability	Direct, explicit focus on environmental regeneration, resilience, and circular economy integration
Human role	Primarily viewed as system operators or overseers of automated processes	Central, active co-creators and decision-makers in collaboration with intelligent systems
Sustainability orientation	Efficiency-driven resource optimization; indirect contribution to sustainability	Direct, explicit focus on environmental regeneration, resilience, and CE integration
Innovation model	Technology-push innovation led by automation and connectivity	Human-centric innovation integrating ethical design, inclusivity, and adaptability
Social impact	Potential risk of job displacement; limited social integration	Focus on job enrichment, worker well-being, inclusiveness, and participatory design
Goal	Maximize productivity and efficiency through automation	Create sustainable, resilient, and human-centered industrial systems

Source: Authors' compilation.

13.1.3 I5.0–CE Integration for Sustainable Development

Integrating I5.0 with the CE principles is vital in accomplishing the United Nations' (UN's) Sustainable Development Goals (SDGs) (Gazi et al., 2024). The overlap between them allows the introduction of smart production systems that are fully cost-competitive and sustainable. This integration reduces environmental impact and increases the economic value in closed-loop systems (Galindo & Herrera, 2025). Combining lifecycle assessment, waste valorization, and human-in-the-loop AI technologies assures that sustainability is built throughout the entire product lifecycle (Raghav et al., 2025). This is a revolutionary aggregation aiming to help the transition to inclusive, resilient, and zero-waste industrial ecosystems.

13.1.4 Unique Contributions of This Chapter

This chapter draws several novel contributions to advance both academic and practical domains beyond the existing literature:

1. Bridging Conceptual Gaps: While prior research has focused on I5.0 and CE individually, this study carefully explores their conceptual synergy,

demonstrating how human-centered innovation and I5.0 can directly boost CE strategies.
2. Policy-Focused Framework: Existing studies often remain descriptive; this study goes further by proposing a comprehensive framework that guides policymakers on harnessing the I5.0–CE nexus to promote sustainability, resilience, and inclusive growth.
3. Cross-Industry Application: By examining how I5.0 technologies enable circular practices across multiple sectors (e.g., manufacturing, textiles, and electronics), the chapter provides practical, transferable insights relevant to a wide range of industries.
4. Balanced Analysis of Barriers and Opportunities: Unlike many studies that emphasize either challenges or benefits, this study critically evaluates both, offering a nuanced roadmap for academics, industry practitioners, and policymakers to anticipate risks and leverage opportunities.

Through these contributions, the chapter advances current academic debates by integrating technological, human, and circular perspectives into a cohesive narrative and equips practitioners with concrete strategies to implement sustainable, human-centered, and circular industrial models.

13.1.5 Theoretical Underpinnings

To rigorously frame the discussion of the transformative nexus between I5.0 and CE, this chapter draws on several established theoretical perspectives:

1. Sustainable Innovation Theory: This body of work explains how firms and industries can embed sustainability objectives into technological and organizational innovation processes (Schot & Geels, 2008). It helps conceptualize how I5.0's human-centric technologies can perform as effective enablers of CE business models.
2. Resource-Based View: Resource-based view (RBV) emphasizes that firms achieve competitive advantage by leveraging unique internal resources and capabilities (Malhotra et al., 2025)—including human expertise, advanced technologies, and learning processes—essential for adopting circular strategies supported by I5.0.
3. Institutional Theory: This concept focuses on how regulatory, normative, and cultural-cognitive forces influence organizational behavior. It is relevant for explaining why industries might integrate human-centric and circular practices in response to policy frameworks, social expectations, and market norms.

By anchoring the analysis in these theories, the chapter moves beyond describing trends and offers a theoretically grounded explanation of how and why the nexus of I5.0 and CE can drive sustainable industrial transformation.

13.2 Conceptual Foundations

13.2.1 Industry 5.0: Human-Centric, Resilient, and Sustainable Industry

I5.0 builds upon the innovations introduced in I4.0 but changes the focus by prioritizing the human element, resilience, and sustainability. In contrast to I4.0, which emphasizes automation and productivity, Industry 5.0 envisions a symbiotic relationship between human talent and breakthrough technology such as AI, robotics, and the IoT. Ethical innovation is an approach that can lead to improved worker well-being and system flexibility (Popa et al., 2025). Human-in-the-loop is the cornerstone of I5.0, emphasizing creative participation and universal value generation (Ertz et al., 2024). Furthermore, one of the key enablers of I5.0 is the development of decentralized agile production networks that can respond to global disruption (e.g., pandemics and geopolitical crises) (Sassanelli, 2024). Its sustainably pillar is based on enablers such as eco-design, the integration of renewable energies, and socioeconomic inclusion, enhancing I5.0 as a strategic enabler of regeneration in the industry.

Table 13.2 serves as a foundational element for understanding the transformative nexus of I5.0 and the CE. It meticulously outlines key concepts that underpin this transformative relationship, providing definitions and their respective sources.

The table begins with human-centric manufacturing, a key pillar of I5.0 that emphasizes the integration of human creativity and problem-solving with

Table 13.2 Industry 5.0 and CE Concepts.

Concept	*Definition*	*Sources*
Human-centric manufacturing	Focuses on integrating human creativity, problem-solving, and customization capabilities with advanced technologies in manufacturing	Nahavandi (2019)
Hyper-customization	Produces highly tailored products on-demand using AI and collaborative robotics, reducing waste and increasing customer happiness	Xu et al. (2021)
Resilient manufacturing	Encourages building adaptive and robust systems that can respond to disruptions while minimizing environmental impact and resource usage	Grabowska et al. (2022)
Closed-loop manufacturing	System that reuses waste materials in production, reducing raw material needs and prolonging product lifecycles	Kalmykova et al. (2018)
Smart circular supply chains	Uses digital technologies like IoT and blockchain to monitor, trace, and optimize material flows in circular business models.	Yan et al. (2021)
Collaborative robotics (cobots)	I5.0 focuses on human–robot collaboration for enhanced precision and efficiency while promoting sustainable practices	Shah et al. (2025)
DT in CE	Throughout the product lifespan, digital reproductions of actual products allow for real-time tracking, predictive maintenance, and improved resource management	Thakuri et al. (2024)

Source: Authors' compilation.

advanced technologies to provide tailored solutions. This is strongly aligned with the human-centricity aspect of the transformative nexus. Building on that, hyper-customization emphasizes producing highly personalized products through on-demand manufacturing supported by robotics and AI. This method contributes to reducing waste (a crucial goal of CE) and enhances customer satisfaction, supporting inclusive growth. Table 13.2 presents the I5.0 concepts and CE practices in diverse business domains.

Furthermore, the table introduces smart circular supply chains, using IoT and blockchain to streamline material flows for circular models, representing the nexus's innovative spirit. Additionally, cobots highlights human–robot teamwork for sustainability and efficiency, aligning with human-centric and eco-friendly aims. Furthermore, DT technology provides real-time monitoring and predictive maintenance, which improves resource use. These demonstrate how I5.0's technology and human attention help the CE achieve its efficiency and waste reduction goals.

13.2.2 CE Principles: Waste Minimization and Resource Efficiency

The CE intends to reduce waste and prolong the life of materials and products (Karim et al., 2025a; Roshid et al., 2025b), pursuing the principles of eliminating waste, reusing materials, and regenerating nature (Han et al., 2025; Waaje et al., 2025a). Biomimicry, which draws inspiration from nature's intricate and tested designs, offers innovative pathways like biodegradable products, self-healing materials, and energy-efficient processes (Ellen MacArthur Foundation, 2013; Karim et al., 2025b; Stahel, 1982). CE helps tackle resource scarcity, pollution, and greenhouse gas (GHG) emissions sustainably by transforming from a linear take–make–dispose model to closed-loop systems. In the following, the authors discuss how manufacturability, reuse, recycling, and the product service system help in the operationalization of CE in industrial and consumer contexts (Sassanelli, 2024). Furthermore, CE requires systemic innovation throughout value chains—spanning from design, procurement, production, to end-of-life management—to increase resource productivity and to maintain economic development within the limits of the planet (Popa et al., 2025). The CE model is inherently linked to the SDGs, particularly Goals 12 (Responsible Consumption and Production) and 13 (Climate Action).

13.2.3 Antecedents, Core Attributes, and Consequences of the I5.0 and CE Nexus

Figure 13.1 presents a conceptual framework that links the antecedents, core attributes, and effects of integrating I5.0 with CE principles.

On the left, the antecedents of the nexus are divided into contextual drivers, such as environmental degradation, resource scarcity, and technological advancements like AI, robotics, and IoT, and organizational enablers, including leadership commitment, innovative culture, and investment in smart technologies. These factors collectively create the foundation for this transformative integration.

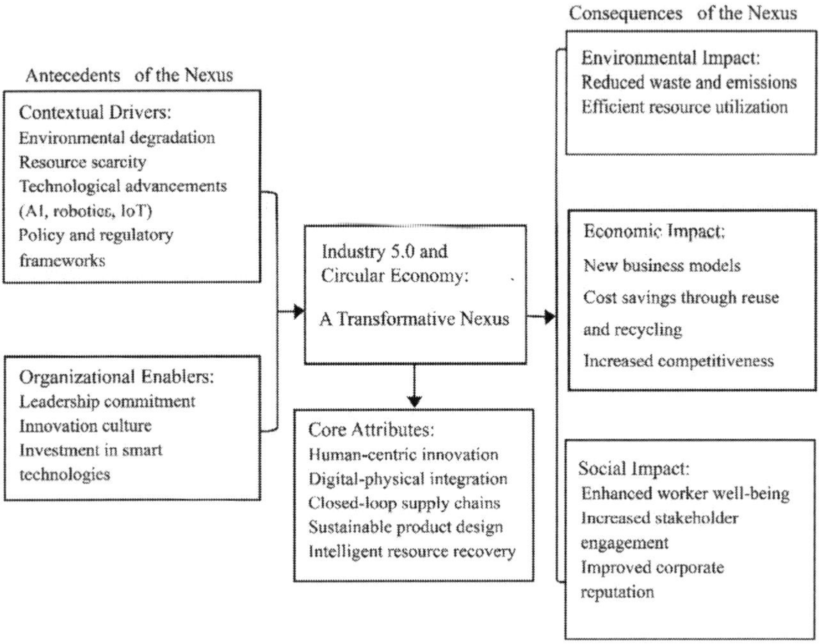

Figure 13.1 Conceptual framework of I5.0–CE integration.

At the center lies the core of the nexus, which characterizes I5.0 and CE through five key attributes: human-centric innovation, digital–physical integration, closed-loop supply chains, sustainable product design, and intelligent resource recovery. These attributes reflect the synergy between technological advancement and sustainability goals.

On the right, the consequences of the nexus are categorized into three impact areas:

- Environmental Impact: Reduced emissions and better resource efficiency
- Economic Impact: New business models, cost savings via reuse/recycling, and enhanced competitiveness
- Social Impact: Improved worker well-being, stronger stakeholder engagement, and better corporate reputation

Overall, the figure encapsulates how strategic enablers and contextual pressures drive a transformative shift, resulting in holistic environmental, economic, and social benefits.

13.2.4 Overlapping Goals: Regenerative Processes and Economic Resilience

The integration of I5.0 and CE is mainly based on the acknowledgment of a strong convergence of values and strategic goals. Both Industry 5.0 and the Circular Economy call for a paradigm shift from merely minimizing harm to actively regenerating and restoring natural and social systems through human-centric innovations such as bio-based materials, cradle-to-cradle design, and circular supply networks (Büchler & Hoon, 2024). The ecosystem approach also strengthens economic resilience by making industrial systems stronger and more adaptable. The decentralization of production, demand-activated production, and digitalized feedback loops will help the organizations become more resilient to supply chain disruptions, market instability, and changes (Hossain & Mollik, 2022; Popa et al., 2025). This also brings about innovation by incorporating digital into smart resource tracking, predictive maintenance, and lifecycle optimization, all of which are crucial to continue operations while protecting the environment. Altogether, this nexus is the guiding framework to a path of transformation for economies looking for longevity, social justice, and technological power.

13.3 Technological Enablers in I5.0 for CE Implications

13.3.1 Artificial Intelligence and Machine Learning

AI and ML are critical enablers for I5.0 in establishing more sustainable and efficient CEs. In the production and supply chain, AI supports smart decision-making concerning resource allocation, predictive maintenance, or reducing waste (Thürer et al., 2024). ML algorithms process large quantities of data fed from IoT devices to track material flow and realize adaptive manufacturing following CE principles. For instance, predictive analytics based on AI allows manufacturers to adopt proactive (as opposed to reactive) maintenance, achieving far longer product lifecycles and minimizing resource usage (Keenavinna & Wickramarachchi, 2025). Moreover, AI can help give dynamic product remanufacturing and recycling processes added flexibility by sorting materials and automating quality control, which is key to high-value reuse in circular loops.

13.3.2 Robotics and Cobots

The role of robotics, in particular the adoption of cobots, encapsulates the human-centric mentality of I5.0 and supports CE solutions. Cobots collaborate with human workers carrying out precise, repetitive, or dangerous tasks, i.e., disassembly tasks for recycling or reuse of components (Thürer et al., 2024). Being highly flexible, they facilitate the cost-effective processing of various waste streams, which in turn helps in recovering material and reducing the dependence on landfills. The ability of cobots to enhance the provision of customization and repair services is also seen to assist in prolonging the life of products and reducing obsolescence (Keenavinna & Wickramarachchi, 2025). Its use is also propagated for work safety, ergonomic aspects, and shaping the development of technology following social sustainability, a key Industry 5.0 postulate.

13.3.3 Digital Twins and Simulation Technologies

DT technologies, virtual models of physical assets or processes, are central to making the CE work effectively in I5.0 (Keenavinna & Wickramarachchi, 2025). They support real-time monitoring, scenario simulation, and lifecycle assessment, helping manufacturers reduce waste and energy use by predicting and avoiding inefficiencies. When combined with AI and IoT, DTs become even more powerful. AI employs real-time data from IoT devices to help with predictive maintenance, process improvement, and automated decision-making. This interaction allows industries to model circular workflows, forecast environmental impacts, and adjust production instantly to minimize waste and energy demands.

Evidence from the automotive and electronics sectors shows that linking carbon reduction and economic performance through DTs boosts sustainability (Thürer et al., 2024). This enables companies to identify inefficiencies and apprehend the environmental impacts of their design choices before implementing them. Such "virtual testing" encourages circular design strategies like modularity, repairability, and material recyclability. Additionally, DTs support closed-loop feedback systems that help businesses continuously improve toward sustainability targets.

13.3.4 Blockchain and Transparency for Circularity

Blockchain technology supports transparency and traceability, which are essential in verifying circularity practices. This decentralized ledger system ensures immutability of information about the material's history of origin, usage, and recycling, resulting in building trust between the involved parties and end-users (Keenavinna & Wickramarachchi, 2025). Blockchain enables the creation of circular supply chains, as it guarantees that raw materials are sustainable and certified, and facilitates secondary material markets (Thürer et al., 2024). Smart contracts enable automated transactions for resource sharing and reuse, such as product renting or return rewards. Moreover, the integration of blockchain fights against fraud and fake items, supporting the real adoption of the CE.

Figure 13.2 presents the key technical enablers that support the integration of I5.0 and the CE. At the heart of the CE is a regenerative model of production and consumption that attempts to eliminate waste and keep products, materials, and resources in use for as long as feasible. This is the ultimate goal and fundamental value toward which all enabling technologies are directed. Surrounding the CE are successive technological layers. The first enabling layer is DT technology, which is critical for simulating, monitoring, and optimizing the product and system lifecycles in real time.

DTs help diminish waste and boost efficiency by precisely exposing product behavior and resource use. Additive manufacturing, also known as 3D printing, enables on-demand production, eliminating waste through customized, efficient manufacturing processes, operating in closed-loop cycles, an essential attribute of the CE.

Blockchain enhances traceability, transparency, and security of supply chains, ensuring the materials are sustainably sourced and contain verifiable recycled content. AI and IoT function together to be the digital infrastructure's core, with AI supporting predictive maintenance and smart forecasting, while IoT collects real-time data, optimizing resource use. Taken together, these technologies

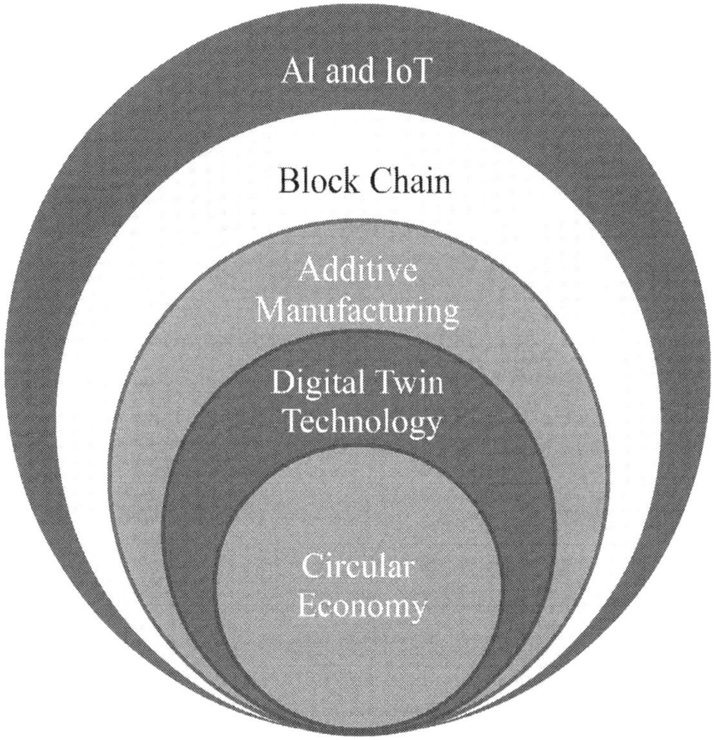

Figure 13.2 Technological enablers in I5.0 and the CE.

construct a synergistic framework transitioning industries from linear to circular models, driving a smarter, more sustainable, and human-centered I5.0 in alignment with CE principles.

13.4 Methodology: Data Collection Process

This chapter pursues a qualitative research perspective with an analysis of themes and concept mapping using ATLAS.ti 2025 for systematic coding and analysis of data. The elaboration on the methodology to ensure transparency and rigor is presented below.

13.4.1 Selection Criteria

The authors analyzed three types of data sources:

1. Academic Literature: The authors selected peer-reviewed articles, systematic literature reviews, and seminal works published between 2016 and 2025, emphasizing I5.0, CE, and sustainable industrial transformation. Databases searched

included Scopus, Web of Science, and Google Scholar using keywords such as "Industry 5.0," "Circular Economy," "human-centric innovation," and "sustainable manufacturing."
2. Case Studies: Empirical cases published in journals, reports, and industry white papers that explicitly demonstrate the integration of I5.0 technologies with CE practices, e.g., Siemens, IKEA, and others, as listed in Table 13.3.
3. Policy Documents and Strategic Reports: Reports from the European Commission, Ellen MacArthur Foundation, and UN SDGs documentation were selected for their relevance to policy frameworks guiding human-centric and circular innovation.

The inclusion criteria required sources to

- address Industry 5.0 and/or CE in an industrial context;
- be published in English; and
- offer conceptual frameworks, empirical data, or policy guidance relevant to sustainable industrial development.

13.4.2 Coding Process and Theme Derivation

Data imported into ATLAS.ti were first open-coded to identify initial codes reflecting recurrent concepts such as human-centricity, DTs, closed-loop manufacturing, and resilience. Next, the codes were grouped into axial codes (e.g., technological enablers, policy implications, barriers, and opportunities). Finally, selective coding identified higher-level themes that structured the narrative of this chapter:

- The synergy of I5.0 and CE
- Human-centric innovation as an enabler of circularity
- Policy frameworks supporting sustainable transformation
- Challenges and strategic opportunities

Concept mapping tools within ATLAS.ti were used to visualize connections among codes and to generate figures such as the word cloud (Figure 13.3) and thematic frameworks (Figures 13.5–13.7).

13.4.3 Validation Techniques

To ensure robustness:

1. A second researcher independently coded a portion (~20%) of the data to evaluate inter-coder reliability. Cohen's kappa was calculated and found to be acceptable (>0.75), suggesting strong agreement.
2. Regular team discussions resolved discrepancies and refined code definitions to enhance reliability.
3. Triangulation of data from academic, industry, and policy sources further increased validity.

13.4.4 Empirical Support

While the study is primarily qualitative, empirical evidence was incorporated where possible:

1. Adoption examples (e.g., Siemens' use of human–robot collaboration and DTs; IKEA's use of AI and product passports).
2. Reference to published quantitative data, such as estimates from Xu et al. (2021) showing increasing I5.0 pilot projects in Europe and Asia.
3. Industry reports forecast that by 2026, over 30% of advanced manufacturing firms will implement human-in-the-loop AI systems (source: Keenavinna & Wickramarachchi, 2025).

Although large-scale quantitative datasets were beyond the scope of this study, integrating available statistics and concrete case outcomes strengthens the discussion and grounds conceptual insights on empirical trends.

13.5 I5.0–CE Synergies and Mutual Reinforcement

13.5.1 How CE Implementation Gets Support from I5.0

I5.0 (human-centric philosophy) embodies a significant value addition to the pragmatic realization of the CE. Achieving the objectives of the Circular Economy (CE) is supported by Industry 5.0, which emphasizes the realignment of humans and machines in manufacturing to create more sustainable, efficient, and human-centric production systems (Rejeb et al., 2025). Industry 5.0 enables the production of adaptable, decentralized, sustainable manufacturing systems that perfectly fit with the CE principles by combining high-tech solutions with human inventiveness (Nechyporuk & Kocherhina, 2025). Such a system enables regenerative production loops, optimizes efficient resource flow, and minimizes waste through advanced automation and real-time feedback. The combination also creates opportunities for industries to establish closed-loop supply chains, where materials and products retain their value throughout multiple lifecycles, resulting in significantly reduced environmental impacts (Ünal, 2024) and promoting economic circularity. In addition, the collaborative aspect of I5.0 fosters the co-design of circular offerings, making them available for longer life, simple maintenance, reuse, and recycling. Thus, I5.0 incorporates CE principles at the upstream stages of the product lifecycle.

13.5.2 CE Integration in the I5.0 Domain: Model and Contrary Cases with Outcomes

Table 13.3 provides concrete examples that vividly illustrate the "transformative nexus" between I5.0 and CE, while additionally displaying the challenges and contradictions.

The "Model Cases" show successful examples of combining I5.0 with CE practices. Model case 1 features Siemens, which uses human–robot collaboration and

Table 13.3 Model and Contrary Cases of CE Integration in I5.0.

Types	*Cases*
Model case 1	Siemens applies I5.0 through human–robot collaboration and AI in decentralized manufacturing systems. It also promotes circularity by utilizing 3D printing with recycled materials and implementing closed-loop systems to minimize raw material consumption and energy waste (Xu et al., 2021)
Model case 2	IKEA integrates CE principles by designing furniture for disassembly and repair. Under I5.0 principles, they utilize digital product passports and AI to tailor user needs, enhance lifecycle tracking, and minimize environmental impact while promoting inclusive design (Geissdoerfer et al., 2022)
Contrary case 1	The fast fashion industry (e.g., Shein and Boohoo) demonstrates a contradiction to CE and I5.0. Despite automation, these companies continue to promote disposable products, exploit labor, and discard unsold inventory, resulting in excessive landfill waste and increased energy consumption (Ellen MacArthur Foundation, 2021)
Contrary case 2	Electronic waste from tech giants: Some large tech firms still produce devices with limited repairability or are intentionally designed to become obsolete. This contradicts CE ideals. While automation exists, the lack of human-centric product lifecycle design contributes to the rising global e-waste problem (Bressanelli et al., 2019)

Source: Authors' compilation.

AI in decentralized manufacturing. Siemens also supports circularity by using 3D printing with recycled materials and closed-loop systems, cutting down on raw material use and energy waste. This shows how I5.0 technology can directly support CE, boosting both efficiency and sustainability. Model case 2 highlights IKEA, which designs products for easy disassembly and repair. IKEA uses digital product passports and AI for customization and lifecycle tracking, reducing environmental impact and promoting inclusive design. These cases show how the mix of human-centered technology and circular thinking can reshape production and consumption.

In contrast, the "contrary cases" show where this integration fails. "Contrary case 1' looks at the fast fashion industry (e.g., Shein and Boohoo), which uses automation but promotes cheap, disposable products; poor labor practices; and high waste. "Contrary case 2" focuses on electronic waste from major tech companies, where products are hard to repair and are designed to become obsolete. This causes rising e-waste and shows that technology alone isn't enough without a human- and environment-focused mindset.

Overall, Table 13.2 makes it clear: true progress needs ethical design, strong commitment to CE, and people-focused innovation, not only better technology.

13.5.3 Additional Global Case Examples to Enhance Relevance

To broaden the chapter's global perspective, this section integrates selected empirical cases and industry reports from different regions and sectors:

- **Asia (Textiles and Apparel):** Bangladesh's garment sector is piloting AI-powered quality inspection tools in SMEs to reduce fabric waste and rework (Islam et al., 2025).
- **Africa (Agri-Processing):** South African wineries are using IoT sensors and DTs to monitor water usage and support circular resource management (Festa et al., 2025).
- **Latin America (Construction):** Brazil's civil engineering firms are adopting blockchain for tracking recycled construction materials, enhancing transparency and compliance with CE regulations (Pinto Varela Alberte & de Oliveira Novelli, 2024).
- **Europe (Automotive):** Volkswagen uses DTs and human–robot collaboration to remanufacture components and reduce carbon footprint (Singh et al., 2022).
- **SMEs in Multiple Regions:** Pilot projects leverage cloud-based AI tools to implement predictive maintenance without major capital investment (Keenavinna & Wickramarachchi, 2025).

These diverse cases illustrate that I5.0 and CE principles can be adapted to local contexts, resource constraints, and sector-specific challenges. They also demonstrate scalability pathways, especially when solutions are modular, cloud-based, or open-source, making them more feasible for SMEs and developing economies.

13.5.4 Critical Reflection on Sectoral Failures and Trade-Offs

While "model cases" illustrate successful integration of I5.0 and CE principles, it is equally important to critically examine why certain sectors—such as fast fashion—struggle to adopt these paradigms effectively, despite having access to advanced technologies.

Failure to integrate I5.0 and CE in fast fashion and similar sectors can be attributed to several factors:

- The Business Model Lock-In: Fast fashion remains driven by high-volume, short lifecycle products, which fundamentally contradicts CE's goals of durability, reuse, and waste minimization.
- Economic Pressure: Cost competitiveness discourages investment in human-centric technologies that slow down production or require retraining staff.
- Cultural and Consumer Demand: Fast-evolving trends often drive linear production, weakening circular and sustainable practices despite available technology.
- Misaligned Incentives: Technologies such as AI and robotics might be leveraged to elevate production rates instead of lowering environmental impact.

Furthermore, there are several inherent trade-offs and conflicts between technological advancement and social or environmental goals, such as the following:

- Rebound Effects: The automation can ensure greater efficiency, resulting in increased production and consumption, offsetting sustainability gains.
- Social Tensions: Then, automation and AI might displace personnel, challenging I5.0's human-centric promise if not managed carefully.
- Environmental Limits: Advanced technologies exert resource and energy costs, which often undermine CE goals if not balanced by lifecycle thinking.

In examining these contradictions, the chapter demonstrates that technology alone is insufficient; genuine change necessitates new business models, policies, and cultural shifts to achieve meaningful social and environmental outcomes.

13.5.5 Concept Mapping

Figure 13.3 (concept analysis) presents a word cloud in which the most common and important words in the literature about I5.0 are visually represented. The central positioning of the word "technology" suggests its role as the bridge between human-centric innovation and sustainable practices. Subsequent dominant terms, such as "industry," "research," "economy," and "system," emphasize the interdisciplinary and systematic character of such a nexus.

Furthermore, the words "chain," "resource," "value," "waste," and "model" indicate that key CE concepts are present, reinforcing its focus on resource and

Figure 13.3 Word cloud of the I5.0 and CE nexus.

waste efficiency and business model change. The embodiment of "AI," "IoT," and "design" reflects the technology drivers that are critical for implementing I5.0 principles. More importantly, the academic breadth is emphasized by designations such as "study," "analyses," "reviews," and "journal" as evidence of the increasing body of scholarly interest and evidence-based investigation in this transformative space.

Throughout, the concept analysis encompasses the emerging dialog among technology, sustainability, and human-centered industrial transformation, reflecting how the dominant ideas converge to delineate the research agenda about I5.0 and CE. It posits a visual integration of established and other new interests, platforms, and lines of scholarly inquiry in this integrative domain.

13.5.6 Human-Centric Innovation Driving Sustainable Practices

At its core, I5.0 brings together human ingenuity and modern technologies to generate long-term innovation. Different from the emphasis of I4.0 on automation, I5.0 emphasizes human–machine collaboration, which enhances ethics inclusivity with a sustainability transformation (Festa et al., 2025). This approach fosters participative design and the integration of indigenous knowledge to accomplish diverse social, economic, and environmental needs (Santos et al., 2024; Roshid et al., 2024). Focusing on social innovation, I5.0 involves communities and stakeholders in CE initiatives, ensuring local relevance and acceptance. Co-creation builds trust, promotes responsibilities, and supports fair, safer workplaces, skilled employment, and empowerment, anchoring circularity in human values and broader societal well-being (David et al., 2024).

13.5.7 Contribution to Sustainable Development Goals (Environmental, Social, and Economic)

Figure 13.4 shows the "transformative role of I5.0 and CE in sustainable development" through a cyclical model of four key, interconnected transformations. I5.0 and CE, at the center of the framework, function as catalysts for transforming from linear economic models to regenerative, human-centric systems. All interconnected transformations form the backbone of sustainable development, and they are discussed as follows.

1. Social transformation emphasizes human-centricity and inclusive growth, ensuring that industrial progress benefits all and promotes fairness.
2. Environmental transformation, "under regeneration over exploitation," strengthens restorative practices that protect natural systems and diminish waste.
3. Economic transformation, marked by "resilience and innovation-driven growth," conceives an economy strengthened by adaptability and ongoing innovation, generating sustainable value.
4. Systematic transformation, defined by "redefining success and progress," broadens success metrics to contain social and environmental well-being, moving beyond mere financial outcomes.

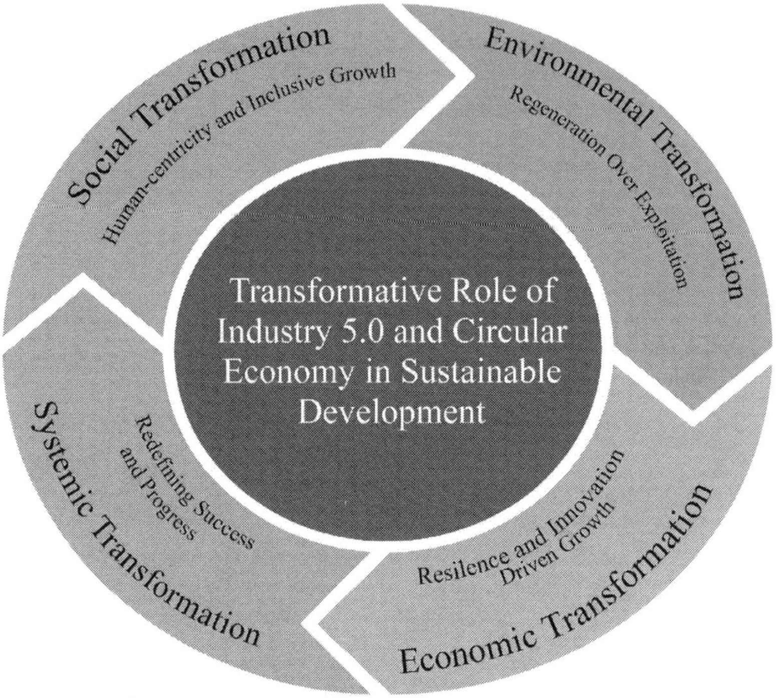

Figure 13.4 Transformative role of I5.0 and CE for sustainable development.

Furthermore, Figure 13.4 presents how I5.0 and CE support the UN SDGs: socially by fostering a decent work environment, minimizing inequalities, and empowering communities (Karim et al., 2024; Mavri et al., 2024); environmentally by cutting waste, promoting recycling, and encouraging renewables (Khan et al., 2025); and economically by improving resource efficiency and global competitiveness (Majoul, 2024). Simultaneously, they drive resilient, regenerative systems aligned with global sustainability goals.

13.6 Challenges and Opportunities

13.6.1 Barriers in Technology Integration and Workforce Adaptation

The integration of advanced I5.0 technologies with the CE is hindered by several factors, particularly related to technological deployment and workforce preparedness (see Figure 13.5). Technological challenges include the costs of advanced AI, robotics, and blockchain technology, and the challenges in integrating them with legacy systems (Rahaman et al., 2025). First, the shift is hampered by the fact that many SMEs lack the requisite capital or technological capabilities to ensure a smooth conversion. Second, workforce adjustment is crucial. It needs new

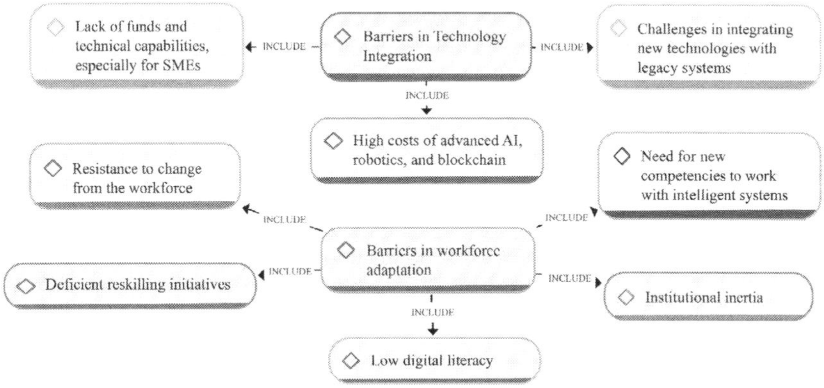

Figure 13.5 Barriers in technology integration and workforce adaptation.

competencies to work effectively with intelligent systems; however, reskilling initiatives are often inadequate (Shabur et al., 2025). Additionally, low digital literacy, resistance to change, and institutional inertia add to the challenge of integration. Bridging these human–technology gaps is key to realizing and achieving the potential benefits of I5.0 in circular initiatives.

The barriers to integrating the technology are as follows:

1. Initial High Costs: The costs of advanced AI, robotics, and blockchain technology.
2. Integration with Legacy Systems: Barriers in seamlessly connecting new technologies with existing, older systems.
3. Lack of Funds/Technical Capabilities: Many entities, especially SMEs, lack the required financial resources and technical expertise for a smooth transition.

Furthermore, the high cost associated with adopting new technologies like AI, robotics, blockchains, and DTs has been cited as one of the fundamental challenges in implementing I5.0 and CE in business. All these technologies often demand substantial capital investment, which is highly burdensome for SMEs (Villarreal & Calvo, 2021). Besides, many firms operate on outdated infrastructure, lacking compatibility with advanced technologies, resulting in making seamless integration more difficult and costly (Bag et al., 2021). Similarly, many organizations with limited financial resources and inadequate technical capabilities in developing economies face complexities in deploying advanced digital solutions effectively (Islam et al., 2025). In such situations, employees need new digital and cognitive skills to work with AI and automation, as human-focused technologies have advanced (Longo et al., 2020; Sung, 2018).

The obstacles to workforce adaptation are highlighted as follows:

1. New Competency Requirements: The workforce highly needs advanced technological skills to effectively interact with intelligent systems.

2. Lacking Reskilling Initiatives: The programs and opportunities for reskilling employees are insufficient.
3. Resistance to Accept Changes: Often, employees are reluctant to accept and adopt new technologies and work methods.
4. Low Digital Literacy: The workforce lacks foundational skills in the digital know-how.
5. Institutional Inertia: Several organizations have a resistance to adopting new practices and technologies, resulting in slow, limited, or no change.

Additionally, employees often resist changes due to fear of job loss or difficulties in adjusting to new digital tools (Rachinger et al., 2019). Limited digital literacy among employees further hinders the adoption of new technologies (Schwab, 2017), while organizational inertia slows adaptation and suppresses innovation (Birkel et al., 2019).

13.6.2 Policy, Regulatory, and Economic Challenges

Since policies and laws are not updated with modern technological development, they are creating burdens for I5.0 and CE adoption. The absence of unified global policies and standards (see Figure 13.6) is a key barrier here. Furthermore, regulatory differences between countries create barriers to scaling circular technologies, underlining the need for international cooperation to advance sustainability goals. Global organizations and partnerships are pivotal to aligning relevant policies and supporting the CE–I5.0 integration.

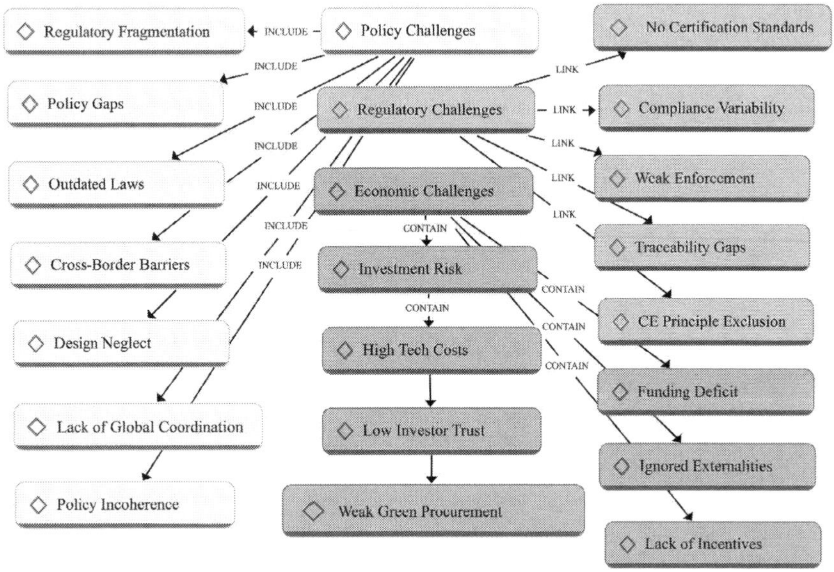

Figure 13.6 Challenges in the I5.0 and CE nexus.

Existing regulations often fail to promote circular design, remanufacturing, and product-as-a-service models, thus preventing business model innovation (Betti et al., 2025). Economic concerns have prevailed in the form of market volatility and investment risk, causing companies to hesitate from investing in circular technologies. There is a further absence of standardized metrics and certification for circularity and sustainability performance (Oro et al., 2025), adding complexity to compliance and consumer confidence. Creating an enabling environment for innovation, circular supply chains, and fair resource usage requires a combination of policy coherence and economic incentives.

13.6.3 Additional Barriers and Unintended Consequences

The integration of I5.0 and CE presents several additional barriers (see Table 13.4)., which in turn generate negative consequences that limit the effectiveness of I5.0 and CE integration.

Firms and policymakers are facing the above-mentioned additional barriers and unintended consequences. This chapter presents a more holistic view of the complex realities by recognizing these additional barriers. A deeper investigation signifies the need for careful policy design, stakeholder engagement, and systemic thinking to ensure that I5.0 and CE truly convey the promises of inclusivity, resilience, and sustainability.

13.6.4 Emerging Opportunities for Academia, Industry, and Policymakers

Despite the significant challenges, the cross-sector opportunities from integrating I5.0 and CE are numerous. In academia, interdisciplinarity is expected to fuel

Table 13.4 Additional Barriers of I5.0 and CE Integration.

Financial constraints in SMEs	Unlike large corporations, SMEs lack sufficient budget to afford AI, robotics, or DT technologies
Technological inequalities	Advanced tools disproportionately benefit technologically advanced firms and developed economies, thereby widening the gap between resource-constrained regions and creating new forms of digital exclusion
Cultural resistance	The resistance of organizations or employees toward new technologies hinders the adoption of new technologies
Overdependence on technology	Overdependence on automation and AI often decreases human creativity and judgment, paradoxically contradicting I5.0's human-centric foundation
Environmental costs of new technologies	Developing and running AI, quantum systems, and data centers consumes substantial resources, potentially negating environmental benefits if not properly managed
Ethical dilemmas	Data-driven design and increased personalization can increase privacy concerns and ethical questions about data ownership and surveillance

Source: Authors' compilation.

research in the domains of human–machine collaboration, sustainable materials, and lifecycle analytics, fostering innovation (Rahaman et al., 2025). Industries can benefit from increased efficiency in operations, customization of products, brand distinction through closed-loop approaches, and new markets and revenue opportunities (Thaha et al., 2025). This momentum may be used by policymakers in developing enabling frameworks to achieve circular procurement, green finance, and workforce development initiatives (Shabur et al., 2025; Roshid et al., 2025a). Collaborative approaches between all stakeholders are essential to co-evolve resilient ecosystems that use digitalization to achieve environmental, social, and economic sustainability.

The sector-wise opportunities are as follows:

Opportunities for Academia:

1. Interdisciplinary Research: Promoting cross-field collaboration of human–machine interaction, lifecycle analysis, and sustainable materials (Rahaman et al., 2025).
2. Innovation Catalysis: Academic contribution to the CE and I5.0 innovation through creating and disseminating advanced knowledge.

Opportunities for Industry:

1. Operational Efficiency: Improving productivity through smart, circular technologies (Thaha et al., 2025).
2. Product Customization: Enabling personalized manufacturing at scale.
3. Brand Differentiation: Enhancing reputation via circular practices.
4. New Market Access: Unlocking value from recycled resources and service-based models.
5. Revenue Expansion: Generating profits from circular economy loops.

Opportunities for Policymakers:

1. Policy Frameworks: Developing enabling laws for circular procurement and green technology (Shabur et al., 2025).
2. Green Financing: Mobilizing investments into circular and sustainable ventures.
3. Workforce Upskilling: Supporting education and training aligned with future jobs.
4. Ecosystem Building: Facilitating cross-sector collaboration for sustainable innovation.

Cross-Sectoral Opportunity:

1. Stakeholder Collaboration: Co-creating resilient, digitalized ecosystems focusing on social, environmental, and economic sustainability.

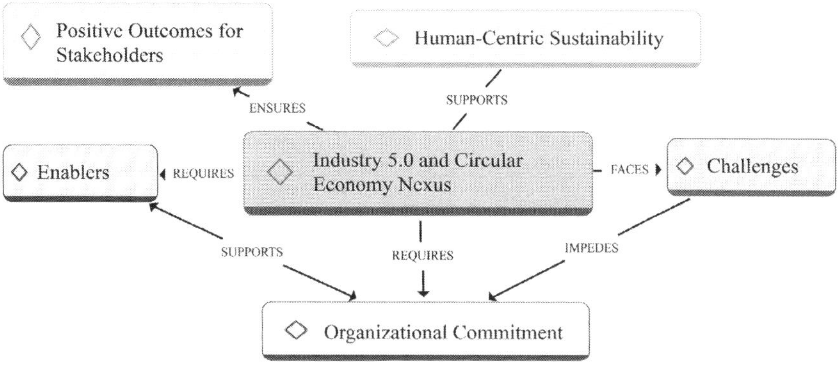

Figure 13.7 Thematic analysis of the I5.0 and CE nexus.

13.6.5 Thematic Analysis

Thematic analysis (Figure 13.7) can be applied within a multi-element framework to study sustainability and innovation. Primarily, the nexus between I5.0 and CE requires organizational commitments and enablers, such as advanced technologies, policies, and stakeholder engagement. However, challenges such as change resistance, limited resources, and regulatory barriers can hinder organizational commitment and slow down the expected progress.

In contrast, the effectively aligned nexus between I5.0 and CE drives positive outcomes for stakeholders, with economic value, social equity, and environmental benefits. Furthermore, it supports human-centric sustainability, a foundational goal of I5.0, emphasizing technological progress. On the whole, this interconnected model shows that a successful transition to a circular and human-centered I5.0 landscape is contingent upon strategic alignment of enablers, strong commitment, and overcoming systematic barriers.

13.7 Strategic Implications and Future Research Directions

13.7.1 Roadmap for Industry, Academia, and Policymakers

As the integration of I5.0 and CE moves forward with a strategic roadmap, industry leaders, researchers, and policymakers must work together to ensure coordinated efforts. To enhance sustainability and resilience, industries should prioritize human-centered automation, circular product design, and smart resource management (Dabija et al., 2024). Academia has a role to play in the evolution of cross-disciplinary research into the ethical, technical, and socioeconomic dimensions of this transformation. Policymakers have a unique role in enabling regulations, incentives, and standards that encourage innovation and promote an inclusive and friendly environment (Nazari & Ansari, 2024). Collaboration groups, sharing networks, and pilot projects can accelerate learning and uptake. The roadmap highlights agile frameworks that can adapt to the evolution of technology and societal needs.

13.7.2 Proposed Roadmap: Coordinating Roles for Industry, Academia, and Policymakers

Achieving the transformative integration of I5.0 and CE requires collaboration across sectors. Figure 13.8 presents a roadmap outlining the complementary roles:

- Industry: Focusing on adopting human-centric technologies (AI, DTs, and cobots) specifically to enable circular practices like reuse, remanufacturing, and closed-loop systems; investing in training of hybrid human–technology teams; and embedding circular design early in product development.
- Policymakers: Developing incentives (grants and tax credits) to promote circularity-enabling technologies; enacting regulations mandating transparent product data (e.g., digital product passports); and supporting national strategies linking digital transformation to sustainability goals.
- Academia: Driving interdisciplinary research among engineering, design, and sustainability; developing new metrics for measuring the social and environmental impact of I5.0 adoption; and working with industries to experiment with and refine circular business models.

The combined roadmap shows how all involved players can participate in unique ways, while simultaneously reinforcing each other to scale sustainable, resilient industrial transformation.

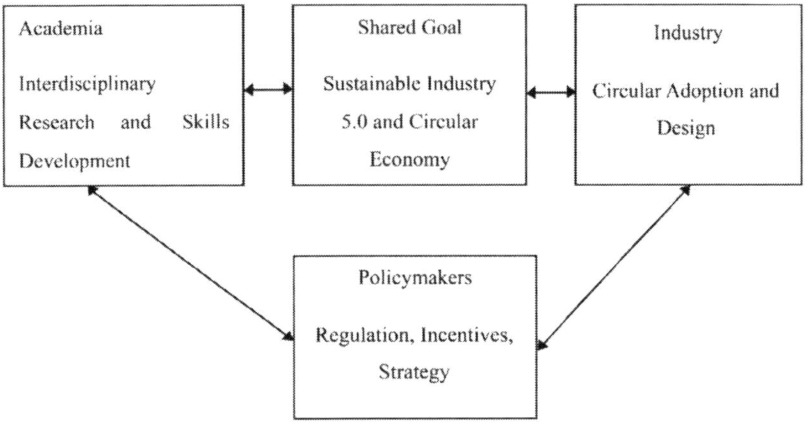

Figure 13.8 Coordinated roadmap of the roles of industry, academia, and policymakers in enabling sustainable I5.0 and CE transformation.

13.7.3 Practical Recommendations

This section provides recommendations to governments, industries, and researchers about the implementation of I5.0 with CE principles, based on the strategic implications indicated in the previous section. The recommendations are as follows:

For policymakers:

- Experimenting with a wide range of incentives (tax cuts, grants, etc.) to promote investment in human-centered technologies by SMEs.
- Developing regulatory frameworks to harmonize DT targets with CE.
- Prioritizing reskilling programs that combine technical literacy with sustainability and ethical design to foster hybrid skills.

For Industry Leaders and Practitioners:

- Bringing "human-in-the-loop" concepts to production planning, whereby people are no longer surrounded by new systems at work but rather are the main actors in decision-making.
- Creating legislative frameworks that interrelate the digitalization objectives and the CE milestones.
- Considering a product's lifecycle in product design and using data analytics in real time to predict end-of-life situations and remanufacturing or reuse.

For Researchers and Educators:

- Investigating multidisciplinary connections among engineering, social sciences, and environmental studies to create new human-centered circular models.
- Performing empirical studies to identify enabling variables and impediments to the fruitful nexus of I5.0 and CE in diverse sectors or countries.
- Developing and testing systems to assess the social impact of Industry 5.0 adoption, in addition to standard productivity metrics.

13.7.4 Advancing Human-Centric and Circular Industrial Ecosystems

The composite objective is to plan, design, and establish sustainable facilities in factories that not only contribute to workers' well-being, nurturing skills, and social justice but also facilitate the restoration of the environment (Leal Filho et al., 2024). AI, IoT, and blockchain solutions can enable transparency, flexibility, and closed-loop principles; reduce waste; and increase resource usage. Nurturing business ecosystems is essential for shaping and executing a company's overall strategy, particularly in advancing circularity, lifecycle thinking, and effective stakeholder communication. This paradigmatic shift is aimed at creating a healthier and more sustainable industry for addressing contemporary global issues, such as climate change and social disparity.

13.7.5 Scalability Considerations for SMEs and Developing Economies

While integrating the concepts of I5.0 and CE is a vibrant approach, scaling them up is not easy for SMEs and for firms in developing countries. The high initial cost of high-tech tools like robots and AI is a major barrier that many SMEs cannot afford. Shared services, open-software solutions, and digital innovation hubs can help reduce these costs and share the financial risk.

People with enough skills in new technologies are insufficient. Training programs, partnerships with different authorities and stakeholders, and other public–private projects can be established to support the development of a technology-focused talent pool.

There are also infrastructure challenges, such as poor Internet and limited digital networks, that are obstructing progress. Governments can assist with expanding digital infrastructure, with a focus on the regions where the smallest businesses operate.

Finally, some of the more sophisticated solutions built for larger companies may have to be tailored down for smaller ones, perhaps relying on cheap sensors or more basic digital tools rather than on full automation. This way of crafting approaches ensures that benefits are much wider and inclusive.

13.7.6 Areas for Further Study

There are still several research gaps in the area, such as developing human–machine partnerships toward strengthening the sustainability of circular practices, the development of AI to predict lifecycles of products, and the application of cobots in the closed-loop systems (Khan et al., 2025). The ethical and social consequences of the integration of I5.0 and CE must be further investigated among the aspects of privacy, data security, job substitution, and equity in access to new technology. This requires ethical frameworks and governance models that safeguard those at risk while also upholding human rights and social justice.

Interdisciplinary work with collaboration of social sciences, engineering, and environmental fields might be able to provide paths toward inclusive and transparent routes for industrial transformation. A further necessity is to create some common indicators to measure both the circularity outcomes and human-centered impacts of integrating I5.0 and CE. Policymakers should especially scrutinize market mechanisms and regulations that promote circular business models.

Equally important is examining social innovation that connects community engagement and technological change. In future work, long-term studies in various sectors could be conducted to explore deeper insights into the sustainability and resiliency implications of the technologies and products reported in this research, thereby informing policy and business decisions and potentially examining additional areas.

13.8 Conclusion

The integration of I5.0 and the CE is a major step forward for industries. It aims to make industrial growth more focused on being sustainable, strong, and beneficial for people. This combination incorporates advanced technologies, such as AI, robotics, DTs, and blockchain, and human intelligence to foster creativity and ethics. The aim is to create circular production systems that minimize waste and ensure the reuse of resources. This holistic approach can lead to industrial transformation, which reconciles economic growth, environmental care, and social inclusion. Crucially, it demonstrates how addressing industrialization and inequality is crucial to achieving the UN's SDGs and realizing an inclusive, sustainable future.

The integration demonstrates that true progress is not only to be found in embracing new technology but also in how it affects people and society. For it to succeed, industries will require cross-sector collaboration, further research, and strong policy backing. Such efforts can help industries transition toward a circular, sustainable, and socially responsible future.

References

Aiduang, W., Jatuwong, K., Ratanapong, K., Promjaidee, T., Xayyavong, O., Hongsanan, S., & Lumyong, S. (2025). Breaking down linear low-density polyethylene (LLDPE) using fungal mycelium (part A): A path towards sustainable waste management and its possible economic impacts. *Life*, *15*(5), 755. https://doi.org/10.3390/life15050755

Awewomom, J., Dzeble, F., Takyi, Y. D., Ashie, W. B., Ettey, E. N. Y. O., Afua, P. E., Sackey, L. N. A., Opoku, F., & Akoto, O. (2024). Addressing global environmental pollution using environmental control techniques: A focus on environmental policy and preventive environmental management. *Discover Environment*, *2*(1). https://doi.org/10.1007/s44274-024-00033-5

Bag, S., Pretorius, J. H. C., & Gupta, S. (2021). Role of institutional pressures and resources in the adoption of big data analytics powered artificial intelligence, sustainable manufacturing practices and circular economy capabilities. *Technological Forecasting and Social Change*, *163*, 120420. https://doi.org/10.1016/j.techfore.2020.120420

Betti, G., Gagliardi, F., Riccaboni, A., Mecca, A., & Tozzi, C. (2025). Sustainability and circularity of the agri-food systems: How to measure it? *Sustainability*, *17*(7), 3169. www.mdpi.com/2071-1050/17/7/3169

Birkel, H. S., Veile, J. W., Müller, J. M., Hartmann, E., & Voigt, K. I. (2019). Development of a risk framework for Industry 4.0 in the context of sustainability for established manufacturers. *Sustainability*, *11*(2), 384. https://doi.org/10.3390/su11020384

Bocken, N. M., Short, S. W., Rana, P., & Evans, S. (2014). A literature and practice review to develop sustainable business model archetypes. *Journal of Cleaner Production*, *65*, 42–56. https://doi.org/10.1016/j.jclepro.2013.11.039

Bressanelli, G., Perona, M., & Saccani, N. (2019). Challenges in supply chain redesign for the Circular Economy: A literature review and a multiple case study. *International Journal of Production Research*, *57*(23), 7395–7422. https://doi.org/10.1080/00207543.2018.1542176

Büchler, J. P., & Hoon, C. (2024). *Hidden Champions Case Compendium*. Springer.

Dabija, D., Nastase, C. E., Chetrariu, A., & Dabija, A. (2024). Multi-criteria analysis in circular economy principles: Using AHP model for risk assessment in sustainable whisky production. *Computation*, *12*(10), 206. www.mdpi.com/2079-3197/12/10/206

David, L. O., Aigbavboa, C., Nwulu, N., & Adepoju, O. O. (2024). Synergistic frameworks for water–energy–food (WEF) nexus. In *Sustainable Synergy: A Digital Framework for the Water–Energy–Food Nexus Project Delivery in Developing Economies. Green Energy and Technology*. Cham: Springer. https://doi.org/10.1007/978-3-031-72833-4_12

Ellen MacArthur Foundation. (2013). Towards the circular economy. *Journal of Industrial Ecology*, *2*(1), 23–44. https://ellenmacarthurfoundation.org/towards-the-circular-economy-vol-1-an-economic-and-business-rationale-for-an

Ellen MacArthur Foundation. (2021). *The Circular Economy: A Transformative Pathway for a Sustainable Fashion Industry*. Ellen MacArthur Foundation. https://ellenmacarthurfoundation.org

Ertz, M., Tandon, U., Sun, S., Sarigöllü, E., & Torrent-Sellens, J. (2024). Introduction to the handbook. In *The Palgrave Handbook of Sustainable Digitalization for Business, Industry, and Society* (pp. 1–14). Springer International Publishing. https://doi.org/10.1007/978-3-031-58795-5

European Commission. (2021). *Industry 5.0: Towards a Sustainable, Human-Centric and Resilient European Industry*. https://research-and-innovation.ec.europa.eu/knowledge-publications-tools-and-data/publications_en

Festa, G., D'Amato, A., Palladino, R., Papa, A., & Cuomo, M. T. (2025). Digital transformation in wine business–from Marketing 5.0 to Industry 5.0 in the world of wine adopting artificial intelligence. *European Journal of Innovation Management*. https://doi.org/10.1108/ejim-04-2024-0465

Galindo, D. F., & Herrera, R. J. G. (2025). Roadmap towards achieving strategic development goals from lean manufacturing to circular economy. *Revista de Gestão Social e Ambiental*. https://doi.org/10.24857/rgsa.v19n4-055

Gazi, M. A. I., Islam, H., Islam, M. A., Karim, R., Momo, S. M., & Senathirajah, A. R. B. S. (2024). Unlocking sustainable development in East Asia Pacific and South Asia: An econometric exploration of ESG initiatives. *Sustainable Environment*, *10*(1). https://doi.org/10.1080/27658511.2024.2366558

Geissdoerfer, M., Santa-Maria, T., Kirchherr, J., & Pelzeter, C. (2022). Drivers and barriers for circular business model innovation. *Business Strategy and the Environment*, *32*(6), 3814–3832. https://doi.org/10.1002/bse.3339

Grabowska, S., Saniuk, S., & Gajdzik, B. (2022). Industry 5.0: Improving humanization and sustainability of Industry 4.0. *Scientometrics*, *127*(6), 3117–3144. https://doi.org/10.1007/s11192-022-04370-1

Han, J., Zuo, J., Zillante, G., Chang, R., & Du, L. (2025). A systematic review of PET circularity technologies and management strategies: Challenges and future directions. *Resources, Conservation and Recycling*, *219*, 108280. https://doi.org/10.1016/j.resconrec.2025.108280

Hashim, M. A. M., & Mason-Jones, R. (2025). *Industry 5.0 and Circular Supply Chain 5.0: Search for Operationalization of Sustainable Development*. Cardiff Metropolitan University. https://doi.org/10.25401/cardiffmet.29099294.v1

Hossain, M. K., & Mollik, M. S. (2022). Exploring the impact of social media platforms on instigating tourist's emotions and behavioral intentions to co-create and visit tourist spots – Bangladesh perspective. *Journal of Business Studies*, *3*(1), 217–236. https://doi.org/10.58753/jbspust.3.1.2022.13

Islam, M. M., Emon, J. I., Ng, K. Y., Asadpour, A., Aziz, M. R. A., Baptista, M. L., & Kim, J. M. (2025). Artificial intelligence in smart manufacturing: Emerging opportunities and prospects. In *Artificial Intelligence for Smart Manufacturing and Industry X.0* (pp. 9–36). Cham, Switzerland: Springer Nature. https://link.springer.com/book/10.1007/978-3-031-80154-9

Kalmykova, Y., Sadagopan, M., & Rosado, L. (2018). Circular economy – From review of theories and practices to development of implementation tools. *Resources Conservation and Recycling*, *135*, 190–201. https://doi.org/10.1016/j.resconrec.2017.10.034

Karim, R., Roshid, M., & Waaje, A. (2024). Circular economy and supply chain sustainability. In *Strategic Innovations for Dynamic Supply Chains* (pp. 1–30). IGI Global. https://doi.org/10.4018/979-8-3693-3575-8.ch001

Karim, R., Waaje, A., & Roshid, M. M. (2025a). Digital circular economy for attaining sustainable development goals: Technologies for global sustainability. In *Effects of*

Digitalization and Circular Economy on Sustainable Policy and Climate Change Prevention (pp. 211–240). IGI Global Scientific Publishing. https://doi.org/10.4018/979-8-3693-9909-5.ch008

Karim, R., Waaje, A., Roshid, M. M., & Yeamin, M. B. (2025b). Turning the waste into wealth: Progressing toward global sustainability through the circular economy in waste management. In *Sustainable Waste Management in the Tourism and Hospitality Sectors* (pp. 507–552). IGI Global Scientific Publishing. https://doi.org/10.4018/979-8-3693-6110-8.ch019

Keenavinna, H., & Wickramarachchi, R. (2025, February). Industry 5.0 readiness in the Sri Lankan apparel industry: A systematic literature review. In *2025 5th International Conference on Advanced Research in Computing (ICARC)* (pp. 1–6). IEEE. https://doi.org/10.1109/ICARC64760.2025.10963306

Khan, A. H., Firoozi, A. A., Alam, P., & Agwa, I. S. (2025). Supplementary cementitious materials: Recent developments, performance insights and potential applicability in circular economy. *Frontiers in Materials*. https://doi.org/10.3389/fmats.2025.1608358

Leal Filho, W., Kautish, S., Wall, T., Rewhorn, S., & Paul, S. K. (2024). *Digital Technologies to Implement the UN Sustainable Development Goals*. Springer. https://doi.org/10.1007/978-3-031-68427-2

Li, L., & Duan, L. (2025). Human centric innovation at the heart of Industry 5.0–exploring research challenges and opportunities. *International Journal of Production Research*, 1–33. https://doi.org/10.1080/00207543.2025.2462657

Longo, F., Padovano, A., & Umbrello, S. (2020). Value-oriented and ethical technology engineering in Industry 5.0: A human-centric perspective for the design of the factory of the future. *Applied Sciences*, *10*(12), 4182. https://doi.org/10.3390/app10124182

Majoul, B. (2024). *Proceedings of the 2024 9th International Conference on Social Sciences and Economic Development*. https://books.google.com/books?id=H4YVEQAAQBAJ

Malhotra, G., Dandotiya, G., Shaiwalini, S., Khan, A., & Homechaudhuri, S. (2025). Benchmarking for organisational competitiveness: A resource-based view perspective. *Benchmarking: An International Journal*, *32*(3), 943–964. https://doi.org/10.1108/BIJ-09-2023-0668

Mavri, M., Mazza, P. I., Karasavvoglou, A., & Polychronidou, P. (2024). *Digital Economy and Green Growth*. Springer. https://doi.org/10.1007/978-3-031-66669-8

Mollik, M. S., Rahman, S. M., Rahat, M. R., Kulsum, C. U., & Sagir, S. A. M. (2024). Does trustworthiness influence travel service use intentions at an online travel agency? A study on the digitalization of the tourism sector in Bangladesh. *Geo Journal of Tourism and Geosites*, *52*(1), 30–40. https://doi.org/10.30892/gtg.52103-1180

Nahavandi, S. (2019). Industry 5.0—A human-centric solution. *Sustainability*, *11*(16), 4371. https://doi.org/10.3390/su11164371

Nazari, M., & Ansari, M. (2024). Bibliometric analysis and presenting the future research path in the field of sustainable business model innovation. *Scientometrics Research*. https://rsci.shahed.ac.ir/article_4664.html?lang=en

Nechyporuk, L., & Kocherhina, O. (2025). *Global Economy: Impact of Industry 5.0 for Sustainable Development*. Baltija Publishing. www.baltijapublishing.lv/omp/index.php/bp/catalog/download/572/15597/32904-1

Noor, Z. Z. (2025). Sustainable production management in circular economy supply chains. *International Journal of Industrial Engineering and Management*. https://doi.org/10.24867/IJIEM-384

Oro, C. E. D., Puton, B. M. S., Bernardi, J. L., Exenberger Finkler, D., Venquiaruto, L. D., Dallago, R. M., & Tres, M. V. (2025). Sustainable valorization of plant residues through enzymatic hydrolysis for the extraction of bioactive compounds: Applications as functional ingredients. *Processes*, *13*(5), 1314. www.mdpi.com/2227-9717/13/5/1314

Pinto Varela Alberte, E., & de Oliveira Novelli, G. (2024). Potential use of blockchain technology in construction supply chain management of Brazil. *Construction Innovation*. https://doi.org/10.1108/ci-08-2022-0208

Popa, I., Corboş, R. A., Ştefan, S. C., & Bunea, O. I. (2025). Circular economy practices, procurement strategies, and organizational competitiveness. *Systems*, *13*(5), 344. https://doi.org/10.3390/systems13050344

Rachinger, M., Rauter, R., Müller, C., Vorraber, W., & Schirgi, E. (2019). Digitalization and its influence on business model innovation. *Journal of Manufacturing Technology Management*, *30*(8), 1143–1160. https://doi.org/10.1108/JMTM-01-2018-0020

Raghav, A., Singh, B., Wongmahesak, K., & Chandra, S. (2025). Towards a greener future: Integrating life cycle assessment for sustainability in Industry 5.0. In *Sustainable Futures with Life Cycle Assessment in Industry 5.0* (pp. 1–20). IGI Global Scientific Publishing. https://doi.org/10.4018/979-8-3693-9346-8.ch001

Rahaman, M., Pranta, A. D., & Ahmed, S. (2025). Transitioning from Industry 4.0 to Industry 5.0 for sustainable and additive manufacturing of clothing: Framework, case studies, recent advances, and future prospects. *Materials Circular Economy*, *7*(1). https://doi.org/10.1007/s42824-025-00176-7

Rejeb, A., Rejeb, K., Keogh, J. G., & Süle, E. (2025). When Industry 5.0 meets the circular economy: A systematic literature review. *Circular Economy and Sustainability*, 1–32. https://doi.org/10.1007/s43615-025-00570-y

Roshid, M. M., Karim, R., Waaje, A., Islam, S., & Bhowmik, R. C. (2025a). Mapping the carbon finance research: A bibliometric exploration of innovation in sustainable finance. *Development and Sustainability in Economics and Finance*, *5*, 100048. https://doi.org/10.1016/j.dsef.2025.100048

Roshid, M. M., Saiyed, S., Karim, R., Haque, T., Siddique, M. S. R., Haque, M. A., Kumar, V., & Habib, S. B. (2025b). The role of green entrepreneurship in driving sustainable innovation in Bangladesh's startup ecosystem: Insights from a belt and road emerging market. *Environment, Innovation and Management*, *1*, 2550010. https://doi.org/10.1142/S3060901125500103

Roshid, M. M., Waaje, A., & Karim, R. (2024). Industry 6.0 as an emerging research field: A systematic and bibliometric analysis. In *EMIDWORLD 2nd International Congress on Economics Public Finance Business & Social Sciences* (pp. 784–802). Altınbas University. www.emidworld.com

Roshid, M. M., Waaje, A., Karim, R., Irtisham, M. R., & Kaswan, M. S. (2025c). Industry 5.0 and organizational sustainability: Integrating cutting-edge technologies and human ingenuity. In *Industry 5.0* (pp. 50–61). CRC Press. https://doi.org/10.1201/9781003535188-7

Santos, E. S., Afif, E., & Forján, R. (2024). By-product amendments for remediation aligned with circular economy principles. *Frontiers in Environmental Science*. www.frontiersin.org/articles/10.3389/fenvs.2024.1498172/full

Sassanelli, C. (2024). Case studies in circular economy. *Frontiers in Sustainability*. www.frontiersin.org/articles/10.3389/frsus.2024.1436950/full

Schot, J., & Geels, F. W. (2008). Strategic niche management and sustainable innovation journeys: Theory, findings, research agenda, and policy. *Technology Analysis and Strategic Management*, *20*(5), 537–554. https://doi.org/10.1080/09537320802292651

Schwab, K. (2017). *The Fourth Industrial Revolution*. Currency.

Shabur, M. A., Shahriar, A., & Ara, M. (2025). From automation to collaboration: Exploring the impact of Industry 5.0 on sustainable manufacturing. *Discover Sustainability*, *6*(1), 12. https://doi.org/10.1007/s43621-025-01201-0

Shah, R., Doss, A. S. A., & Lakshmaiya, N. (2025). Advancements in AI-enhanced collaborative robotics: Towards safer, smarter, and human-centric industrial automation. *Results in Engineering*, 105704. https://doi.org/10.1016/j.rineng.2025.105704

Singh, M., Srivastava, R., Fuenmayor, E., Kuts, V., Qiao, Y., Murray, N., & Devine, D. (2022). Applications of digital twin across industries: A review. *Applied Sciences*, *12*(11), 5727. https://doi.org/10.3390/app12115727

Stahel, W. R. (1982). The product life factor. In *An Inquiry into the Nature of Sustainable Societies* (pp. 72–105). Houston: The Role of the Private Sector, Houston Area Research Centre.

Stahl, B. (2012). Morality, ethics, and reflection: A categorization of normative IS research. *Journal of the Association for Information Systems*, *13*(8), 636–656. https://doi.org/10.17705/1jais.00304

Sung, T. K. (2018). Industry 4.0: A Korea perspective. *Technological Forecasting and Social Change*, *132*, 40–45. https://doi.org/10.1016/j.techfore.2017.11.005

Thaha, A. N., Ghamari, M., Jothiprakash, G., Velusamy, S., Karthikeyan, S., Ramesh, D., & Sundaram, S. (2025). High impact biomass valorization for second generation biorefineries in India: Recent developments and future strategies for sustainable circular economy. *Biomass*, *5*(1), 16. https://doi.org/10.3390/biomass5010016

Thakuri, P. K., Alkki, L., & Aarikka-Stenroos, L. (2024). Digital technologies enabling component reuse in circular value chains: Using digital twin, internet of things and robots in construction and manufacturing sectors. *R and D Management*. https://doi.org/10.1111/radm.12744

Thürer, M., Riedel, R., Von Cieminski, G., & Romero, D. (2024). Advances in production management systems for volatile, uncertain, complex, and ambiguous environments. In *Production Management Systems*. Springer. https://doi.org/10.1007/978-3-031-65894-5

Ünal, E. (2024). *Circular Industry 5.0: Regenerative-Dominant Logic of Value Creation and Manufacturing*. Remade Institute. https://remadeinstitute.org/wp-content/uploads/2025/03/74_Circular-Industry_Unal.pdf

Villarreal, P., & Calvo, R. A. (2021). The cost of AI adoption for SMEs: Barriers and enablers. *AI and Society*, *36*(3), 835–848. https://doi.org/10.1007/s00146-020-00974-0

Waaje, A., Karim, R., Roshid, M. M., & Meem, T. N. (2025a). Minimizing waste in the hospitality industry: The opportunities and challenges of lean tourism. In *Sustainable Waste Management in the Tourism and Hospitality Sectors* (pp. 125–166). IGI Global Scientific Publishing. https://doi.org/10.4018/979-8-3693-6110-8.ch006

Waaje, A., Karim, R., Roshid, M. M., Yeamin, B., & Meem, T. N. (2025b). Integration of Industry 4.0 with personalized medicine: A shift of the paradigm in healthcare. In Chemingui, H., & Lamloumi, M. (Eds.), *AI-Driven Personalized Healthcare Solutions* (pp. 27–69). IGI Global. https://doi.org/10.4018/979-8-3693-7858-8.ch002

Wuyts, W. (2025). Circular economy and entangled water infrastructures: Hydrofeminist perspectives. *Architecture, Structures and Construction*, *5*(1), 36. https://doi.org/10.1007/s44150-025-00154-2

Xu, X., Lu, Y., Vogel-Heuser, B., & Wang, L. (2021). Industry 4.0 and Industry 5.0—inception, conception and perception. *Journal of Manufacturing Systems*, *61*, 530–535. https://doi.org/10.1016/j.jmsy.2021.10.006

Yan, X., Liu, W., Lim, M. K., Lin, Y., & Wei, W. (2021). Exploring the factors to promote circular supply chain implementation in the smart logistics ecological chain. *Industrial Marketing Management*, *101*, 57–70. https://doi.org/10.1016/j.indmarman.2021.11.015

Zong, Z., & Guan, Y. (2024). AI-driven intelligent data analytics and predictive analysis in Industry 4.0: Transforming knowledge, innovation, and efficiency. *Journal of the Knowledge Economy*, 1–40. https://doi.org/10.1007/s13132-024-02001-z

14 Harnessing Industry 5.0 for Sustainable Industrial Transformation in Uganda

Policy, Practice, and Pathways

Mahadih Kyambade and Afulah Namatovu

14.1 Introduction

There is a radical shift taking place in the global business landscape to a model that is less about automation or robots or Industry 4.0 and more about human or sustainable models of Industry 5.0 (Golovianko et al., 2023; Mourtzis et al., 2022; Zizic et al., 2022). In contrast to its predecessor, Industry 5.0 focuses on the coordinated partnership between human beings and smart technologies, such as artificial intelligence (AI), Internet of Things (IoT), robotics, and machine learning (ML), and on inclusive, ethical, and environmentally sustainable industrial results (Alves et al., 2023; Nahavandi, 2019; Kyambade & Namatovu, 2025a). This change of perspective is particularly relevant to developing countries, which are struggling to turn to economic development for sustainable development. Industry 5.0 presents considerable opportunities for Africa, particularly given the region's priorities around industrial growth, employment for its youth, and environmental management. This next phase of industrialization is not only about adopting advanced technologies, but it is also about reimagining industry with sustainability, social inclusion, and resilience at the core (see Javaid et al., 2020; Ghobakhloo & Iranmanesh, 2021; Kyambade & Namatovu, 2025a). Still, it is worth noting that Africa faces some tough structural challenges: inadequate infrastructure, uneven digital access, and a lag in innovation ecosystems (Kayanja et al., 2025). For any meaningful industrial transformation to happen, the solutions must address these foundational issues in addition to technological adoption. Otherwise, the potential of Industry 5.0 will remain largely theoretical.

Key to understanding the transformative potential of Industry 5.0 is having a thorough understanding of its conceptual foundations and interrelated components. Industry 5.0 refers to the fusion of human intelligence with smart machines, aimed at co-creating value in ways that are personalized, ethical, and sustainable (Zizic et al., 2022; Kyambade et al., 2025a). In contrast to the efficiency-driven goals of Industry 4.0, Industry 5.0 places people and the planet at the heart of industrial innovation, sustainable industrial transformation. Our study is aimed at restructuring industrial processes and systems to reduce carbon emissions, adopt circular economy models, and prioritize human well-being (Santiago et al., 2025). Interconnected enablers such as policy alignment, public–private partnerships,

DOI: 10.4324/9781003647621-17

technological readiness, and stakeholder participation have been shown to influence this transformation positively (Sangarathas & Shanmugathas, 2025; França et al., 2017; Rosati et al., 2023). However, literature also highlights persistent barriers, including regulatory inertia, low technology absorption rates, and mismatched skills, particularly in low-resource settings (Ali et al., 2025; Namatovu & Kyambade, 2025a; Kyambade et al., 2024). These barriers must be systematically addressed for meaningful uptake of Industry 5.0 practices in the Global South.

This study draws on concepts from socio-technical transitions theory and multilevel perspective (MLP) to understand how Industry 5.0 can be realized in low-resource settings like Uganda. Socio-technical transitions (Geels, 2002) emphasize that industrial change is not only technological but also requires coordinated shifts in institutions, user practices, policy frameworks, and cultural norms. This perspective aligns with Industry 5.0's emphasis on human-centered and sustainable innovation. In addition, MLP (Ullman & Kittner, 2024) provides a useful lens for analyzing the interdependencies among governments, private sectors, academia, and civil society in fostering inclusive industrial transformation. These frameworks help conceptualize Industry 5.0 not as a linear technological upgrade but as a complex, multi-actor, and value-driven transition that requires alignment of systemic capabilities, policies, and partnerships. Together, they offer a robust conceptual foundation for exploring how Uganda can navigate the institutional, infrastructural, and social dimensions of Industry 5.0 adoption.

Uganda offers a pertinent case study for examining Industry 5.0 implementation in sub-Saharan Africa. The country's industrial policy frameworks, particularly Vision 2040 and the Third National Development Plan (NDP III), emphasize the importance of inclusive and innovation-led industrial growth (MoFPED, 2021). Nevertheless, the country's manufacturing sector remains underdeveloped, contributing just 8.4% to the national GDP, far below the 20% benchmark targeted under NDP III (Uganda Bureau of Statistics, 2023). Despite recent investments in digital skilling, ICT hubs, and agro-industrialization, Uganda continues to face significant challenges, including underutilized production capacity, youth unemployment, and environmental degradation (World Bank, 2023; UNCTAD, 2022). Technological infrastructure, especially outside urban areas, remains insufficient, limiting the country's readiness to adopt advanced manufacturing solutions. Furthermore, while initiatives in smart farming and e-government are underway, there is a lack of coherent strategies to operationalize Industry 5.0 across sectors. These shortcomings underscore the need for a systemic and inclusive approach to industrial modernization that aligns with global sustainability targets and local development realities.

Despite growing scholarly interest in Industry 5.0, most studies remain focused on high-income countries, often neglecting the contextual challenges and opportunities faced by developing economies like Uganda. There is a clear research gap in understanding how human–machine collaboration, ethical AI governance, and sustainability transitions can be realized in low-resource environments with fragile industrial ecosystems (David, 2024; Ogunode et al., 2023). Furthermore, little empirical research exists on how Africa can leapfrog traditional industrialization barriers by adopting Industry 5.0 principles in sectors like agro-processing, green manufacturing, and energy-efficient construction. This chapter seeks to bridge this gap by offering an in-depth exploration of Uganda's policy and practical readiness for Industry 5.0. Through qualitative methods such as stakeholder interviews, policy analysis, and sectoral case studies, the study aims to identify context-appropriate pathways for enabling sustainable, inclusive, and technologically enhanced industrial transformation.

This study contributes to emerging global debates on sustainable digital industrialization by localizing Industry 5.0 within Uganda's development agenda. It demonstrates that Industry 5.0 is not only a technological upgrade but also a socio-technical transition that requires ethical considerations, institutional reforms, and human-centered design. The findings will provide actionable insights for policymakers, industrial actors, educators, and civil society organizations on how to co-create a resilient and inclusive industrial future that places sustainability and human well-being at its core.

14.2 Literature Review

The transition from Industry 4.0 to Industry 5.0 has gained increasing scholarly attention as economies seek to align industrial transformation with broader sustainability and human development goals. While Industry 4.0 emphasized automation, data exchange, and efficiency, Industry 5.0 places human well-being, social responsibility, and environmental consciousness at the center of innovation (Alves et al., 2023). It introduces a paradigm where humans collaborate with intelligent machines, supported by AI, robotics, the IoT, and blockchain to create inclusive, ethical, and resilient production systems (Nahavandi, 2019; Kyambade & Namatovu, 2025a).

Several recent studies have emphasized the integrative value of Industry 5.0 in advancing sustainable development. For instance, Javaid et al. (2020) assert that Industry 5.0 promotes not only technological innovation but also societal resilience by encouraging collaborative intelligence and human–machine co-evolution. This is particularly relevant in the face of global disruptions such as pandemics, climate change, and labor market volatility. Similarly, Ghobakhloo and Iranmanesh (2021) argue that the alignment of digital transformation with human-centered values is essential for industrial sectors aiming to balance competitiveness with sustainability.

The key pillars of Industry 5.0 include sustainable production, circular economy practices, responsible AI governance, and skills development for the future workforce. França et al. (2017), Mpaata et al. (2025), and Rosati et al. (2023) demonstrate how business model innovation driven by circular principles can foster resource efficiency and value retention across supply chains. Santiago et al. (2025) and Kyambade et al. (2025c) highlight the role of green technologies and digital platforms in creating resilient supply networks and reducing environmental footprints. However, implementing such models in developing countries remains a challenge due to infrastructural, regulatory, and human capital constraints (Ali et al., 2025).

In the African context, the discourse on Industry 5.0 is still emerging. David (2024) call for a decolonized approach to sustainability transitions, one that recognizes local knowledge systems and community participation. Sangarathas and Shanmugathas (2025) and Kyambade et al. (2025d) emphasize that successful human-centered digital transformation in Africa requires robust institutional support, inclusive policy frameworks, and public–private partnerships. Despite these insights, empirical studies focusing on low-resource environments remain limited, especially those exploring practical pathways to operationalize Industry 5.0 in national development strategies.

A growing number of empirical studies are beginning to shed light on how African countries are adapting to the evolving demands of digital and sustainable industrialization. For example, David et al. (2024) examine Nigeria's use of AI and IoT in agro-processing clusters, revealing both the potential and gaps in regulatory support and digital skills. In Kenya, Yin et al. (2022) assess the integration of smart technologies in green manufacturing firms, noting that while digital readiness is improving, inclusive policy frameworks and financing remain major hurdles. In Rwanda, government-led programs such as the Smart Rwanda Master Plan have demonstrated how coordinated digital infrastructure and innovation hubs can accelerate industrial modernization in low-resource contexts (Maurya et al., 2025). These cases illustrate both the challenges and adaptive strategies that can inform Uganda's own pathway. They also underscore the importance of regional knowledge-sharing and context-sensitive policies for successfully localizing Industry 5.0 principles in sub-Saharan Africa (Kyambade et al., 2023b).

Uganda, like many sub-Saharan African countries, faces unique challenges in digital industrialization. Although policy frameworks such as Uganda Vision 2040 and the Third National Development Plan (NDP III) advocate for inclusive industrial growth, real-world implementation lags due to infrastructural deficits, digital divides, and misalignment between education systems and labor market needs (MoFPED, 2021; Uganda Bureau of Statistics, 2023). Yet, nascent initiatives in smart agriculture, green energy, and digital entrepreneurship suggest potential entry points for Industry 5.0 integration. The literature indicates that a coordinated and context-specific approach is needed to harness the transformative power of Industry 5.0 in such environments.

14.2.1 Theoretical Framework

This study is anchored in two interrelated theoretical perspectives: the socio-technical systems (STS) theory and the MLP on sustainability transitions.

STS theory posits that successful technological innovation is contingent on the alignment of social and technical subsystems (Appelbaum, 1997). In the context of Industry 5.0, this framework underscores the importance of human–machine collaboration, organizational culture, skills development, and ethical governance in achieving sustainable industrial transformation (Kyambade & Namatovu, 2025b). STS provide a lens to assess how technological adoption in Uganda's industrial sector is mediated by workforce readiness, regulatory structures, and societal values.

The MLP, as articulated by Geels (2002), provides a dynamic model for understanding transitions within complex STS. It identifies three analytical levels including niche innovations, socio-technical regimes, and the broader socio-technical landscape. In this study, niche innovations refer to pilot initiatives in AI, green manufacturing, and smart agriculture. The socio-technical regime encompasses Uganda's current industrial norms, practices, and institutions, while the socio-technical landscape includes macro-trends such as globalization, climate change, and demographic shifts. The MLP allows for a nuanced examination of how Industry 5.0 initiatives can disrupt or align with existing industrial structures in Uganda.

By combining STS theory and the MLP, this study offers a holistic understanding of the enablers and barriers to Industry 5.0 adoption. It allows for the exploration not only of technical feasibility but also of sociocultural compatibility, institutional readiness, and long-term sustainability. This dual-theoretical grounding enhances the explanatory power of the research and provides a robust foundation for policy and practice recommendations tailored to Uganda's development context.

14.3 Methodology

This study employed a qualitative exploratory case study design to examine how Uganda can leverage Industry 5.0 for sustainable industrial transformation (Kyambade et al., 2025b). The choice of a qualitative approach was guided by the study's aim to understand complex, context-specific dynamics involving emerging technologies, human capital, policy frameworks, and sustainability challenges factors that are best explored through in-depth, interpretive inquiry (Garcia & Gluesing, 2013). A case study strategy allowed the researchers to investigate real-world practices and stakeholder perceptions within their natural contexts, particularly across Uganda's manufacturing and agricultural sectors, which were chosen due to their centrality in national development priorities and potential alignment with Industry 5.0 elements.

Data were collected using three interrelated methods including semi-structured interviews, policy and document analysis, and nonparticipant observations. The interview component involved 28 key informants selected purposively for their expertise in digital technology, sustainability, and industrial operations. Respondents included government officials from the Ministry of Trade, Industry, and Cooperatives; senior managers in manufacturing and agri-tech firms; university researchers in AI and innovation studies; and representatives from NGOs and international development agencies. Interview questions were designed around four thematic areas: awareness of Industry 5.0; perceived opportunities and risks; institutional and infrastructural readiness; and strategic pathways for sustainable adoption. All interviews were conducted in English, lasted between 30 and 60 minutes, and were audio-recorded and transcribed with participant consent.

To ensure the credibility and consistency of data collection, interview protocols were pilot-tested with two respondents and refined for clarity, relevance, and thematic alignment. All interviews followed a flexible guide that allowed participants to elaborate on context-specific experiences while ensuring comparability across responses. Interviewers maintained detailed field notes alongside recordings to capture nuances and nonverbal cues. Coding consistency was enhanced through a collaborative coding process involving two independent researchers who reviewed the same transcripts and compared emerging themes. Where discrepancies in interpretation arose, they were resolved through discussion and consensus-building, supported by revisiting raw data and relevant theoretical constructs. An audit trail was maintained documenting coding decisions, theme development, and category refinement, thereby improving the dependability and confirmability of the findings.

To complement the interviews, policy and document reviews were conducted using over 20 national and regional frameworks. These included Uganda's NDP III, the Industrial Policy (2020), the Green Growth Development Strategy, and the National Science, Technology, and Innovation Roadmap. Regional documents such as the African Union's Agenda 2063 and the EAC Digital Transformation Strategy were also analyzed. The goal was to examine how existing policy environments align with or hinder the operationalization of Industry 5.0 principles, especially those related to human-centric automation, circular economy, and ethical AI deployment.

Nonparticipant observations were carried out at three manufacturing plants and two agri-processing startups located in Kampala, Jinja, and Mbarara. During these site visits, the researchers observed production lines, employee–machine interactions, sustainability practices, and digital tool usage. Field notes and photographs were used to capture the observations where permissions were granted. These visits enabled the researchers to validate the claims made during interviews and to understand the degree of practical integration of Industry 5.0 technologies and principles.

A purposive sampling technique was used to ensure the inclusion of diverse yet relevant stakeholders. To identify harder-to-reach but critical actors in agri-tech and logistics startups, snowball sampling was also used. Data saturation was achieved when new interviews no longer yielded novel insights. To enhance the study's trustworthiness, several strategies were employed, including triangulation across methods, member checking of findings with selected respondents, and peer debriefing to ensure interpretive accuracy. Data was analyzed thematically using Braun and Clarke's six-step approach, and NVivo 14 software was used for coding and organizing qualitative data. Initial codes were derived both deductively from theoretical constructs related to Industry 5.0 (e.g., human–machine collaboration, resilience, and sustainability) and inductively from the field data. All participants received clear explanations about the purpose and confidentiality of the study and voluntarily consented to participate. Anonymity was maintained through coded identifiers, and all digital data were stored on encrypted, password-protected devices.

Despite its strengths, the study faced limitations. The geographic and sectoral focus while adequate for depth limits generalizability to the broader Ugandan economy. Additionally, the rapidly evolving nature of Industry 5.0 means that current findings may need to be revisited as new technologies and policies emerge. Future studies could use mixed-methods or longitudinal designs to investigate how Industry 5.0 adoption unfolds over time in different regions and industries.

14.4 Findings

This section presents the major findings from the study, thematically organized around the key domains that emerged from interviews, document reviews, and field observations. Four central themes were identified: (1) awareness and understanding of Industry 5.0; (2) institutional readiness and capacity; (3) human-centric technology integration; and (4) policy alignment and implementation gaps. These findings offer a multilevel understanding of the opportunities, challenges, and pathways for leveraging Industry 5.0 in Uganda's industrial transformation.

14.4.1 Awareness and Understanding of Industry 5.0

The study found that awareness and understanding of Industry 5.0 among stakeholders in Uganda remain limited and fragmented, with substantial variations across sectors. While some actors in academia and international development organizations have started engaging with the concept, the majority of stakeholders including those in manufacturing, agriculture, and public policy were unfamiliar with its terminology and principles.

Most respondents were conversant with Industry 4.0, particularly its emphasis on automation, digitization, the IoT, and AI. These technologies have received notable attention through training programs, donor-funded projects, and government digitalization strategies. However, the evolution toward Industry 5.0, which prioritizes human–machine collaboration, sustainable development, and societal well-being, was largely unknown or misunderstood.

> "To be honest, I only heard of Industry 5.0 from this interview. We have been training for digital systems but not this new version that focuses on people and sustainability."
>
> (Training Officer, AgriTech Firm—Interview P07)

This response typifies the knowledge gap that exists, even among technically trained professionals. The notion of Industry 5.0 as a holistic framework, one that goes beyond efficiency and productivity to emphasize resilience, inclusiveness, and environmental consciousness, is not yet mainstreamed in Uganda's industrial discourse.

Yet, once the core values of Industry 5.0 were explained, many participants quickly identified with its relevance and expressed optimism about its potential to address real-world challenges in their sectors. This was particularly evident in manufacturing and agriculture, where human judgment, contextual knowledge, and localized problem-solving are still essential.

> "This is what we need, technology that works with people, not replaces them. Our farmers won't trust robots alone, but with human input, it becomes useful."
>
> (Ministry Official—Interview P03)

This quote reflects a cultural and contextual readiness to embrace a hybrid approach, blending technological efficiency with human values and social trust, especially in rural and semi-urban settings. Respondents emphasized that the success of any innovation, be it automation, robotics, or AI, depends on how well it accommodates local realities, from infrastructure limitations to labor market dynamics and user literacy levels.

The findings resonate with recent global literature that emphasizes the need to decolonize and localize Industry 5.0 frameworks for the Global South (Matenga & Mpofu, 2023; Atobishi & Mansur, 2025). Scholars argue that the dominant narratives of smart industries must evolve beyond Eurocentric blueprints and account for social equity, digital inclusion, and co-creation (collaborative process where local communities, stakeholders, and relevant actors actively participate in designing, developing, and shaping Industry 5.0 frameworks and innovations) with local communities.

Furthermore, participants indicated that language and branding may pose barriers. The term "Industry 5.0" was perceived as abstract or overly futuristic by some, yet its core values, human-centric design, sustainability, ethical AI, and circular economies were welcomed when explained in more practical, sector-specific terms. This suggests that policy communication and knowledge translation will be crucial in fostering a wider understanding.

Several respondents in education and training institutions also highlighted the need for curriculum reform, noting that most technical programs still focus on Industry 4.0 or general digital literacy, without integrating the broader socio-technical and sustainability dimensions central to Industry 5.0.

While current awareness of Industry 5.0 is low, there is strong latent receptivity to its principles once explained in relatable language. This gap between conceptual knowledge and practical alignment presents both a challenge and an opportunity for Uganda's transition toward a more inclusive, resilient, and sustainable industrial future.

From a theoretical standpoint, these findings align closely with the STS theory, which emphasizes the interplay among human actors, technological artifacts, and organizational structures in shaping system performance and innovation adoption. The limited awareness of Industry 5.0 among Ugandan stakeholders reflects a disjointed socio-technical landscape in which human-centered values and sustainability considerations are not yet integrated with the country's industrial discourse and practice. This advocates a need for systemic redesign that balances technological advancement with social, ethical, and cultural appropriateness.

Furthermore, when viewed through the lens of the MLP on sustainability transitions, Uganda's low awareness of Industry 5.0 can be understood as a reflection of niche-level innovations struggling to break through entrenched socio-technical regimes. While some awareness exists in academia and among international actors (niches), it has not yet diffused into dominant policy, industrial, or public sector regimes. The readiness of stakeholders to embrace these ideas when framed appropriately underscores an important leverage point for transition: effective knowledge translation and policy framing tailored to Uganda's development context.

This presents a novel insight in contrast to Eurocentric narratives that assume linear transitions and homogeneous diffusion of industrial paradigms.

14.4.2 Institutional Readiness and Capacity Gaps

The study revealed that Uganda's institutional landscape is not yet fully equipped to harness the potential of Industry 5.0, primarily due to persistent infrastructural deficits, misaligned human capital development, and weak interinstitutional linkages. While isolated initiatives to promote digitalization and automation exist, the systemic foundation needed for human-centric and sustainable industrial transformation remains inadequate.

One of the most frequently cited challenges was the obsolescence of industrial equipment and ICT infrastructure in many local firms, particularly in manufacturing hubs such as Jinja and Kampala. As one factory manager put it:

> "We are stuck with machinery from the 1990s. Before we even talk about collaborative robots, we need to digitize the basics."
>
> (Factory Manager, Jinja—Interview P11)

This statement underscores a crucial insight: many Ugandan industries have not yet fully transitioned into Industry 4.0, which is a prerequisite for the hybrid intelligence and human–machine collaboration envisioned in Industry 5.0. Without fundamental digital infrastructure (e.g., IoT devices, real-time data collection systems, and smart sensors), the leap to a more integrated, intelligent, and sustainable industrial model becomes difficult to realize (Chander et al., 2022).

Another significant barrier identified was the misalignment between education systems and industrial needs. Although universities and tertiary institutions offer computer science and engineering programs, these were often described as overly theoretical and disconnected from real-world applications:

> "We still teach programming as theory. Our graduates have little experience in applying AI to solve real industrial problems."
>
> (University Researcher—Interview P18)

This points to a fundamental skills gap in preparing future workers not only to operate but also to collaborate with advanced technologies in ethically sound and socially beneficial ways. Industry 5.0 emphasizes human-centricity, ethical design, and resilience, which demand new skill sets such as systems thinking, interdisciplinary collaboration, and digital sustainability management (European Commission, 2021). The absence of such capacities within current training pipelines was widely seen as a strategic bottleneck.

Technical and vocational education and training (TVET) institutions, which are ideally placed to provide mid-level, hands-on training, were also flagged as underfunded, outdated, and underutilized in Industry 5.0 preparedness. Despite their potential to produce technicians capable of maintaining and coworking with collaborative robotics, AI systems, and green industrial technologies, most are constrained by lack of resources, qualified instructors, and partnerships with local industries.

Moreover, the institutional ecosystem remains fragmented. Respondents repeatedly cited the absence of effective collaboration among universities, industries, government agencies, and innovation hubs. There is limited support for co-creation, co-investment, or knowledge transfer between academia and industry. Unlike in countries where triple-helix models (innovation as the result of dynamic collaboration among universities, industry, and government) drive innovation (Adel, 2024), Uganda's siloed institutional arrangements reduce agility in responding to global industrial trends.

> "Most research remains on paper. We rarely see prototypes being commercialized or scaled up through partnerships with firms."
>
> (Innovation Hub Coordinator—Interview P04)

Furthermore, funding mechanisms for innovation are inconsistent. Many stakeholders highlighted that reliance on donor-driven digital projects, rather than locally owned and sustained investments, creates short-term gains without building systemic capacity. A lack of domestic industrial research funding was viewed as a significant barrier to developing technologies tailored to Ugandan realities whether in agro-processing, green manufacturing, or energy management.

Another critical insight was the absence of a national innovation framework aligned with Industry 5.0 values. While Uganda's NDP III and STI Policy 2020 emphasize innovation and industrialization, they fall short in defining strategic priorities, performance metrics, or incentives for inclusive, human-centered industrial upgrading.

These institutional readiness gaps align closely with the STS theory, which emphasizes the need for balanced co-evolution among social, technical, and organizational elements. The study demonstrates that Uganda's technological ambitions, whether in AI, digital agriculture, or green energy, are constrained by outdated educational systems, fragmented institutional partnerships, and a lack of systemic investment in innovation infrastructure. This imbalance indicates that Uganda's industrial ecosystem is technologically aspirational but socio-institutionally underprepared.

Applying the MLP, these findings suggest Uganda is caught between a regime that has not fully transitioned into Industry 4.0 and emerging niche activities that reflect Industry 5.0 ideals (e.g., inclusive technology incubators or ethical AI pilots). This dual reality presents both friction and opportunity. A unique insight from Uganda is that informal innovation hubs and localized practices (e.g., youth technology startups and green agri-tech) may be more agile and value-aligned than legacy institutions, suggesting that bottom-up innovation could accelerate regime shifts if supported with proper governance and interinstitutional collaboration.

14.4.3 Human-Centric Technology Integration

One of the defining pillars of Industry 5.0 is placing humans at the center of technological transformation, rather than displacing them through automation. The study revealed cautious optimism among Ugandan industrial stakeholders toward human–machine collaboration, especially when digital tools were implemented in a context-sensitive, inclusive, and participatory manner.

Respondents working in manufacturing and agro-processing firms piloting AI-enabled systems reported tangible benefits in productivity, quality control, and workplace safety. However, they were equally clear that technology cannot replace human insight, particularly in nuanced production processes where contextual knowledge is crucial:

> "We've started using AI-enabled sensors for quality control, but we still need people on the line. Their eyes and judgment catch things machines don't."
>
> (Quality Assurance Officer—Interview P09)

This observation reflects what scholars describe as symbiotic automation where human intuition complements machine precision rather than being displaced by it (Emmanouilidis & Waschull, 2021). In environments where quality control requires flexibility, cultural understanding, or ethical judgment (e.g., food production, healthcare, and education), machines alone cannot replicate human adaptability. Thus, Industry 5.0 is not anti-automation but rather pro-collaboration, embedding human values, ethics, and creativity at the heart of smart systems (Demir et al., 2019).

Evidence of early Industry 5.0 traits was also visible in some agri-businesses and green enterprises where environmentally conscious technologies such as solar-powered systems and IoT-enabled irrigation were deployed. Yet, these technologies did not eliminate the need for human expertise; instead, they shifted the skill requirements toward technical interpretation, remote monitoring, and preventive maintenance:

> "We installed solar systems that we control remotely. It helps us save costs and reduce emissions, but we still need trained technicians to interpret the data."
>
> (Agro-processing Startup Founder—Interview P14)

This highlights a critical lesson: digital sustainability innovations must be accompanied by investments in human capacity, particularly in data interpretation, ethical AI application, and troubleshooting of intelligent systems. The success of Industry 5.0 lies not only in deploying advanced tools but also in co-creating value with workers, empowering them to make informed decisions in partnership with machines (European Commission, 2021).

A prominent concern raised by multiple participants was the risk of excluding marginalized groups, especially women, rural communities, and non-tech-savvy workers, from the benefits of digital industrialization. For example, gender disparities in access to technical education and digital tools were seen as a barrier to equitable participation in emerging Industry 5.0 ecosystems. Participants emphasized the need for inclusive design principles, echoing similar findings in global studies (Mukherjee & Mandal, 2025; Tushabe et al., 2025):

> "We talk about smart factories, but what about the rural cooperatives? Or the women who can't afford smartphones? Inclusion must be part of the system."
>
> (NGO Officer—Interview P06)

Another respondent from a youth innovation hub stressed the importance of participatory co-design, especially when implementing technologies in rural or low-literacy communities:

> "We piloted an AI-driven tool for sorting coffee beans, but the first version was rejected because it didn't match how local farmers classified quality. We had to rework it with their input."
>
> (Technology Incubator Coordinator—Interview P12)

This underscores that local knowledge, user feedback, and cultural values must be embedded into system design, a key tenet of Industry 5.0's human-centric ethos (Longo et al., 2020; Kyambade et al., 2023a). Top-down technology rollouts, even if efficient, risk failure if they ignore end-user realities and social acceptance.

Ethical considerations also emerged as a recurrent theme. Stakeholders expressed a strong desire for responsible AI governance frameworks that protect privacy, ensure transparency in decision-making algorithms, and prevent labor exploitation through digital surveillance. Without such safeguards, the integration of smart technologies could undermine trust, especially among workers unfamiliar with automated decision systems.

> "We're okay with sensors and tracking, but only if there's transparency. People must know what data is collected and how it's used."
>
> (Labor Union Representative—Interview P19)

This reflects global calls for ethical-by-design industrial systems, where technologies uphold fairness, accountability, and user autonomy, core principles of the Industry 5.0 vision.

These findings reinforce the STS perspective that technological change is not only about machines but also about the reconfiguration of human roles, ethics, and institutional practices. The Ugandan experience highlights how people-centered values such as trust, contextual judgment, and inclusive design are central to the success of smart technologies in real-world settings. Technology alone is insufficient; its social embedding, acceptability, and cultural alignment determine its viability.

From an MLP standpoint, the early examples of Industry 5.0 integration in Uganda's agri-tech and green energy sectors can be viewed as niche innovations that could eventually disrupt dominant production models. However, these niche practices require protective spaces and enabling environments such as inclusive funding, ethical guidelines, and participatory design norms to mature. A key insight is that co-design and contextual customization are not only ethical imperatives but also strategic necessities for low-resource settings where user misalignment can stall or derail innovation diffusion.

14.4.4 Policy Alignment and Implementation Gaps

While Uganda has made notable strides in crafting development strategies that reference digital transformation, innovation, and green growth, this study revealed a persistent disconnect between policy intentions and coordinated implementation, particularly in the context of Industry 5.0. Despite the presence of high-level frameworks such as NDP III; the Science, Technology, and Innovation (STI) Policy (2020); and the Digital Uganda Vision, stakeholders widely agreed that these documents lack a coherent, actionable framework to guide human-centric and sustainable industrial transformation.

A policy analyst interviewed in the study summarized the dilemma succinctly:

> "Our policies mention technology and green growth separately. There's no framework that brings them together under something like Industry 5.0."
>
> (Policy Analyst—Interview P05)

This quote reflects the fragmentation in Uganda's policy architecture, where digitalization, sustainability, and industrial development are treated in sectoral silos rather than through a convergent systems approach. As emphasized in emerging Industry 5.0 scholarship, realizing the potential of human–machine collaboration, ethical AI, and green innovation requires interministerial coordination, multilevel governance, and integrated investment pathways (Demir et al., 2019; European Commission, 2021).

Several participants pointed to the absence of a dedicated national roadmap for Industry 5.0 adoption as a significant policy blind spot. Unlike regions such as the European Union or East Asia, which have begun drafting frameworks for next-generation industrial policy (e.g., the EU's Industry 5.0 Strategy), Uganda remains in the early stages of conceptualizing what Industry 5.0 means within its local context.

> "We need a national vision that tells ministries, investors, and academia how to collaborate on future industries. Right now, it's guesswork."
>
> (Innovation Hub Director—Interview P10)

This lack of policy clarity hampers cross-sectoral coordination and limits the ability of stakeholders, especially small and medium-sized enterprises (SMEs), universities, and local governments, to align their efforts. It also discourages long-term private sector investment in AI, robotics, and renewable energy integration, all of which are key enablers of sustainable Industry 5.0 ecosystems (Ghobakhloo et al., 2023; Mujabi et al., 2025).

Another policy concern raised was Uganda's dependency on donor-led pilot projects to drive innovation. While development partner collaborations have supported digital agriculture, e-mobility, and ICT infrastructure pilots, there was concern that these efforts lacked sustainability or systemic scaling:

> "Many of the so-called digital transformation programs are donor-funded. Once the funding ends, the project dies. There's little integration into national systems."
>
> (Development Economist—Interview P07)

This critique echoes findings in development policy literature warning of "projectization," the phenomenon where externally funded innovations are not institutionalized within local systems due to poor domestic policy ownership, weak feedback loops, or over-centralization of decision-making (Companie & Windapo, 2024). In the case of Uganda, this approach risks reducing Industry 5.0 to a collection of disconnected experiments, rather than fostering a resilient, inclusive, and scalable industrial revolution.

Despite these challenges, some participants noted positive steps, such as increased government interest in smart manufacturing zones, digital skilling, and collaboration with institutions like the Uganda National Council for Science and Technology (UNCST). However, they emphasized that intentional coordination is still lacking, particularly in integrating ethics, gender equity, and sustainability into the digital industrial agenda.

> "We need to embed social and environmental standards in every innovation fund or tech park. Otherwise, it's not Industry 5.0, it's just another wave of automation."
>
> (Policy Advisor—Interview P16)

This concern reflects Industry 5.0's emphasis on value-driven innovation, where equity, inclusion, and ecological sustainability are not add-ons but core design principles (Mukherjee & Mandal, 2025; Nahavandi, 2019). Uganda's current innovation frameworks largely prioritize efficiency and productivity, with less attention to ethical governance, worker well-being, or rural inclusion, all of which are critical for building trust and resilience in emerging industrial systems.

The findings regarding fragmented policy frameworks and weak implementation echo the STS theory's assertion that institutional coherence is essential for successful socio-technical transitions. Uganda's policy ecosystem, although rhetorically committed to digital and sustainable transformation, lacks the integrative mechanisms necessary to operationalize Industry 5.0 principles across sectors. This disjointed policy architecture hampers the coordination and scaling of human-centric, ethical, and environmentally sustainable innovation.

Using the MLP, these gaps illustrate the absence of regime-level support structures that could allow niche Industry 5.0 experiments to scale. Without a national vision or roadmap explicitly framing Industry 5.0 priorities, Uganda's stakeholders operate in silos, limiting cross-sectoral learning and systemic innovation. A unique insight here is the overreliance on donor-led initiatives, which often remain decoupled from domestic policy frameworks, thereby reducing institutional ownership and long-term sustainability. In contrast to global contexts where Industry 5.0 strategies are domestically funded and politically championed, Uganda's transition remains externally influenced and sporadically integrated. This underscores the need for a homegrown policy agenda that aligns industrial strategy with social values and technological foresight.

Table 14.1 provides a thematic summary of the study's key findings on Uganda's readiness for Industry 5.0. Under Awareness and Understanding, stakeholders showed limited knowledge and often confused Industry 5.0 with Industry 4.0. However, once explained, many expressed strong interest, suggesting that awareness campaigns tailored to local contexts could be effective. The theme of Institutional Readiness and Capacity highlights major gaps in infrastructure, outdated equipment, and weak links between education and industry. These challenges

Table 14.1 Summary of Key Findings by Thematic Area.

Theme	*Key Insights*	*Challenges Identified*	*Opportunities/ Recommendations*
Awareness and understanding of Industry 5.0	-Limited awareness among stakeholders -Misconceptions with Industry 4.0	-Low conceptual clarity -Terminology perceived as abstract	-Strong latent interest once explained -Embed Industry 5.0 in sector-specific narratives
Institutional readiness and capacity	-Infrastructure and skills gaps -Misaligned education and training -Weak academia–industry linkages	-Obsolete machinery -Fragmented policy support -Poor curriculum relevance	-Invest in TVET and university reform -Develop co-creation platforms -Establish industry–academia–government consortia
Human-centric technology integration	-Early examples in agri-tech, energy -Human roles still critical in smart tech	-Risk of marginalizing rural/women stakeholders -Limited ethical AI frameworks	-Promote participatory design -Draft responsible AI policies -Incentivize inclusive innovation
Policy alignment and implementation gaps	-Strong policy rhetoric but poor integration -Siloed strategies and reliance on donors	-Lack of a national Industry 5.0 strategy -Weak coordination and ownership	-Create a centralized Industry 5.0 roadmap -Link NDP III, STI Policy, and Digital Vision -Institutionalize multi-stakeholder platforms

Source: Author's own work.

can be addressed through investment in technical and vocational education, curriculum reform, and stronger collaboration among universities, government, and industry. Human-Centric Technology Integration is already emerging in sectors like agri-tech and renewable energy, but risks remain, especially the exclusion of rural communities and women. Promoting participatory design and establishing ethical AI guidelines can make technology adoption more inclusive and socially aligned. Finally, Policy Alignment and Implementation Gaps reveal a disconnect between Uganda's strong policy rhetoric and actual implementation. The absence of a unified Industry 5.0 strategy and overreliance on donor projects limit progress. A centralized roadmap and coordinated national vision linking existing policies (like NDP III and the STI Policy) are essential to guide systemic change.

14.5 Discussion

The study's findings provide critical insights into the contextual realities shaping Uganda's readiness and capacity to harness Industry 5.0 for sustainable industrial transformation. These findings underscore both the latent potential and the systemic constraints facing Uganda as it seeks to transition from digital adoption (Industry 4.0) to a more integrated, human-centric, and sustainability-oriented industrial paradigm (Industry 5.0). The discussion integrates empirical findings with recent literature from global and African contexts, highlighting policy, capacity-building, and socio-technical implications.

14.5.1 Awareness and Understanding of Industry 5.0

A major theme that emerged from the findings is the limited and fragmented awareness of Industry 5.0 among Ugandan stakeholders. This mirrors global observations that Industry 5.0 remains conceptually vague and under-communicated, especially outside Europe and Asia (Demir et al., 2019; Nahavandi, 2019). In Uganda, although Industry 4.0 has gained traction through donor-funded digitalization programs and national digital strategies (e.g., Digital Uganda Vision), the broader values of Industry 5.0 emphasizing societal well-being, ethical AI, and human–machine collaboration are poorly understood or entirely unknown. This knowledge gap is particularly concerning given that successful implementation of Industry 5.0 requires a foundational understanding across industries, policy sectors, and educational institutions (European Commission, 2021). Studies by Longo et al. (2020) and Mukherjee and Mandal (2025) emphasize that awareness is the first step toward institutional and technical preparedness. The strong latent receptivity observed among Ugandan stakeholders upon learning about Industry 5.0 supports these insights, suggesting that targeted awareness campaigns, sector-specific translations, and curriculum redesign could bridge this gap.

14.5.2 Institutional Readiness and Capacity Gaps

The readiness of Uganda's institutions to implement Industry 5.0 is undermined by outdated technical infrastructure, a misalignment between education systems and industrial needs, and fragmented innovation ecosystems. These findings align with studies in similar low- and middle-income contexts, such as South Africa, Kenya, and Nigeria, which show that structural constraints in digital infrastructure and skills ecosystems inhibit transition to smart and sustainable industrial models (Atobishi & Mansur, 2025; Chander et al., 2022). The data also confirm the presence of a "two-speed" industrial system in Uganda, where some sectors (notably fintech and ICT) are advancing rapidly, while manufacturing and agro-processing still operate with analog machinery and outdated production systems. This imbalance complicates the adoption of collaborative robotics, cyber–physical systems, and real-time AI tools, the core pillars of Industry 5.0 (Emmanouilidis & Waschull, 2021).

More importantly, the study echoes global findings that highlight the crucial role of TVET institutions and mid-level professionals in implementing Industry 5.0 (Adel, 2024). Uganda's underutilized and under-resourced TVET institutions represent a missed opportunity. Strengthening these institutions could serve as a bridge between high-level research and on-the-ground application of human-centric technologies. The observed disconnect between research institutions and local industry also supports existing critiques of Africa's weak innovation systems. Research by Matenga and Mpofu (2023) and Companie and Windapo (2024) stress that unless academic research is aligned with local industrial problems and supported by commercialization strategies, technological advancement will remain slow and fragmented.

14.5.3 Human-Centric Technology Integration

A significant and encouraging finding is the cautious optimism around human–machine collaboration, particularly when technologies are contextually adapted and inclusively implemented. This mirrors the principles of "symbiotic automation," where human judgment complements algorithmic precision (Demir et al., 2019; Emmanouilidis & Waschull, 2021). The study highlights that successful integration of technologies like IoT-enabled irrigation and solar-powered processing relies not only on infrastructure but also on trained personnel who can interpret, maintain, and adapt these tools to local conditions. This reflects the global insight that Industry 5.0 is less about technology per se and more about the design of STS that empower users (Longo et al., 2020; Namatovu & Kyambade, 2025b).

However, ethical and equity considerations were central to participants' concerns, particularly regarding surveillance, algorithmic transparency, and digital exclusion. This aligns with the European Commission's Industry 5.0 framework, which places emphasis on ethical AI, fairness, and transparency. Studies by Mukherjee and Mandal (2025) and Nahavandi (2019) also caution that without strong ethical guardrails, the deployment of Industry 5.0 technologies can exacerbate social inequality. The findings on gender and rural exclusion echo similar concerns in other sub-Saharan African studies. For example, Kasinathan et al. (2022) emphasize that Industry 5.0 must prioritize rural digital inclusion, while Matseke (2023) and Tushabe et al. (2023) call for gender-sensitive innovation policies. Uganda's emphasis on "smart urbanization" must therefore be recalibrated to include grassroots innovation and marginalized communities.

14.5.4 Policy Alignment and Implementation Gaps

The study uncovered critical gaps in Uganda's policy framework with regard to coherent and actionable strategies for Industry 5.0. Despite having several national policies that address digitalization and sustainability, there is no integrated roadmap that aligns these ambitions within a human-centric industrial strategy. This fragmentation mirrors what is described in policy literature as "policy misalignment," where overlapping mandates and siloed implementation inhibit effective governance (Ghobakhloo et al., 2023). Similar to findings in Kenya and Rwanda, Uganda's overreliance on donor-driven digital transformation projects risks undermining local ownership, sustainability, and scalability (Companie & Windapo, 2024; Atobishi & Mansur, 2025). The absence of a clear policy framework for human-centric innovation also deters private sector investment in frontier technologies, as uncertainty breeds risk aversion. The study supports global calls for national governments to articulate dedicated Industry 5.0 strategies that bridge digital transformation, green growth, and social inclusion. Examples from the European Union and Japan suggest that having a centralized vision accelerates cross-sectoral coordination, investment attraction, and public trust in new technologies (European Commission, 2021).

Overall, the study confirms that Uganda possesses a strong contextual and cultural foundation for Industry 5.0 adoption, particularly through its receptivity to inclusive, people-centered innovation. However, realizing this potential will require systemic changes in awareness-building, institutional reform, ethical governance, and policy alignment. These findings contribute to a growing body of knowledge advocating for a localized, decolonized, and socially grounded approach to Industry 5.0 in the Global South.

14.6 Implications of the Study

The findings of this study present critical implications for policymakers, industry actors, educational institutions, and development partners seeking to harness Industry 5.0 for sustainable industrial transformation in Uganda. First, the limited awareness and conceptual clarity surrounding Industry 5.0 call for targeted knowledge translation strategies. Government agencies, innovation hubs, and higher education institutions must develop localized communication tools, curricula, and training programs that demystify Industry 5.0 and align it with Uganda's socioeconomic realities. Second, the study highlights the urgent need to strengthen institutional readiness through investment in technical infrastructure, digital skills development, and policy coherence. This implies that national development frameworks such as Uganda's NDP III should be revisited to explicitly incorporate human-centric, sustainable, and collaborative innovation goals aligned with Industry 5.0 principles.

Furthermore, the results underscore the importance of inclusive innovation ecosystems. For Industry 5.0 to be transformative rather than extractive, its implementation must actively involve marginalized groups, including women, rural communities, and informal workers. Firms must adopt inclusive design principles and promote ethical AI, transparent data practices, and worker co-creation models. The study also highlights the necessity of aligning technical education, especially in TVET institutions, with practical Industry 5.0 competencies such as human–machine collaboration, data ethics, and sustainable production to prepare Uganda's workforce for next-generation industries. Lastly, Uganda's overreliance on donor-led pilot projects suggests a need to reorient innovation funding toward locally owned, scalable, and systems-oriented models that can institutionalize Industry 5.0 beyond fragmented initiatives.

The government should also introduce targeted fiscal incentives such as tax breaks or matching grants for firms investing in ethical AI, circular manufacturing technologies, or collaborative robotics. The establishment of a national Industry 5.0 coordination unit within the Ministry of Trade or Science and Technology could promote cross-sectoral alignment, develop regulatory sandboxes for emerging technologies, and ensure ethical governance. Additionally, regulatory frameworks must be updated to support responsible innovation. This includes drafting an AI ethics policy, updating national industrial standards to include sustainability metrics, and mandating inclusive stakeholder consultations in digital transformation projects. These concrete steps will enable policymakers and practitioners to translate Industry 5.0 from concept to systemic practice.

Table 14.2 summarizes Uganda's readiness for Industry 5.0 across six key domains, revealing significant gaps and opportunities. Awareness and understanding are low, with fragmented knowledge outside academia, calling for targeted public education. Institutional infrastructure is outdated and poorly coordinated, requiring investment and stronger cross-sector collaboration. Human capital readiness is medium-low; current training remains theoretical, highlighting the need for curriculum reform and skills development in ethics and human–machine collaboration. Policy integration is weak, with disconnected strategies and no unified Industry 5.0 roadmap. The innovation ecosystem shows moderate potential but lacks scale-up support, requiring inclusive funding mechanisms. Lastly, ethical and inclusive design is underdeveloped, with no formal governance frameworks, stressing the need for ethical AI policies and stakeholder inclusion. Together, these findings highlight priority areas for intervention to enable a sustainable, human-centered industrial transition in Uganda.

14.7 Study Limitations and Future Research Directions

While this study offers valuable qualitative insights into Uganda's preparedness and pathways for Industry 5.0 adoption, several limitations should be acknowledged, which also point toward promising avenues for future research. First, the use of a qualitative research design and purposive sampling limits the generalizability of the findings. Data was gathered primarily from stakeholders in urban

Table 14.2 Readiness Indicators for Industry 5.0 in Uganda.

Indicator Domain	*Current Status*	*Readiness Level (Low/Medium/ High)*	*Needed Actions*
Awareness & Understanding	Fragmented knowledge, sector-dependent	Low	Knowledge translation strategies, public awareness campaigns
Institutional Infrastructure	Legacy machinery, siloed governance	Low	Invest in infrastructure, promote cross-sectoral R&D
Human Capital & Skills	Theoretical training, underutilized TVETs	Medium-low	Reform curricula, build capacity in systems thinking and digital ethics
Policy Integration	Multiple disconnected policies	Low	Draft Industry 5.0 strategy, establish coordination unit
Innovation Ecosystem	Active hubs but weak scale-up	Medium	Create funding pipelines for inclusive, ethical, and green technology
Ethical and Inclusive Design	Emerging concerns but no frameworks	Low	Develop ethical AI guidelines, mandate stakeholder inclusion

Source: Author's own work (adapted from findings).

and peri-urban regions such as Kampala, Jinja, and Mbarara, potentially excluding the perspectives of rural and marginalized industrial actors who may face different infrastructural and technological constraints. Furthermore, the study concentrated on the manufacturing and agro-processing sectors, leaving out other critical areas such as construction, healthcare, education, and mining, which could reveal additional challenges and opportunities for Industry 5.0 integration. The study also relied heavily on self-reported data through interviews, which may be influenced by social desirability bias or inaccuracies in retrospective accounts. Additionally, the absence of longitudinal data limited the ability to evaluate the sustained impact of emerging Industry 5.0 initiatives on outcomes such as productivity, sustainability, or workforce development. The lack of comprehensive national datasets on Industry 5.0 readiness further constrained the potential for quantitative benchmarking or comparative policy analysis.

While this study provides valuable insights into Uganda's readiness for Industry 5.0, it is important to acknowledge a key limitation of the use of purposive and snowball sampling methods primarily engaging urban and peri-urban stakeholders. As a result, perspectives from rural communities, informal enterprises, and hard-to-reach regions may be underrepresented. This sampling bias limits the generalizability of the findings, particularly in understanding how Industry 5.0 transitions may unfold in decentralized, non-urban contexts where infrastructural and technological constraints are more pronounced. Therefore, findings should be interpreted with caution when applied to rural and informal sectors, where the dynamics of digital adoption, workforce readiness, and policy engagement may differ significantly.

There is also a need to develop and propose preliminary quantitative indicators to benchmark Uganda's Industry 5.0 readiness. Suggested metrics may include AI and IoT adoption rates across key sectors, digital literacy and skills indices, public–private R&D investment levels, and the proportion of green technology usage in production. Integrating such indicators into future mixed-method frameworks would not only enhance empirical depth but also support evidence-based policymaking and cross-country comparisons. Longitudinal case studies tracking the design, deployment, and adaptation of pilot technologies could provide evidence of long-term sustainability, resilience, and social equity outcomes.

Furthermore, future research should explore the sociocultural and ethical dimensions of Industry 5.0, including local trust in AI systems, gender dynamics in digital adoption, and the influence of indigenous knowledge systems on technology co-creation. Comparative analyses across sub-Saharan African countries could uncover context-specific best practices, policy innovations, and models of regional cooperation. As Industry 5.0 continues to evolve globally, further studies should also investigate its intersection with broader agendas such as climate resilience, inclusive digital economies, and post-pandemic recovery, particularly in resource-constrained or vulnerable contexts like Uganda.

References

Adel, A. (2024). The convergence of intelligent tutoring, robotics, and IoT in smart education for the transition from Industry 4.0 to 5.0. *Smart Cities*, *7*(1), 325–369.

Ali, I., Nguyen, K., & Oh, I. (2025). Systematic literature review on Industry 5.0: Current status and future research directions with insights for the Asia Pacific countries. *Asia Pacific Business Review*, 1–28.

Alves, J., Lima, T. M., & Gaspar, P. D. (2023). Is industry 5.0 a human-centred approach? A systematic review. *Processes*, *11*(1), 193.

Appelbaum, S. H. (1997). Socio-technical systems theory: An intervention strategy for organizational development. *Management Decision*, *35*(6), 452–463.

Atobishi, T., & Mansur, H. (2025). Bridging digital divides: Validating government ICT investments accelerating sustainable development goals. *Sustainability*, *17*(5), 2191.

Chander, B., Pal, S., De, D., & Buyya, R. (2022). Artificial intelligence-based Internet of Things for Industry 5.0. In *Artificial Intelligence-Based Internet of Things Systems* (pp. 3–45). Cham: Springer International Publishing.

Companie, F., & Windapo, A. (2024, August). Project portfolio and construction management practices impacting urban infrastructure performance: Theoretical perspectives. In *International Conference on Management Science and Engineering Management* (pp. 1400–1414). Singapore: Springer Nature.

David, J. O. (2024). Decolonizing climate change response: African indigenous knowledge and sustainable development. *Frontiers in Sociology*, *9*, 1456871.

David, L. O., Aigbavboa, C., Nwulu, N., & Adepoju, O. O. (2024). Application of digital technologies of the Fourth Industrial Revolution (4IR) in water, energy, and food (WEF) nexus. In *Sustainable Synergy: A Digital Framework for the Water–Energy–Food Nexus Project Delivery in Developing Economies* (pp. 287–317). Cham, Switzerland: Springer Nature.

Demir, K. A., Döven, G., & Sezen, B. (2019). Industry 5.0 and human–robot co-working. *Procedia Computer Science*, *158*, 688–695.

Emmanouilidis, C., & Waschull, S. (2021). Human in the loop of AI systems in manufacturing. In *Trusted Artificial Intelligence in Manufacturing: A Review of the Emerging Wave of Ethical and Human-Centric AI Technologies for Smart Production* (pp. 158–172). Now Publishers Inc.

European Commission. (2021). *Industry 5.0: Towards a Sustainable, Human-Centric and Resilient European Industry*. Publications Office of the European Union. https://research-and-innovation.ec.europa.eu

França, C. L., Broman, G., Robert, K. H., Basile, G., & Trygg, L. (2017). An approach to business model innovation and design for strategic sustainable development. *Journal of Cleaner Production*, *140*, 155–166.

Garcia, D., & Gluesing, J. C. (2013). Qualitative research methods in international organizational change research. *Journal of Organizational Change Management*, *26*(2), 423–444.

Geels, F. W. (2002). Technological transitions as evolutionary reconfiguration processes: A multi-level perspective and a case-study. *Research Policy*, *31*(8–9), 1257–1274.

Ghobakhloo, M., & Iranmanesh, M. (2021). Digital transformation success under Industry 4.0: A strategic guideline for manufacturing SMEs. *Journal of Manufacturing Technology Management*, *32*(8), 1533–1556.

Ghobakhloo, M., Iranmanesh, M., Morales, M. E., Nilashi, M., & Amran, A. (2023). Actions and approaches for enabling Industry 5.0-driven sustainable industrial transformation: A strategy roadmap. *Corporate Social Responsibility and Environmental Management*, *30*(3), 1473–1494.

Golovianko, M., Terziyan, V., Branytskyi, V., & Malyk, D. (2023). Industry 4.0 vs. Industry 5.0: Co-existence, transition, or a hybrid. *Procedia Computer Science*, *217*, 102–113.

Javaid, M., Haleem, A., Singh, R. P., Haq, M. I. U., Raina, A., & Suman, R. (2020). Industry 5.0: Potential applications in COVID-19. *Journal of Industrial Integration and Management*, *5*(4), 507–530.

Kasinathan, P., Pugazhendhi, R., Elavarasan, R. M., Ramachandaramurthy, V. K., Ramanathan, V., Subramanian, S., . . . Alsharif, M. H. (2022). Realization of sustainable development goals with disruptive technologies by integrating Industry 5.0, Society 5.0, smart cities and villages. *Sustainability*, *14*(22), 15258.

Kayanja, W., Kyambade, M., & Kiggundu, T. (2025). Exploring digital transformation in higher education setting: The shift to fully automated and paperless systems. *Cogent Education*, *12*(1), 2489800.

Kyambade, M., Abdul, S. M., & Namatovu, A. (2023a). Adoption theories for internet banking in Uganda. *Advances in Research*, *24*(6), 136–145.

Kyambade, M., Kagere, R., Namatovu, A., Ssentumbwe, A. M., Tushabe, M., & Oyella, S. (2024). Technological innovation and commercial banks' financial performance: A mediation analysis of risk management practices. *Seisense Business Review*, *4*(1), 98–117.

Kyambade, M., Mugambwa, J., Namuddu, R., Kiggundu, T., Sewante, L., & Namatovu, A. (2023b). Socially responsible leadership: A drive to Uganda's institutions of higher learning performance. *Journal of Economics, Business and Management*, *2*(9), 13–20.

Kyambade, M., & Namatovu, A. (2025a). Health-care leaders' perspectives on AI implementation in Uganda: Overcoming barriers, driving innovation and strategic considerations. *Leadership in Health Services*, *38*(3), 442–463.

Kyambade, M., & Namatovu, A. (2025b). The impact of ethical leadership and code of ethics on employee integrity, mediated by job satisfaction in Uganda's public sector. *International Journal of Public Leadership*, *21*(1), 72–89.

Kyambade, M., Namatovu, A., & Male Ssentumbwe, A. (2025a). Exploring the evolution of artificial intelligence in education: From AI-guided learning to learner-personalized paradigms. *Cogent Education*, *12*(1), 2505297.

Kyambade, M., Namatovu, A., Namubiru, B., & Muganzi, C. (2025b). Socially responsible leadership practices in university context: A developing nation perspective. *International Journal of Educational Management*, *39*(1), 275–293.

Kyambade, M., Namatovu, A., Ssentumbwe, A. M., & Tushabe, M. (2025c). Work from home can be distracting: Exploring the moderation role of transformational leadership on teleworking and cyber-slacking. *Cogent Education*, *12*(1), 2450115.

Kyambade, M., Tushabe, M., Mugerwa, G. W., & Namatovu, A. (2025d). Am not contented at work: Unraveling socially responsible leadership and psychological safety on job satisfaction in Ugandan public universities. *Management Research Review*, *48*(6), 863–882.

Longo, F., Padovano, A., & Umbrello, S. (2020). Value-oriented and ethical technology engineering in industry 5.0: A human-centric perspective for the design of the factory of the future. *Applied Sciences*, *10*(12), 4182.

Matenga, A. E., & Mpofu, K. (2023). Africa and industry 5.0: Challenges and opportunities in the future of manufacturing. *Procedia CIRP*, *120*, 1256–1261.

Matseke, C. (2023). China and the BRI in Africa: A focus on Rwanda. In *Perspectives on Africa–China Infrastructural and Industrial Cooperation: Empirical Findings and Conceptual Implications* (pp. 189–207). Cham: Springer International Publishing.

Maurya, R. K., Maurya, S. R., & Yewale, V. (2025). Closing the digital gap: Rural and urban communities' E-government options. In *Startup-Driven E-Government: Digital Innovation for Sustainable Ecosystems* (pp. 147–182). IGI Global Scientific Publishing.

MoFPED. (2021). *Third National Development Plan (NDP III) 2020/21–2024/25*. Ministry of Finance, Planning and Economic Development, Republic of Uganda.

Mourtzis, D., Angelopoulos, J., & Panopoulos, N. (2022). A literature review of the challenges and opportunities of the transition from Industry 4.0 to Society 5.0. *Energies*, *15*(17), 6276.

Mpaata, E., Kyambade, M., Matovu, A., & Naigwe, J. (2025). Impact of social influence, financial literacy, and self-control on saving behavior among micro and small enterprise owners in Uganda. *Cogent Psychology*, *12*(1), 2471703.

Mujabi, S., Kyambade, M., Waiswa, Y., Manyindo, E. K., & Nabasirye, M. (2025). When borrowers choose the worst credit sources; borrower cognitive ability and credit decision rationality in the context of informal credit sector in Uganda. *SN Business & Economics*, *5*(6), 69.

Mukherjee, D., & Mandal, B. K. (2025). Introduction to industry 4.0. In *Industry 4.0, Smart Manufacturing, and Industrial Engineering* (pp. 1–23). CRC Press.

Nahavandi, S. (2019). Industry 5.0—A human-centric solution. *Sustainability*, *11*(16), 4371.

Namatovu, A., & Kyambade, M. (2025a). Artificial intelligence and employee performance in Uganda's healthcare institutions: Exploring the mediation effects of perceived ease of use and skills enhancement. *Journal of Health Organization and Management*. https://doi.org/10.1108/JHOM-01-2025-0045

Namatovu, A., & Kyambade, M. (2025b). Leveraging AI in academia: University students' adoption of ChatGPT for writing coursework (take home) assignments through the lens of UTAUT2. *Cogent Education*, *12*(1), 2485522.

Ogunode, N. J., Edinoh, K., & Okolie, R. C. (2023). Public private partnership model and implementation of tertiary education program in Nigeria. *American Journal of Science on Integration and Human Development*, *1*(6), 1–12.

Rosati, F., Rodrigues, V. P., Cosenz, F., & Li-Ying, J. (2023). Business model innovation for the sustainable development goals. *Business Strategy and the Environment*, *32*(6), 3752–3765.

Sangarathas, D., & Shanmugathas, S. (2025). Human-centric digital transformation in Industry 5.0: A phenomenological study of family-owned SME in Sri Lanka. *The Electronic Journal of Information Systems in Developing Countries*, *91*(4), e70026.

Santiago, B. D. S., Scavarda, L. F., Gusmão Caiado, R. G., Santos, R. S., & Mattos Nascimento, D. L. D. (2025). Corporate social responsibility and circular economy integration framework within sustainable supply chain management: Building blocks for industry 5.0. *Corporate Social Responsibility and Environmental Management, 32*(1), 269–290.

Tushabe, M., Kyambade, M., Kalisa, G., & Birungi, F. (2023). Work–life balance, gender role beliefs, gender participation and women's representation in leadership positions in public universities: A case of Makerere university. *Asian Research Journal of Arts & Social Sciences*, *21*(4), 9–21.

Tushabe, M., Kyambade, M., Namatovu, A., & Oyella, S. (2025). Advancing gender diversity in leadership: The role of work–life balance and gender participation in Ugandan public universities. *Cogent Social Sciences*, *11*(1), 2486553.

Uganda Bureau of Statistics. (2023). *Annual Statistical Abstract 2023*. Kampala, Uganda: Uganda Bureau of Statistics. https://www.ubos.org/2023-statistical-abstract/

Ullman, A. N., & Kittner, N. (2024). Are global efforts coordinated for a Just Transition? A review of civil society, financial, government, and academic Just Transition frameworks. *Energy Research & Social Science*, *108*, 103371.

UNCTAD. (2022). *The Least Developed Countries Report: Structural Transformation for Sustainable Development*. United Nations Conference on Trade and Development.

World Bank. (2023). *Uganda Economic Update: Digital Transformation for Inclusive Growth*. Washington, DC: The World Bank.

Yin, S., Zhang, N., Ullah, K., & Gao, S. (2022). Enhancing digital innovation for the sustainable transformation of manufacturing industry: A pressure-state-response system framework to perceptions of digital green innovation and its performance for green and intelligent manufacturing. *Systems*, *10*(3), 72.

Zizic, M. C., Mladineo, M., Gjeldum, N., & Celent, L. (2022). From Industry 4.0 towards Industry 5.0: A review and analysis of paradigm shift for the people, organization and technology. *Energies*, *15*(14), 5221.

Index

Note: numbers in **bold** indicate a table. Numbers in *italics* indicate a figure.

Made in the USA
Monee, IL
12 November 2026

7d198e65-4493-487b-a173-83f5c8b6d01fR01